SURVEY OF INDUSTRIAL CHEMISTRY

2nd
Revised Edition

● ● ●

Philip J. Chenier

VCH

To my chemistry instructors,
who taught me a fascination for the subject,
and to all of my students, past, present, and future,
with the hope that they will learn this too.

Philip J. Chenier
Department of Chemistry
University of Wisconsin-Eau Claire
Eau Claire, WI 54702-4004
Library of Congress Cataloging-in-Publication Data

Chenier, Philip J.
    Survey of industrial chemistry / Philip J. Chenier
        p.   cm.
    Includes bibliographical references and indexes.
    ISBN 1-56081-082-3
    1. Chemistry, Technical. I. Title
TP145.C44 1992
660--dc20   92-19693
                                             CIP

Printed in the United States of America.

ISBN 1-56081-082-3 VCH Publishers
ISBN 1-56081-622-8 VCH Publishers (paper cover)

Published jointly by:

VCH Publishers, Inc.	VCH Verlagsgesellschaft mbH	VCH Publishers (UK) Ltd.
220 East 23rd Street	P.O. Box 10 11 16	8 Wellington Court
Suite 909	D-6940 Weinheim	Cambridge CB1 1HZ
New York, New York 10010	Federal Republic of Germany	United Kingdom

Preface

This book arose from a need for a basic text dealing with industrial chemistry for use in a one-semester, three-credit senior level course taught at the University of Wisconsin-Eau Claire. The course was added in 1981 as a requirement for our B.S. degree in Chemistry/with Business Emphasis and is strongly recommended as an elective in our other chemistry majors, including our ACS-accredited program. There are some good extensive texts and valuable reference works dealing with applied chemistry. What was needed for our course, and what I believe will be useful for similar courses, is a basic text of introductory material, sufficient to cover all important areas of the chemical industry, yet limited in scope so as to be a reasonable goal to complete in 40-45 hours of lecture.

Industrial chemistry means different things to different people. Most will agree that the phrase includes the practical, applied chemistry that bridges the gap between basic research and development and at least two other disciplines, chemical engineering and chemical marketing. The present text attempts to lessen the lack of knowledge that most graduates have in both of these areas. Some attempt is made to instill in chemists an appreciation for both the manufacturing and economic problems facing the chemical industry on a day-to-day basis, as well as introduce them to the chemistry used by our industry everyday. Although some space is devoted to economics and engineering, this is largely a chemistry book, and chemical reactions and processes, even mechanisms of reactions, are given full coverage.

In developing such a book the toughest job is always deciding what to include. I have tried to cover a little of everything, since the text is meant to be a survey of important sectors of industrial chemistry. The manufacture and uses of the top 100 basic chemicals are covered in detail. The chemistry of all important industrial polymers are included and their applications are discussed. Finally certain selected specific technologies, the most important of the many areas that Chemical and Allied Products (SIC 28) covers, are given a chapter each. If one measures treatment in terms of the value of shipments, the book covers some 94% of the chemical industry, as well as information on other industries separate from Chemicals and Allied Products that do contain interesting chemistry and do employ many chemists, such as Paper and Allied Products (SIC 26), Petroleum and Coal Products (SIC 29), and Rubber and Miscellaneous Plastic Products (SIC 30).

Granted there are many areas of applied chemistry that contribute to the other 6% of industrial shipments. Practice has shown that in our course these cannot be covered in any great detail because of time limitations. However, the requirement of having students give a short talk and write on one of these other subjects (see Appendix), and study three extra topics to enable them to pass a short quiz, gives students some choice in material to be covered and further completes their study of the most important divisions of the chemical industry.

Perhaps the most challenging part of teaching this course and of writing a good text is to keep the important economic data current. In some cases this is done easily; in others it is difficult. Rather than having to revise this material yearly or even monthly with changing economic times, I have made recourse to some references to periodic updates that students can consult for the latest data. Examples of these series are the "Top 50 Chemicals" in *Chemical and Engineering News* and "Chemical Profiles" in *Chemical Marketing Reporter*. Government figures are particularly frustrating, since official numbers for shipments are not available in *Annual Survey of Manufactures* until three years later. But even those numbers give students a general feeling for the economic trends of the industry. It is virtually impossible for a text to remain economically accurate and complete for more than a year or two, and the present text is no exception. In presenting this material I use numerous transparencies (over 800 have now been developed) to which up-to-date numbers for each year can easily be added when necessary. The graphs and charts are on computer disk so that easier updating of information and replotting requires a minimum of effort. The course material is also supplemented with over 200 color slides of various chemical plants, manufacturing sites, and research labs which I have visited in the last few years.

To attempt to thank everyone who has helped me expand my knowledge of this subject would be an impossible task, but certain organizations deserve a special mention. A University of Wisconsin System Undergraduate Teaching Improvement Grant allowed me to plan the course initially during one summer. University of Wisconsin-Eau Claire Faculty Development Grants enabled me to visit chemical plants throughout the United States to get first-hand experience in manufacturing. They also funded some release time for one semester to write a portion of an earlier version of this book. A number of companies must be mentioned for letting me visit their facilities, talk with their personnel, and obtain pictures for use in class: Air Products and Chemicals, Allied Chemical Corporation, Amoco Chemicals Corporation, Amoco Oil Co., E. I. Du Pont de Nemours and Co., Georgia-Pacific Corporation, Occidental Chemical Corporation, Vista Chemical Company, and W.H. Brady Co. I also thank the Department of Chemistry at UW-Eau Claire for allowing me to develop and teach the course, and to the students I have had who have given me valuable feedback on the course and text. Finally, I wish to thank one individual, Dr. Harold Wittcoff, who first got me interested in teaching industrial chemistry when I audited his course during a sabbatical at the University of Minnesota.

Philip J. Chenier

Contents

List of Important References and Their Abbreviations

There are a variety of useful sources for obtaining information on industrial chemistry. The following list is not exhaustive. In the text an attempt has been made to specify appropriate reading for each chapter or in some cases individual sections to supplement the chapter. Some references are given in shortened form in the text by citing the author's name or by use of an abbreviation. These references, as well as some of the others the author has found useful, are given below with the abbreviations used.

Books

Abbreviation	*Reference*
Austin	Austin, G.T. *Shreve's Chemical Process Industries*, 5th ed.; McGraw-Hill: New York, 1984.
B & K	Billmeyer, F.W.; Kelly, R.N. *Entering Industry: A Guide for Young Professionals*; John Wiley & Sons: New York, 1975; Chapters 1 and 2.
BSWB	Büchner, W.; Schliebs, R.; Winter, G.; Büchel, K.H. *Industrial Inorganic Chemistry*; VCH: New York, 1989.
C & T	Chang, R.; Tikkanen, W. *The Top Fifty Industrial Chemicals*; Random House: New York, 1988.
EPSE	Kroschwitz, J.I. *Concise Encyclopedia of Polymer Science and Engineering*; John Wiley & Sons: New York, 1990.
Kent	Kent, J.A. *Riegel's Handbook for Industrial Chemistry*, 8th ed.; Van Nostrand Reinhold: New York, 1983.
Kline	Eveleth, W.; Kollonitsh, V. *The Kline Guide to the Chemical Industry*, 5th ed.; Kline and Company, Inc.: Fairfield, NJ, 1990.

L & M
Lowenheim, F.A.; Moran, M.K. *Faith, Keyes, and Clark's Industrial Chemicals*, 4th ed.; Wiley-Interscience: New York, 1975.

R & B
Reuben, B.G.; Burstall, M.L. *The Chemical Economy*; Longmans: London, 1973; pp 121-136.

S & C
Seymour, R.B.; Carraher, Jr., C.E. *Polymer Chemistry: An Introduction*; Marcel Dekker, Inc.: New York, 1981.

Szmant
Szmant, H.H. *Organic Building Blocks of the Chemical Industry*; Wiley-Interscience: New York, 1989.

Thompson
Thompson, R. *The Modern Inorganic Chemicals Industry*; The Chemical Society: London, 1977.

Ulrich
Ulrich, H. *Raw Materials for Industrial Polymers*; Oxford University Press: New York, 1988.

White
White, H.L. *Introduction to Industrial Chemistry*; Wiley-Interscience: New York, 1986.

W & R I
Wittcoff, H.A.; Reuben, B.G. *Industrial Organic Chemicals in Perspective. Part One: Raw Materials and Manufacture*; Wiley-Interscience: New York, 1980.

W & R II
Wittcoff, H.A.; Reuben, B.G. *Industrial Organic Chemicals in Perspective. Part Two: Technology, Formulation, and Use*; Wiley-Interscience: New York, 1980.

Wiseman
Wiseman, P. *Petrochemicals*; Ellis Horwood Limited: Chichester, 1986.

Multi-Volume Works

CEH
Chemical Economics Handbook; Stanford Research Institute International: Menlo Park, CA, 1991.

KO
Kirk-Othmer's *Encyclopedia of Chemical Technology*, 3rd ed.; Wiley-Interscience: New York, 1983.

Ullmann
Ullmann's *Encyclopedia of Industrial Chemistry*; VCH: New York, 1991.

Periodicals and Government Documents

AS
Annual Survey of Manufactures; U.S. Department of Commerce, Bureau of the Census, each year.

C & E News
Chemical and Engineering News; American Chemical Society: Washington, DC, 1991. Contains many interesting articles each week and valuable annual series including "Facts and Figures for the Chemical Industry," "Facts and Figures for Chemical R&D," "Salary Survey," "Employment Outlook," "Top 50 Chemicals," and "Top 100 Chemical Companies."

CMR
Chemical Marketing Reporter; Schnell Publishing Co.: New York, 1991. Contains many informative articles in each issue and up-to-date chemical prices each week.

CP
"Chemical Profiles," a weekly series in *Chemical Marketing Reporter*.

KC
"Key Chemicals," a series published in *Chemical and Engineering News*.

KP
"Key Polymers," a series published in *Chemical and Engineering News*.

IO
U.S. Industrial Outlook; U.S. Department of Commerce, International Trade Commission, each year.

SA   *Statistical Abstract of the United States*; U.S. Department of Commerce, Bureau
            of the Census, each year.
SOC   *Synthetic Organic Chemicals*; U.S. Department of Commerce, International
            Trade Commission, each year.

Audio Courses

J   Jonnard, A. *Business Aspects of Chemistry*; American Chemical Society Audio
            Course: Washington, DC, 1974.
W   Wittcoff, H. *Industrial Organic Chemistry*; American Chemical Society Audio
            Course: Washington, DC, 1979.

Top 50 Chemicals

Rank 1990	Chemical	Production Billion lb 1990	Average % Annual Growth 1980-90	Average Price ¢/lb 1991
1	Sulfuric acid	88.56	0	3.8
2	Nitrogen	57.32	5.1	na
3	Oxygen	38.99	0.9	na
4	Ethylene	37.48	2.7	30
5	Lime	34.80	-0.9	2.5
6	Ammonia	33.92	-1.5	7.4
7	Phosphoric acid	24.35	1.2	3.1
8	Sodium hydroxide	23.38	0.1	15
9	Propylene	22.12	4.9	23
10	Chlorine	21.88	-0.4	9.5
11	Sodium carbonate	19.85	1.8	7.3
12	Urea	15.81	0.1	11
13	Nitric acid	15.50	-1.7	8.8
14	Ammonium nitrate	14.21	-2.5	6.3
15	Ethylene dichloride	13.30	1.8	16
16	Benzene	11.86	-2.2	20
17	Carbon dioxide	10.98	6.2	na
18	Vinyl chloride	10.65	5.1	21
19	Ethylbenzene	8.99	1.6	18
20	Styrene	8.02	1.6	40
21	Methanol	7.99	1.1	8.7
22	Terephthalic acid	7.69	2.4	na
23	Formaldehyde	6.41	1.4	12
24	Methyl t- butyl ether	6.30	na	na
25	Toluene	6.10	-1.9	12
26	Xylene	5.70	-1.4	13
27	Ethylene oxide	5.58	0.7	51
28	p- Xylene	5.20	2.1	25
29	Ethylene gylcol	5.03	1.4	30
30	Ammonium sulfate	4.99	1.6	6.5
31	Hydrochloric acid	4.68	-2.1	3.3
32	Cumene	4.31	2.2	24
33	Acetic acid	3.76	2.4	36
34	Potash	3.62	-3.1	4.0
35	Phenol	3.51	3.2	33
36	Propylene oxide	3.20	6.1	58
37	Butadiene	3.16	1.2	20
38	Acrylonitrile	3.03	5.2	39
39	Carbon black	2.87	1.2	22
40	Vinyl acetate	2.55	2.8	44
41	Cyclohexane	2.47	2.3	22
42	Aluminum sulfate	2.42	-0.6	12
43	Acetone	2.22	0.7	38
44	Titanium dioxide	2.16	4.2	99
45	Sodium silicate	1.76	0.9	20
46	Adipic acid	1.64	1.1	60
47	Sodium sulfate	1.47	-4.3	5.7
48	Calcium chloride	1.38	-3.5	8.3
48	Isopropyl alcohol	1.38	-2.8	40
48	Caprolactam	1.38	4.3	89

na = not available

The Second 50 Chemicals

Rank	Chemical	Production Billion lb	Average % Annual Growth	Average Price c/lb
51	Phosgene	2.28	4.2	61
52	Butyraldehyde	1.78	12.8	20
53	Acetic anhydride	1.75	1.1	46
54	Linear alpha olefins (LAO)	1.54	11.0	54
55	Chlorofluorocarbons (CFC)	1.44	2.9	94
56	Acetone cyanohydrin	1.31	4.2	na
57	n-Butyl alcohol	1.28	5.1	40
58	Methyl methacrylate (MMA)	1.24	5.0	71
59	Hexamethylenediamine (HMDA)	1.19	3.8	111
60	Hydrogen cyanide	1.18	4.8	60
61	Isobutylene	1.16	1.1	24
62	Bisphenol A	1.13	8.0	93
63	Cyclohexanone	1.06	8.3	77
64	Aniline	1.03	5.0	61
65	Sodium tripolyphosphate (STPP)	1.03	-3.5	36
66	Acrylic acid	1.01	5.5	67
67	2,4- and 2,6-Dinitrotoluene	0.98	4.8	37
68	Isobutane	0.97	-1.9	8
69	Methylene diphenyl diisocyanate (MDI)	0.96	7.2	120
70	Nitrobenzene	0.96	1.3	34
71	Phthalic anhydride	0.95	0.1	42
72	Propylene glycol	0.86	6.2	58
73	o-Xylene	0.86	-1.5	17
74	n-Paraffins	0.82	-0.4	28
75	Toluene diisocyanate (TDI)	0.78	2.6	129
76	Lignosulfonic acid, salts	0.74	1.0	10
77	Acetaldehyde	0.73	4.0	46
78	1,1,1-Trichloroethane	0.71	-0.3	41
79	Phosphorus	0.70	-2.1	91
80	Hexanes	0.68	5.5	10
81	Linear alkylbenzenes (LAB)	0.68	0.9	49
82	Sodium bicarbonate	0.66	3.0	19
83	Carbon tetrachloride	0.65	-0.7	24
84	Ethanolamines	0.65	4.6	63
85	2-Ethylhexanol	0.65	5.5	48
86	Linear alcohols, ethoxylated	0.64	3.3	49
87	Hydrofluoric acid	0.63	2.0	69
88	Potassium hydroxide	0.59	1.9	14
89	Nonene	0.58	3.5	22
90	1-Butene	0.57	16	30
91	Ethanol (synthetic)	0.56	1.2	31
92	Sodium chlorate	0.56	6.0	22
93	Butyl acrylate	0.55	9.6	53
94	Methyl chloride	0.55	1.7	26
95	Chloroform	0.53	3.8	35
96	Diethylene glycol	0.53	1.4	30
97	Perchloroethylene	0.50	-4.9	31
98	Sulfur dioxide	0.50	1.0	12
99	Hydrogen peroxide	0.48	7.2	48
100	Methylene chloride	0.48	-2.7	28

na = not available

Top Polymers

Polymer	Production Billion lb 1990	Average % Annual Growth 1980-90	Average Price ¢/lb
PLASTICS			
Thermosetting Resins	**6.36**	**4.5**	
Phenol/Formaldehyde	2.95	7.0	26-96
Urea/Formaldehyde	1.50	2.5	55-80
Unsaturated Polyesters	1.22	2.6	55-150
Epoxies	0.50	4.7	128-227
Melamine	0.20	1.9	107-108
Thermoplastic Resins	**41.93**	**5.6**	
Low-Density Polyethylene	11.18	4.4	52
Poly (Vinyl Chloride)	9.09	5.2	39
High-Density Polyethylene	8.33	6.6	51
Polystyrene	5.01	3.6	69
Polypropylene	8.32	8.6	55
Total	**48.29**	**5.4**	
SYNTHETIC FIBERS			
Cellulosics	**0.50**	**-4.6**	
Rayon	0.30	-4.9	85
Acetate	0.21	-4.2	160
Noncellulosics	**8.18**	**0.4**	
Polyester	3.19	-2.2	72
Nylon	2.66	1.2	125
Olefin	1.82	9.3	72
Acrylic	0.51	-4.2	78
Total	**8.68**	**0**	
SYNTHETIC RUBBER			
Styrene-Butadiene	1.88	-2.3	64
Polybutadiene	0.89	2.6	74
Ethylene-Propylene	0.56	5.9	138
Nitrile	0.12	-1.2	122
Other	1.20	2.6	
Total	**4.66**	**0.5**	

Shipments of Selected Sectors of Chemicals and Allied Products and Other Chemical Process Industries

SIC Code	Name	Shipments $ Billion	%
281	Industrial Inorganic Chemicals	22.110	8.5
286	Industrial Organic Chemicals	59.972	23.1
2821	Plastics	33.823	13.0
2823	Cellulosic Fibers	1.346⎤	4.2
2824	Organic Fibers, Noncellulosic	9.512⎦	
2822	Synthetic Rubber	3.916	1.5
2851	Paints and Allied Products	12.851	4.9
2891	Adhesives and Sealants	4.845	1.9
2873	Nitrogenous Fertilizers	3.281⎤	
2874	Phosphatic Fertilizers	4.150⎬	3.4
2875	Fertilizers, Mixed Only	1.506⎦	
2879	Agricultural Chemicals (Pesticides)	6.360	2.4
283	Drugs	43.987	16.9
284	Soaps, Cleaners, and Toilet Goods	37.856	14.6
	Other	14.184	5.6
28	Chemicals and Allied Products	259.699	100.0
26	Paper and Allied Products	122.560	
29	Petroleum and Coal Products	131.415	
30	Rubber and Miscellaneous Plastic Products	94.200	

1

Introduction to the Chemical Industry: An Overview

References
C & E News
W & R. I, pp. 20-31

THE NATIONAL ECONOMY

Before beginning a detailed discussion of the chemical industry, we should have a basic appreciation for the main sectors of a developed economy so that we may understand the role that this industry plays in the overall picture. Table 1.1 gives the major subdivisions of the U.S. economy along with their official designations or Standard Industrial Classification (SIC) by the U.S. Bureau of Census. A similar classification system is used in Western Europe, Japan and other complex societies. These sectors are separate but interdependent. For example, manufacturing draws on mining to buy iron ore for steel manufacture. The manufacturing sector also converts steel to machinery to sell back to mining for its operations.

The third column gives an estimate of the size of these various sectors in terms of value added in billions of dollars. The value added is simply the difference between the output (goods and services) and the input (labor, land, and capital) of the industry. The total value added, $4,526.7 billion in 1987, is the gross national product (GNP) for the entire economy. This is the latest year that official government figures are available at the time of this writing, although the estimated 1990 GNP is over $5.7 trillion.

Although the numbers change each year, percentages of each sector do not change very much. Note that manufacturing is the largest sector in terms of value added and amounts to about 19% or almost one fifth the GNP. The chemical industry is a part of this manufacturing sector.

1

TABLE 1.1 U.S. Gross National Product by Industry

Industry	SIC	Value Added ($ billion)
Agriculture, forestry, and fisheries	01-09	94.9
Mining	10-14	85.4
Construction	15-17	218.5
Manufacturing	20-39	853.6
Transportation, communication and electric, gas, and sanitary services	40-49	408.2
Wholesale and retail trade	50-59	740.4
Finance, insurance, and real estate	60-67	775.4
Other services including medicine, education, social services, and entertainment	70-97	793.5
Governmental and government enterprises	98	535.3
Other	99	21.5
		GNP= $4,526.7

Source: SA

DEFINITION AND DIVISIONS OF THE CHEMICAL INDUSTRY

Chemical Process Industries

Just what exactly do we mean when we refer to "the chemical industry"? This is a general term that may mean different things to different people. A very broad interpretation of this phrase might, according to certain Standard Industrial Classifications, refer to the "chemical process industries" that are divided into the following areas: chemicals and allied products; paper and allied products; petroleum refining and related industries; rubber and miscellaneous plastic products; and stone, clay and glass products. These are some of the subdivisions of manufacturing dealing heavily in chemicals and chemical products, as listed in Table 1.2. However, this broader interpretation for the chemical industry is not commonly used.

Chemicals and Allied Products

Most people, when referring to the chemical industry, really have in mind one specific division of manufacturing which is classified as chemicals and allied products. Note that it is the fourth largest division of manufacturing in terms of manufacturers' shipments, which is the usual dollar amount quoted in the manufacturing sector to estimate division size. Shipment figures are easier to calculate than value added. In 1988, the latest year for which all figures are available, Chemicals and Allied Products had shipments totaling $259.7

TABLE 1.2 U.S. Chemicals and Allied Products Industry Versus Other Manufacturing Industries

SIC	Industry	Value Added[a] ($ billion)	Shipments ($ billion)
20	Food and Kindred Products	128.8	351.5
21	Tobacco Products	17.2	23.8
22	Textile Mill Products	26.3	64.8
23	Apparel, Other Textile Products	32.5	65.0
24	Lumber and Wood Products	29.0	72.1
25	Furniture and Fixtures	20.9	39.2
→ 26	Paper and Allied Products	57.4	122.6
27	Printing and Publishing	94.1	143.9
→ 28	Chemicals and Allied Products	137.9	259.7
→ 29	Petroleum and Coal Products	25.3	131.4
→ 30	Rubber and Misc. Plastic Products	46.6	94.2
31	Leather and Leather Products	4.5	9.7
→ 32	Stone, Clay and Glass Products	34.2	63.1
33	Primary Metal Industries	56.5	149.1
34	Fabricated Metal Products	79.9	158.8
35	Machinery, Except Electric	129.3	243.3
36	Electric, Electronic Equipment	103.5	187.0
37	Transportation Equipment	143.5	354.0
38	Instruments, Related Products	76.1	114.5
39	Miscellaneous Manufacturing	19.0	34.9
	Total	1,262.3	2,682.5

Source: AS

[a] Value added = (shipments + services rendered) - (cost of materials, supplies, containers, fuel, purchased electricity, and contract work). Value added avoids the duplication in shipments resulting from the use of products by other sections of the industry. It is the best measure of the relative economic importance of different segments of the industry.

billion, or about 9.7% of all manufacturing. Unless specified otherwise, when we use the term *chemical industry* we mean this division, Chemicals and Allied Products (SIC 28).

What does Chemical and Allied Products segment include? This is summarized in Table 1.3 in terms of both shipments and value added. Note that organic chemicals is the largest subdivision in percentage for shipments and drugs is the largest for value added. Although there is a similarity for both methods of subdivisions, there are differences for fine chemicals such as drugs. These are high-priced products with specialized markets. Their manufacture requires less capital but is more labor intensive. Their contribution to value added therefore is much larger than to shipments.

TABLE 1.3 Divisions of U. S. Chemical and Allied Products Industry

SIC	Industry	Value Added^a ($ billions)	Percent	Shipments ($ billions)	Percent
281	Industrial Inorganic Chemicals	12.841	9.3	22.110	8.5
282	Plastic Materials, Synthetics	20.755	15.1	48.388	18.6
283	Drugs	31.026	22.5	43.987	16.9
284	Soaps, Cleaners, Toilet Goods	23.602	17.1	37.856	14.6
285	Paints and Allied Products	6.489	4.7	13.532	5.2
286	Industrial Organic Chemicals	26.985	19.6	59.972	23.1
287	Agricultural Chemicals	7.605	5.5	16.077	6.2
289	Misc. Chemical Products	8.576	6.2	17.778	6.9
	Total	137.879		259.699	

Source: AS.

^{a}Value added = (shipments + services rendered) - (cost of materials, supplies, containers, fuel, purchased electricity, and contract work). Value added avoids the duplication in shipments resulting from the use of products by other sectors of the industry. It is the best measure of the relative economic importance of different segments of the industry.

Diversity and Complexity in the Chemical Industry

The chemical industry is actually a set of related industries with many diverse functions and products. Table 1.4 lists some of these different areas and at the same time emphasizes that certain raw materials are used to prepare key chemicals, monomers, and intermediates that may be sold independently or used directly in additional steps to give various polymers and end chemicals. These in turn can be formulated and fabricated into chemical products, which can sometimes be modified into finished products. Hence the term *chemicals and allied products* accurately emphasizes this diversity as well as the flow of materials and products from raw sources to finished formulations. Although the division is approximate, about 60% of the chemical industry manufactures industrial products that are further modified, whereas 40% of their products are sold directly to the consumer. Chemistry may not be a household word, but it should be.

Further proof of diversity in the chemical industry is apparent in other statistics. There are over 12,000 manufacturing plants in operation in the United States. Over 55,000 chemicals are commercially produced, but only 10% of these account for over 99.9% of production and are made in excess of 1 million lb/yr in the United States. Although the top four companies in the chemical industry have about 25% of the sales in the industry, the top ten have only 35% of all sales. This percentage is small compared to other industries like automobiles, airplanes, tires, and glass, where 80–99% of sales are taken by the top eight companies or less. Diversity has increased in the last few years. Before 1940 chemical companies sold nothing but chemicals. Although some are primarily chemical, others have diversified so that it is common to have

TABLE 1.4 The Chemical Business is 30 Businesses

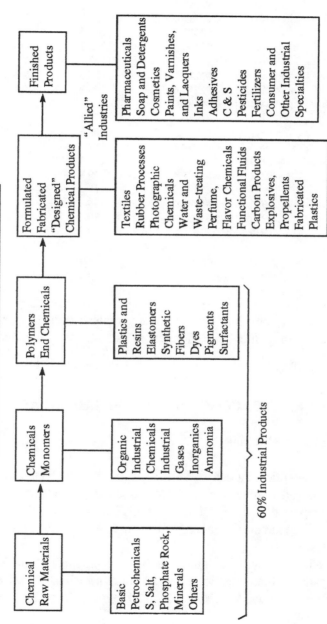

Source: J. Copyright 1974 by the American Chemical Society and reprinted with permission.

TABLE 1.5 U.S. Shipments

	All Manufacturing Industries $ billions	Chemicals and Allied Products $ billions
1990	$2860.3	265.7
1989	$2781.4	255.5
1988	$2611.6	240.5
1987	$2390.0	212.7
1986	$2260.3	197.1
1985	$2279.1	197.3
1984	$2254.4	198.2
1983	$2054.9	183.2
1982	$1960.2	170.7
1981	$2017.5	180.5
1980	$1852.7	162.5
1979	$1727.2	147.7
1978	$1523.4	129.8
1977	$1358.4	118.2
1976	$1185.6	104.1
1975	$1039.1	89.7
1974	$1017.5	83.7
Percent Annual Change		
1989-90	3%	4%
1980-90	4%	5%

Source: C & E News , "Facts and Figures".

chemicals account for only half of the company's sales. Many corporations such as the petroleum companies have chemical sales with a very low percentage of total sales.

SIZE AND CURRENT ECONOMICS OF THE CHEMICAL INDUSTRY

Shipments and Production

How big is the chemical industry? This is a difficult question to answer. What should be the best determining factor? One good measure of size is dollar value of shipments reported. Table 1.5 shows that this industry had 1990 shipments of $265.7 billion compared to all manufacturing at $2860.3 billion, about 9.3% of all manufacturing.

1982 was a bad year and the dollar volume of shipments fell for the first time in over a decade for both chemicals and all manufacturing. In 1986 a similar but minor slowdown was also seen. Fig. 1.1 shows the production index in

Production index,[a] 1987 = 100

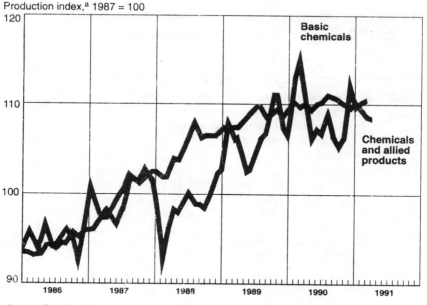

[a] Seasonally adjusted.
Figure 1.1 Production in chemicals and allied products. (*Source*: Reprinted with permission from *Chem. Eng. News* **1991**, 69(25), 39. Copyright 1991 American Chemical Society.)

Chemicals and Allied Products and basic chemicals, a subdivision. Healthy years have been experienced since 1986. Over the decade an average increase in chemicals of 5% per year is still healthy and better than 4% for all manufacturing.

Sales, Earnings, and Profit

The chemical industry weathered the 1981–1982 recession with relatively little damage, especially compared to other segments of the national economy. Figure 1.2 (page 9) shows that sales and earnings were down sharply in these years but bounced back by 1984. A short slowdown in 1985 has been followed by good years until 1989.

Sales Destination of Chemicals

To whom does the chemical industry sell all of its chemicals? It is its own best customer. Table 1.6 shows that 52.1% of industrial chemicals are sold within the Chemicals and Allied Products industry. The industrial inorganic and organic chemicals subdivision buys 16.3% of the chemicals and another 10.4% goes into making polymers. For example, chlorine might be sold to another company

TABLE 1.6 Who Buys Industrial Organic and Inorganic Chemicals According to U. S. Bureau of Census Figures

SIC No.		Percent of Total Sales by Value
	A. Sales to the Chemical and Allied Products Industries	
281 excl. 28195	Industrial inorganic and organic chemicals	16.3
2821	Plastics materials and resins	10.4
2861, 289	Miscellaneous chemical products	4.4
284 excl. 2844	Cleaning preparations	4.0
2851	Coatings	3.3
2873, 2874, 2875	Fertilizers	2.9
2824	Synthetic organic fibers, noncellulosic	2.6
2822	Synthetic rubber	2.5
2879	Agrochemicals	2.1
283	Drugs	1.8
2823	Cellulosic man-made fibers	1.0
2844	Toilet preparations	0.8
		52.1
	B. Exports and Sales to Government and Private Consumers	
	Net exports	9.0
	Federal government for defense	3.3
	Federal government for other purposes	2.3
	Personal consumption	1.2
	State and local government	0.2
		16.0
	C. Sales to Industries Other Than Chemicals and Allied Products	
	Agriculture, forestry, and fishing	4.4
	Petroleum refining and related products	3.3
	Crude petroleum and natural gas	0.8
	Ammunition (except for small arms)	1.2
	Food and kindred products and tobacco manufacturers	1.0
	Textiles	2.7
	Paper and allied products, printing, and publishing	3.2
	Rubber and miscellaneous plastics products	2.2
	Primary iron, steel, and nonferrous metal manufacturing	2.9
	Electrical and electronic equipment	1.4
	Photographic equipment and supplies	1.0
	Wholesale trade	0.7
	Real estate	0.8
	All other industries	6.3
		31.9

Source: W & R II. Reprinted with permission from John Wiley & Sons, Inc.

to make vinyl chloride, which in turn is sold to someone else to make polyvinyl chloride plastic.

$$Cl_2 + CH_2{=}CH_2 \xrightarrow{-HCl} CH_2{=}CHCl \xrightarrow{\Delta} (-CH_2-CHCl-)_n$$

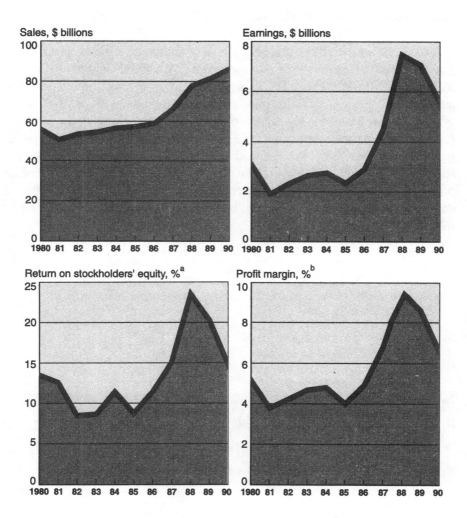

Figure 1.2 Chemical sales, earnings, and profit. (*Source*: Reprinted with permission from *Chem. Eng. News* 1991, 69(25), 30. Copyright 1991 American Chemical Society.)

Note: Based on data for 30 major chemical companies. [a] After-tax earnings as a percentage of year-end stockholders' equity. [b] After-tax earnings as a percentage of sales.

Part B of Table 1.6 shows that a large percentage, 9%, of the chemicals are exported. About 3.3% of all chemicals are sold to the government for defense. Part C shows sales to other industries, especially to agriculture, petroleum refining, and paper products.

LOCATION OF THE CHEMICAL INDUSTRY

Table 1.7 shows a state-by-state breakdown of shipments for the top ten states. The West South Central (Texas and Louisiana), Atlantic (New Jersey and New York), and East North Central regions (Illinois, Ohio, Pennsylvania, Tennessee, and Indiana), along with California, account for the largest share of chemical manufacturing. These ten states have 63% or nearly two thirds of the industry. California is the fastest growing for chemical production. Ohio is also increasing in chemical productivity. It is interesting to note that California, which had the sharpest chemical growth, also has the nation's strictest hazardous waste and air pollution control programs. Research and Development (R & D) technical employment is centered more in the Middle Atlantic and East North Central regions.

TABLE 1.7 Top 10 Chemical-Producing States

Rank	Percent of Industry Total
1 Texas	11.0%
2 New Jersey	10.6
3 Illinois	6.2
4 Ohio	6.2
5 California	5.6
6 Louisiana	5.5
7 New York	5.4
8 Pennsylvania	4.7
9 Tennessee	4.1
10 Indiana	3.4
Total	62.7
U.S. Chemical Industry Total	100.0

Source: C & E News.
Totals represent value added by chemical manufacturers
(the value of industry shipments minus costs of materials
and energy).

EMPLOYMENT IN THE CHEMICAL INDUSTRY

Types of Employment

If you are a chemist you have almost a 2:1 chance of eventually working in the chemical industry. About 58% of chemists are employed by private industry, 25% are in academics, 9% work for the government, and 8% are in other miscellaneous areas. Within this workforce 46% work in R & D, 17% are in management, 12% in teaching, 7% in production and quality control, 4% in marketing, sales, purchasing, and technical service, and 14% are in other fields. These other areas encompass many different jobs, including process development, personnel, public relations, patent literature, library science, and scientific writing. Many chemists start in R & D because it is most like academic chemistry. They progress into managerial positions where greater financial rewards are usually present. Many chemistry majors with some business background may start in marketing and sales. They may also enter management at a later point. It is interesting to note that of chemists age 35 and above, over 50% are in managerial capacities of one type or another.

TABLE 1.8 Chemical Degrees

Academic Year	Degrees awarded in chemistry			Degrees awarded in chemical engineering		
	Bachelor's	Master's	Ph.D.s	Bachelor's	Master's	Ph.D.s
1969	11,807	2070	1941	3557	1136	409
1970	11,617	2146	2208	3720	1045	438
1971	11,183	2284	2160	3615	1100	406
1972	10,721	2259	1971	3663	1154	394
1973	10,226	2230	1882	3636	1051	397
1974	10,525	2138	1828	3454	1045	400
1975	10,649	2006	1824	3142	990	346
1976	11,107	1796	1623	3203	1031	308
1977	11,322	1775	1571	3581	1086	291
1978	11,474	1892	1525	4615	1237	259
1979	11,643	1765	1518	5655	1149	304
1980	11,446	1733	1551	6383	1271	284
1981	11,347	1654	1622	6527	1267	300
1982	11,062	1751	1722	6740	1285	311
1983	10,746	1604	1746	7145	1304	319
1984	10,704	1667	1744	7475	1514	330
1985	10,482	1719	1789	7146	1544	418
1986	10,116	1754	1908	5877	1361	446
1987	9661	1738	1976	4983	1184	497
1988	9052	1708	1995	3917	1088	579
1989	8654	1785	2034	3684	1097	599

Source: Reprinted with permission from *Chem., Eng. News* **1990**, 68 (34), 49. Copyright 1990 American Chemical Society.

Number and Types of Employees in the Chemical Industry

Table 1.8 shows the average number of chemical and chemical engineering degrees granted at the B.S., M.S., and Ph.D. levels in representative years. There are more chemists than chemical engineers each year. In the last ten years the number of chemists has been dropping. B.S. and M.S. engineers have decreased in the same time period, whereas Ph.D. chemists and engineers have increased. Although a fairly high percentage of chemists go on for their Ph.D.s, a lower number of engineers obtain their doctorate.

The breakdown by academic areas of chemistry in which chemists are employed is as follows: analytical, 21%; organic, 14%; polymer, 11%; environmental, 9%; physical, 7%; biochemical, 7%; medicinal and pharmaceutical, 7%; general, 6%; material science, 5%; inorganic, 5%; and other, 8%.

Table 1.9 shows the total employment of all workers, technical and nontechnical, by the chemical industry as well as by all manufacturing.

TABLE 1.9 Total Employment

All Employees (thousands)	1990	1989	1988	Annual Change 1989-90	Annual Change 1980-90
All manufacturing	19,062	19,426	19,350	-2%	-1%
Chemicals and allied products	1,086	1,074	1,059	0	0
Industrial inorganic chemicals	135	133	133	2	-2
Plastics materials and synthetics	185	184	177	0	-1
Drugs	239	233	228	3	2
Soap, cleaners, and toilet goods	159	160	160	0	1
Paints and allied products	63	63	63	0	0
Industrial organic chemicals	153	150	146	2	-1
Agricultural chemicals	53	53	52	0	-2
Miscellaneous chemical products	100	100	101	0	1
Petroleum and coal products	160	157	160	2	-2
Rubber and miscellaneous plastics	867	884	868	-2	1

Source: Reprinted with permission from *Chem. Eng. News* **1991**, 69(25), 58. Copyright 1991 American Chemical Society.

Note that about 19.1 million workers are in all manufacturing, about 1.1 million in chemicals and allied products. Employment in the chemical industry is relatively constant. This is to be contrasted to other major industries—construction and automobile, for example—where employment can be down during a recession. Overall the chemical industry is in good shape. It is believed that about 159,000 chemists and 115,000 chemical engineers are employed in the United States.

Unemployment of chemists has always been low, and in March 1991 it was at 1.6% as compared to nationwide unemployment of 5%.

TABLE 1.10 Salaries of Ph.D. Scientists and
Engineers

All Scientists	$47,800
Chemists	50,500
Physicists and astronomers	53,400
Mathematicians	46,600
Computer/Information specialists	54,400
Earth scientists	50,800
Oceanographers	44,300
Atmospheric scientists	50,000
Biological scientists	44,500
Agricultural scientists	44,300
Psychologists	44,300
Economists	50,800
Sociologists and anthropologists	41,700
All Engineers	58,100
Chemical	58,900
Civil	53,300
Electrical	60,500
Mechanical	55,700
Nuclear	58,900

Source: Reprinted with permission from *Chem. Eng.
News* **1990**, 68(43), 33. Copyright 1990 American
Chemical Society.
[a] Domestic employment

TABLE 1.11 Chemists Median Salaries ($ thousands)[a]

	B.S.	M.S.	Ph.D.	All chemists
1991	$40.3	$47.4	$58.0	$52.0
1990	39.0	45.0	55.0	49.8
1989	37.0	43.0	52.5	47.3
1988	35.4	41.0	50.0	45.0
1987	33.5	39.0	47.7	42.5
1986	33.0	37.9	47.8	42.5
1985	32.0	36.0	44.0	40.0
1984	30.9	34.0	42.0	38.0
1983	30.0	33.0	40.0	36.0
1982	28.5	31.6	37.5	34.7
1981	27.5	30.0	35.0	32.0
Annual Change				
1986-91	4.1%	4.6%	3.9%	4.1%
1981-91	3.9	4.7	5.2	5.0
1990-91	3.3	5.3	5.5	4.4

[a] As of March 1.
Source: Reprinted with permission from *Chem. Eng. News* **1991**,
69(28), 36. Copyright 1991 American Chemical Society.

TABLE 1.12 Chemists' Salaries Versus Experience

Median Salary $ thousands[a]	Years Since B.S. degree									Overall Median
	2 to 4	5 to 9	10 to 14	15 to 19	20 to 24	25 to 29	30 to 34	35 to 39	40 or more	
ALL CHEMISTS	$30.1	$38.0	$47.5	$52.0	$58.0	$61.0	$60.7	$64.0	$66.0	$52.0
By degree										
Bachelor's	30.0	35.0	42.0	46.6	50.7	51.5	52.0	60.0	59.0	40.3
Master's	—	37.0	44.9	49.6	52.0	49.4	53.2	55.0	62.5	47.4
Doctor's	—	47.0	51.0	56.0	61.3	65.0	65.5	68.0	69.9	58.0
By employer										
Industry	31.0	40.0	50.0	56.0	62.3	68.3	70.0	73.8	68.0	54.0
Government	—	35.0	40.6	45.7	52.4	55.5	56.3	57.5	—	50.0
Academia	—	31.6	37.0	40.0	43.4	48.1	54.1	54.6	60.7	46.0

[a] As of March 1991.

Source: Reprinted with permission from Chem. Eng. News 1991, 69 (28), 37. Copyright 1991 American Chemical Society.

Salaries of Chemists

Table 1.10 compares Ph.D. chemists' salaries with other professions. In general, chemists have good salaries as compared to other scientists such as biologists and sociologists. They are usually not paid as high as engineers, computer scientists, or physicists. Table 1.11 shows the average salary of chemists working in all areas, industrial and academic, at the B.S., M.S., and Ph.D. levels. Note that Ph.D. salaries are of course substantially higher than M.S., which in turn are higher than B.S. Fair raises were obtained in 1991 over 1990 salaries, similar to the national average percentage increase.

Table 1.12 summarizes in detail the salary ranges at all degree levels and years of experience for nonacademic chemists. Students will particularly note that the median salary for a fresh B.S. chemist is now $30.0 thousand. A starting Ph.D. chemist is making $47.0 thousand. The truly dedicated academic chemist's salary is substantially lower than that of nonacademic chemists. In academia (Table 1.13) for overall experience only full professors at Ph.D. granting universities compare favorably with nonacademic chemists.

Finally, salaries for chemists vary with the work function of individuals (Table 1.14). At the B.S. level salaries are highest by far in management and marketing, lowest in basic research and production.

TABLE 1.13 Academic Chemists' Salaries

Median salary, $ thousands[a]	9-10 month contracts		11-12 month contracts	
	Non-Ph.D. school	Ph.D. school	Non-Ph.D. school	Ph.D. school
Full Professor	$47.2	$60.0	$69.8	$80.0
Associate Professor	36.8	44.4	54.4	53.0
Assistant Professor	31.5	36.0	40.0	42.9

[a] As of March 1, 1991

Source: Reprinted with permission from *Chem. Eng. News* **1991**, 69 (28), 37. Copyright 1991 American Chemical Society.

TABLE 1.14 Chemists' Salaries by Work Function

Median salary, $ thousands	By degree		
	B.S.	M.S.	Ph.D.
R & D management	$60.3	$65.0	$76.1
General management	53.7	62.0	78.0
Marketing	43.2	53.8	64.5
Applied research	42.0	48.4	58.0
Production	38.0	46.0	56.0
Basic Research	36.0	44.7	60.0

Source: Reprinted with permission from *Chem. Eng. News* **1991**, 69 (28), 37. Copyright 1991 American Chemical Society.
[a] As of March 1991.

GENERAL CHARACTERISTICS OF THE CHEMICAL INDUSTRY

Now that we have some idea about the chemical industry let us focus on a few general characteristics of this important industry. Wittcoff and Reuben define nine important traits that summarize some interesting concepts with regard to the industry, which are listed in Table 1.15.

Rapid Growth

The Chemicals and Allied Products division usually grows faster than all other manufacturing . In the United States it grew 9% per year from 1954-1974 and 13% per year from 1970-80. It has grown at a slower rate, 6% per year, during

TABLE 1.15 Characteristics of the Chemical Industry

1.	Rapid growth
2.	High research and development (R & D) expenditures
3.	Intense competition
4.	Capital intensity and economies of scale
5.	Rapid obsolescence of facilities
6.	Freedom of market entry
7.	"Feast or famine"
8.	International trade
9.	Criticality and pervasiveness

Source: W & RI. Reprinted by permission of John Wiley & Sons, Inc.

the 1980s. Economists' opinions are divided on the future growth of the industry. It may be approaching maturity and the rate of growth may be slowing down because of some things that will be with us for some time: increasing pollution abatement costs, stringent worker safety regulations, new FDA (Food and Drug Administration) laws, the Toxic Chemical Substance Act of 1976 (TOSCA), the Resource Conservation and Recovery Act of 1976 establishing OSHA, the Comprehensive Environmental Response, Compensation, and Liability Act of 1980 establishing the "superfund," and the Bhopal, India tragedy. Despite these increasing costs and more regulation, most observers feel that the chemical industry will be healthy for a long time to come.

High R & D Expenditures

The chemical industry is research intensive. The four industrial sectors spending the largest amounts on R & D are aircraft and missiles, 25%; electrical equipment, 17%; chemicals and allied products, 11%; and motor vehicles and related equipment, 11%. All other industries spend only 36% of the 1989

R & D industrial total. Of the total for chemicals about 10% of chemicals and allied products R & D is federally financed, compared to 76% of aircraft and missiles R & D and 44% of electrical and communications equipment R & D.

The chemicals and allied products industry is the large investor in basic research—the planned search for new knowledge without reference to specific commercial objectives. Applied research is the use of existing knowledge for the creation of new products or processes, and development is commericialization of research and improvements to present products or processes.

Some chemical companies spend a very high percentage of sales on R & D expenditures. Almost all of these companies are pharmaceutical companies, that portion of the industry that is highly competitive technically and requires substantial basic research to remain competitive. These types of companies can spend 10-15% of their sales on R & D. Most general large U.S. chemical companies spend a smaller portion of their sales on R & D: du Pont, 3.9%; Dow, 5.0%; Monsanto, 6.9%; American Cyanamid, 8.4%; and Union Carbide, 2.1%. Figure 1.3 shows the dramatic increase in funding of R & D by 15 selected chemical companies in the past ten years, even in constant 1982 dollars that take into consideration the inflationary pressures. But sales have gone up fast, so that R & D spending as a percentage of sales was increasing until 1985 but has remained nearly constant in the late 1980s.

There are two approaches to R & D and to its funding. There can be *technology push,* where a manufacturer discovers a certain technology through basic research and then creates a market for it. Television, sulfonamides, and lasers are products of this approach, for there was no established market for any of these before they were discovered. The second approach to R & D is *demand pull,* which examines a specific market need and then does R & D to solve the technology required to meet this specific need. Hard water detergents, jets, and automobiles with low exhaust emissions are examples of products derived from this mission-oriented approach. Large companies must have both attitudes toward expenditures for R & D to succeed.

The final proof that the chemical industry has high R & D expenditures is given by the fact that this industry is limited to the developed countries, where a highly technical work force exists. Of the 50 largest chemical companies in the world, 16 are in the United States, 11 in Japan, 5 in Germany, 5 in the United Kingdom, 3 in Switzerland, 2 each in France, Netherlands, and Belgium, and 1 each in Italy, Norway, Finland, and Saudi Arabia. Poorer countries must rely on them for goods and services in the chemical industry.

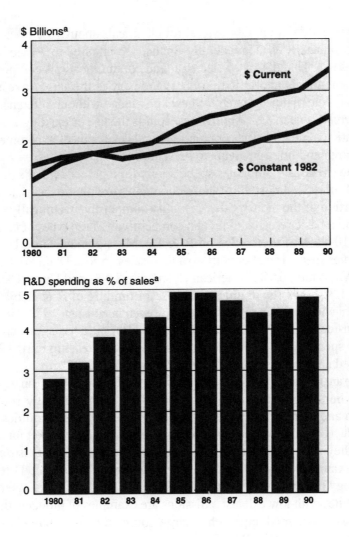

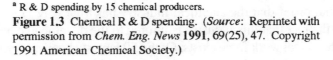

[a] R & D spending by 15 chemical producers.

Figure 1.3 Chemical R & D spending. (*Source*: Reprinted with permission from *Chem. Eng. News* **1991**, 69(25), 47. Copyright 1991 American Chemical Society.)

Intense Competition

There are many reasons why there is intense competition in the chemical industry. First, trade in chemicals is international. As seen in the preceding listing of top chemical companies by country, many developed countries have profiting industries. There is also the tendency of companies to build bigger plants that usually make the manufacturing costs cheaper per unit sold. Larger production volumes then require exporting many products to other countries.

Second, there is much diversity in chemical processes for certain products, inviting intense competition between processes and companies. For example, there are six industrial syntheses of acrylic acid. Different companies specialize in each of these processes and competition is fierce.

The third major reason for competition is the patent system. Although a patent allows a company to have a monopoly on a particular composition or process in reward for its discovery, this exclusive right lasts only 17 years. This may seem like a long time, but this includes many years of research, development, and testing for product safety. The time of testing is becoming longer. Thus many processes can be held exclusively only for a short period of time before they become free game for other companies. There are advantages and disadvantages to this system. Congress has allowed an extension of patent coverage to a longer time when warranted.

Capital Intensity and Economics of Scale

There is a basic rule that applies to production in the chemical industry: Invest huge capital to make a big plant so that there is less overhead and the product can be produced more cheaply on this larger scale. This is the principle of economy of scale. A typical ethylene plant capacity rose from 70 million lb/yr in the 1950s to 2 billion lb/yr in 1991. In 1950 vinyl chloride sold for 14¢/lb and was produced at a rate of 250 million lb/yr. In 1969 it sold for 5¢/lb (in spite of 20 years of inflation) because it was being made at the rate of 3.6 billion lb/yr. In 1950 sulfuric acid, the number one chemical in terms of U.S. production, sold for $20/ton. In 1980 the price was only $40/ton despite many double-digit inflation years in the 1970s. The reason is that the production went from 20 billion lb to 80 billion lb.

The chemical industry has a high investment of current capital for an industry. Other industries invest more but their equipment lasts longer without becoming out-of-date. This high capital investment means that the industry is not so labor intensive. Personpower productivity (sales per employee) is high. Employee salaries are a small percentage (20%) of the cost in the chemical industry as compared to other industries. As a result, labor relations are usually good and pay increases are substantial.

Rapid Obsolence of Facilities

Major R & D advances, along with other factors such as corrosion and patent time limits, make for rapid obsolescence of many processes and plants. An example of this is given by acetic acid. Given below is the synthesis in use during different periods of its history. In the 1930s it was made by a laborious three-step synthesis which also gave a side product, acetic anhydride. In the 1940s

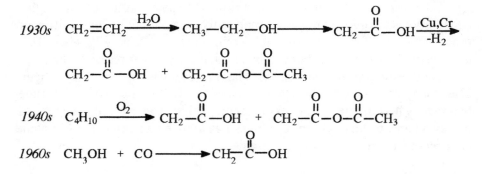

it was made in one step, but methyl ethyl ketone was also formed. In 1969 a one-step synthesis with no side products was achieved. In 1948 94% of the production of sodium carbonate was by the Solvay process. After natural deposits were found in Wyoming, 91% was obtained from this source by 1981. There are many other examples of new processes being so advantageous as to make the rapid switchover very economical.

"Feast or Famine"

The chemical industry is noted for its "ups and downs," its "gluts and shortages," its "feasts and famines." This is due in part to its rule of economy of scale. A bigger than necessary plant is built because it is cheaper in the long run. This floods the market with a product, the price falls, other obsolete plants close, the supply is used up, and the price goes up again. This circle is sometimes continually repeated. Take the case of phthalic anhydride. In 1958 it was 20¢/lb, in 1966 10¢/lb, in 1968 13¢/lb, in 1971 8¢/lb, and in 1974 19¢/lb. Sodium hydroxide was $120/ton in 1970, $180 in 1974, $140 in 1975, $280 in 1982, $150

hydroxide was $120/ton in 1970, $180 in 1974, $140 in 1975, $280 in 1982, $150 in 1984, and $290 in 1990. This is not unusual in the chemical industry, although in recent high-inflation years not many chemicals drop in price anymore. Instead, they vary between large and small increases and are still cyclic.

International Trade

Although some chemicals are transported only with danger and difficulty, many can be transported more easily and cheaply by truck, ship, and pipeline. This ease of transportation creates a large international trade. Table 1.16 shows the U.S. total trade balance as compared to that for chemicals. Although the U.S. total trade balance (exports minus imports) improved two years in a row in 1988 and 1989, is still negative. Oil imports are still the chief culprit in the U.S. trade picture. Were it not for the oil imports the U.S. would have probably enjoyed a trade *surplus*. But other than in agricultural commodities, nowhere is the U.S. export strength more obvious than in chemicals, giving a chemical trade *surplus* of $16.5 billion in 1990. The chemical industry has been setting records for its trade surplus for many years. Figure 1.4 shows this chemical trade balance in graphic form and the percentage of the world chemical exports shared by the United States. Finally, it should be noted that many companies have a large percentage of foreign sales compared to their total sales, sometimes as much as one third, again emphasizing large amounts of international trade.

Criticality and Pervasiveness

The final general characteristic of the chemical industry that we will discuss is its importance in everyday life. It is both critical and pervasive. It is critical to the economy of a developed country. In the first half of this century a nation's industrial development was gauged by its production of sulfuric acid, the chemical with the largest amount of production. It has been called "the grandfather of economic indicators." Lately ethylene, the largest volume organic chemical, is used to judge this development. The chemical industry cannot be replaced by any other industry. If a country does not have one, it must rely on imports. It is critical to the prosperity of a country, as well as pervasive— it is reflected in so many goods and services necessary for modern life as we know it. Finally, many of the problems concerning pollution, energy, and raw materials have been detected and monitored by chemical methods, and chemistry will have an important part to play in their solutions.

TABLE 1.6 U.S. Total Trade Balance

$ Billions	1990	1989	1988	1980
Total exports	$394.0	$364.3	$322.4	$200.6
Total imports	495.0	472.9	440.9	240.8
Trade Balance	-101.0	-108.6	-118.5	-40.2
Chemical exports	$33,983	$36,485	$31,909	$20,740
Chemical imports	22,468	20,752	20,326	8,594
Chemical trade balance	16,515	15,733	11,583	12,146

Source: Reprinted with permission from *Chem.. Eng. News* **1991**, 69(25), 64. Copyright 1991 American Chemical Society.

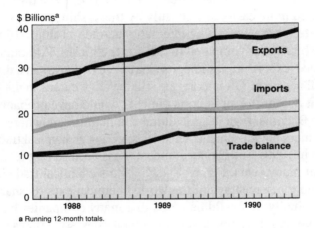

a Running 12-month totals.

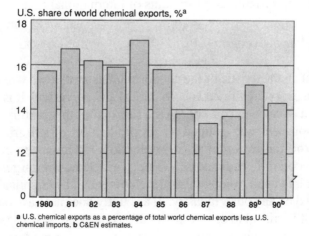

a U.S. chemical exports as a percentage of total world chemical exports less U.S. chemical imports. **b** C&EN estimates.

Figure 1.4 Chemical exports and imports. (*Source:* Reprinted with permission from *Chem. Eng. News* **1991**, 69(25),64. Copyright 1991 American Chemical Society.)

TOP 50 CHEMICALS

As part of an introduction to the chemical industry it is appropriate that we become acquainted with important chemicals, polymers, and chemical companies. *Chemical and Engineering News* publishes yearly lists of the top 50 chemicals and top 100 companies. The top 50 chemicals are ranked (Table 1.17, pp. 23-24) in terms of billions of pounds of chemicals produced in the U.S. for a given year. Sulfuric acid is number 1 by far, with a volume of over 88 billion lb produced yearly. It is almost twice the amount of number 2, nitrogen, which is produced at more the 57 billion lb per year. The highest volume organic chemical is ethylene, the basic petrochemical used to synthesize so many other important organic chemicals. It is the leader of the basic seven organics—ethylene, propylene, butadiene, benzene, toluene, xylene, and methane—from which all other important organic chemicals are derived. Methane does not itself appear in the list because it is not synthesized by a chemical process. However, it is the major constituent in natural gas and is used to make many other chemicals.

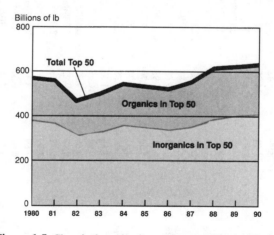

Figure 1.5 Chemical production. (*Source:* Reprinted with permission from *Chem. Eng. News* **1991**, 69(25),16. Copyright 1991 American Chemical Society.)

There are 21 inorganics making up the list, with a total volume of 409 billion lb. There are 29 organics with a substantially smaller production total of 217 billion lb. But the organics have much higher prices, so their overall commercial

TABLE 1.17 Top 50 Chemicals

Rank 1990	Rank 1989[a]		Billions of lb 1990	Billions of lb 1989	Common units[b] 1990	Common units[b] 1989	Avg growth 1989-90	Avg growth 1988-89	Avg growth 1985-90	Avg growth 1980-90
1	1	Sulfuric acid	88.56	86.60	44,281tt	43,301tt	2.3%	2.8%	2.1%	0%
2	2	Nitrogen	57.32	53.91	791 bcf	744bcf	6.3	3.0	3.8	5.1
3	3	Oxygen	38.99	37.42	471 bcf	452bcf	4.2	0	3.7	0.9
4	4	Ethylene	37.48	34.95	37,480 mp	34,953mp	7.2	-6.0	4.7	2.7
5	5	Lime[c]	34.80	34.36	17,400tt	17,178tt	1.3	0.7	1.9	-0.9
6	6	Ammonia	33.92	32.72	16,958tt	16,362tt	3.6	-2.7	-0.4	-1.5
7	7	Phosphoric acid	24.35	23.47	12,175tt	11,735tt	3.7	0.5	3.0	1.2
8	9	Sodium hydroxide	23.38	20.98	11,688tt	10,492tt	11.4	-0.3	1.4	0.1
9	10	Propylene	22.12	20.23	22,117mp	20,229mp	9.3	-4.7	8.2	4.9
10	8	Chlorine	21.88	22.83	10,942tt	11,413tt	-4.1	1.4	1.0	-0.4
11	11	Sodium carbonate[d]	19.85	19.83	9,925tt	9,915tt	0.1	2.9	3.1	1.8
12	13	Urea[e]	15.81	15.93	7,905tt	7,963tt	-0.7	0.6	3.4	0.1
13	12	Nitric acid	15.50	16.70	7,749tt	8,349tt	-7.2	4.5	1.0	-1.7
14	14	Ammonium nitrate[f]	14.21	15.74	7,107tt	7,871tt	-9.7	4.9	1.0	-2.5
15	15	Ethylene dichloride	13.30	13.68	13,301 mp	13,675mp	-2.7	5.0	1.9	1.8
16	16	Benzene	11.86	11.67	1,610mg	1,585mg	1.6	-0.5	4.8	-2.2
17	17	Carbon dioxide[g]	10.98	10.68	5,491tt	5,339tt	2.8	5.6	1.9	6.2
18	18	Vinyl chloride	10.65	9.62	10,652mp	9,618mp	10.8	6.2	2.4	5.1
19	19	Ethylbenzene	8.99	9.22	8,988mp	9,223mp	-2.5	-7.1	4.0	1.6
20	21	Styrene	8.02	8.13	8,018mp	8,129mp	-1.4	-9.5	1.0	1.6
21	22	Methanol	7.99	7.14	7,987mp	7,139mp	11.9	-12.3	9.8	1.1
22	20	Terephthalic acid[h]	7.69	8.31	7,692mp	8,309mp	-7.4	-18.8	3.5	2.4
23	23	Formaldehyde[i]	6.41	6.37	6,413mp	6,370mp	0.7	1.4	2.7	1.4
24	30	Methyl tert-butyl ether[j]	6.30	4.98	6,302mp	4,976mp	26.6	-12.4	27.2	na
25	25	Toluene[k]	6.10	5.84	841mg	805mg	4.5	-8.0	3.8	-1.9
26	26	Xylene	5.70	5.80	791 mg	805 mg	-1.7	5.8	1.4	-1.4
27	29	Ethylene oxide	5.58	5.32	5,581mp	5,322mp	4.9	-10.6	0.6	0.7
28	28	p-Xylene	5.20	5.49	5,201 mp	5,494mp	-5.3	-1.9	1.7	2.1
29	27	Ethylene glycol	5.03	5.50	5,028 mp	5,499 mp	-8.6	-0.3	3.8	1.4
30	31	Ammonium sulfate	4.99	4.69	2,495tt	2,347tt	6.3	0.6	3.6	1.6

TABLE 1.17 (continued)

Rank 1990	Rank 1989[a]		Billions of lb 1990	1989	Common units[b] 1990	1989	Average annual growth 1989-90	1988-89	1985-90	1980-90
31	24	Hydrochloric acid	4.68	6.35	2,341tt	3,177tt	-26.3	20.3	-3.5	-2.1
32	32	Cumene	4.31	4.54	4,312mp	4,535mp	-4.9	1.8	5.2	2.2
33	34	Acetic acid	3.76	3.83	3,756mp	3,826mp	-1.8	21.1	5.3	2.4
34	35	Potash[l]	3.62	3.52	1,640mt	1,595tmt	2.8	4.9	4.8	-3.1
35	33	Phenol[m]	3.51	3.89	3,512mp	3,893mp	-9.8	9.3	4.8	3.2
36	36	Propylene oxide	3.20	3.20	3,200 mp	3,200 mp	0	2.9	5.9	6.1
37	37	Butadiene[n]	3.16	3.09	3,156mp	3,094mp	2.0	-2.4	6.2	1.2
38	39	Acrylonitrile	3.03	2.61	3,029mp	2,608mp	16.1	0	5.2	5.2
39	38	Carbon black	2.87	2.91	2,868mp	2,913mp	-1.5	-0.1	2.2	1.2
40	42	Vinyl acetate	2.55	2.47	2,546 mp	2,470mp	3.1	-3.6	3.8	2.8
41	43	Cyclohexane	2.47	2.39	2,468 mp	2,389 mp	3.3	4.0	8.3	2.3
42	41	Aluminum Sulfate	2.42	2.49	1,208tt	1,244tt	-2.9	1.0	-1.0	-0.6
43	40	Acetone	2.22	2.50	2,221mp	2,497mp	-11.1	8.4	4.4	0.7
44	44	Titanium dioxide	2.16	2.20	1,079tt	1,101 tt	-2.0	7.7	4.6	4.2
45	46	Sodium silicate	1.76	1.67	878tt	834tt	5.3	2.7	4.9	0.9
46	47	Adipic acid	1.64	1.64	1,640mp	1,640mp	0	2.5	2.5	1.1
47	48	Sodium sulfate[o]	1.47	1.51	733tt	755tt	-2.9	-7.6	-2.4	-4.3
48	45	Calcium chloride[p]	1.38	1.92	690tt	962tt	-28.3	32.0	-2.9	-3.5
48	49	Isopropyl alcohol	1.38	1.43	1,380mp	1,432mp	-3.6	3.1	2.2	-2.8
48	50	Caprolactam	1.38	1.31	1,380mp	1,312mp	5.2	4.0	5.0	4.3
		TOTAL ORGANICS	216.82	211.07			2.7%	-3.0%	4.6%	2.1%
		TOTAL INORGANICS	409.07	402.51			1.6%	2.0%	2.1%	0.5%
		GRAND TOTAL	625.89	613.58			2.0%	0.2%	2.9%	1.0%

a Revised. b tt = thousands of tons, bcf = billions of cubic feet, mp = millions of gallons, mg = millions of pounds, mp = millions of pounds, tmt = thousands of metric tons. c Except refractory dolomite. d Natural and synthetic. e 100% basis. f Original solution. g liquid and solid only. h Includes both acid and ester without double counting. i 37% by weight. j Production data for earlier years unavailable. k All grades. K_2O basis. m Synthetic only. n Rubber grade. o High and low purity. p Solid and liquid. na = not available.

Source: Reprinted with permission from *Chem. Eng. News* 1991, 69(14),14. Copyright 1991 American Chemical Society.

25

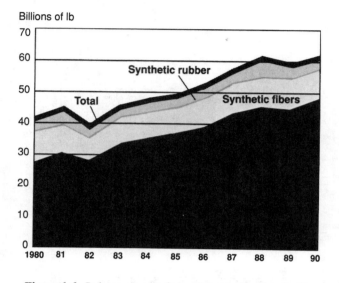

Billions of lb

Figure 1.6 Polymer production. (*Source*: Reprinted with permission from *Chem. Eng. News* **1991**, 69(25), 19. Copyright 1991 American Chemical Society.)

value is high in terms of dollar amounts. The total of 626 billion lb of chemicals represented by the top 50 is approximately one half the weight of all chemicals and polymers produced yearly. Although there are thousands of chemicals commercially produced, much of the approximately 1 trillion lb is centered in these top 50.

An examination of recent trends in production is possible using the average annual change columns of Table 1.17 as well as Fig. 1.5, page 23. The 1980, '82, and '86 recessions required most chemicals to decrease in production for the first time ever, especially compared to the excellent year 1979. However, the rest of the 1980s showed good growth. Overall for the top 50 the average annual growth was 1% for the entire decade, but chemicals increased in production by 2.9% annually for 1985-90. Only 5 chemicals showed a negative average annual change for 1985-90, whereas 14 chemicals showed this for the ten-year period. Note that the last two available yearly changes demonstrate the ups and downs of recent years, with a 2.0% increase in production from 1989-90 but only a 0.2% increase from 1988-89.

TABLE 1.18 Top Polymer Production

	Billions of lb		Common Units					Average annual growth			
	1990	1989	1990	1989	1988	1985	1980	1989-90	1988-89	1985-90	1980-90
PLASTICS (millions of lb)											
Thermosetting resins	6.36	6.41	6,364	6,405	6,588	5,631	4,093	-0.6%	-2.8%	2.5%	4.5%
Phenol and other tar acid resins	2.95	2.88	2,946	2,879	3,066	2,621	1,499	2.3	-6.1	2.4	7.0
Urea resins	1.50	1.48	1,496	1,476	1,425	1,210	1,165	1.4	3.6	4.3	2.5
Polyesters (unsaturated)	1.22	1.32	1,221	1,319	1,404	1,223	947	-7.4	-6.1	0	2.6
Epoxies (unmodified)	0.50	0.51	499	509	486	385	315	-2.0	4.7	5.3	4.7
Melamine resins	0.20	0.22	202	222	207	192	167	-9.0	7.2	1.0	1.9
Thermoplastic resins	41.93	38.62	41,928	38,617	39,608	31,525	24,335	8.6%	-2.5%	5.9%	5.6%
Low-density polyethylene	11.18	9.70	11,176	9,695	10,397	8,889	7,291	15.3	-6.8	4.7	4.4
PVC and copolymers	9.09	8.48	9,088	8,478	8,350	6,772	5,470	7.2	1.5	6.1	5.2
High-density polyethylene	8.33	8.10	8,334	8,102	8,400	6,671	4,405	2.9	-3.5	4.6	6.6
Polystyrene[a]	5.01	5.10	5,012	5,104	5,187	4,054	3,521	-1.8	-1.6	4.3	3.6
Polypropylene	8.32	7.24	8,318	7,238	7,274	5,139	3,648	14.9	-0.5	10.1	8.6
TOTAL[b]	48.29	45.02	48,292	45,022	46,196	37,156	28,428	7.3%	-2.5%	5.4%	5.4%
SYNTHETIC FIBERS (millions of lb)											
Cellulosics	0.50	0.58	502	580	614	558	806	-13.4%	-5.5%	-2.1%	-4.6%
Rayon[c]	0.30	0.36	296	363	400	353	490	-18.5	-9.3	-3.5	-4.9
Acetate[d]	0.21	0.22	206	217	214	205	316	-5.1	1.4	0.1	-4.2
Noncellulosics[b]	8.18	8.52	8,175	8,515	8,526	7,564	7,874	-4.0%	-0.1%	1.6%	0.4%
Polyester	3.19	3.59	3,193	3,594	3,680	3,341	3,989	-11.2	-2.3	-0.9	-2.2
Nylon[e]	2.66	2.74	2,661	2,740	2,670	2,343	2,358	-2.9	2.6	2.6	1.2
Olefin[f]	1.82	1.64	1,815	1,639	1,588	1,249	748	10.7	3.2	7.8	9.3
Acrylic[g]	0.51	0.54	506	542	588	631	779	-6.6	-7.8	-4.3	-4.2
TOTAL[b]	8.68	9.10	8,677	9,095	9,140	8,122	8,680	-4.6%	-0.5%	1.3%	0%

27

TABLE 1.18 Top Polymer Production (continued)

SYNTHETIC RUBBER (thousands of metric tons)

Styrene-butadiene[h]	1.88	1.93	853	874	909	735	1,074	-2.4	-3.9	3.0	-2.3
Polybutadiene	0.89	0.91	403	411	407	330	311	-1.9	1.0	4.1	2.6
Ethylene-propylene	0.56	0.57	256	260	263	215	144	-1.5	-1.1	3.6	5.9
Nitrile	0.12	0.15	56	69	76	53	63	-18.8	-9.2	1.1	-1.2
Other[i]	1.20	1.43	546	648	679	505	423	-15.7	-4.6	1.6	2.6
TOTAL[b]	**4.66**	**4.99**	**2,114**	**2,262**	**2,334**	**1,838**	**2,015**	**-6.5%**	**-3.1%**	**2.8%**	**0.5%**

[a] No longer includes acrylonitrile-butadiene-styrene, or styrene acrylonitrile resins; historical data are restated. [b] Totals are for those products listed and may not add because of rounding. [c] Based on C&EN estimates. [d] Includes diacetate and triacetate yarn, but does not include cigarette tow. Beginning with 1985, includes rayon yarn. [e] Excludes aramid after 1982. [f] Includes olefin film, olefin fiber, polypropylene, and vinyon. Includes modacrylic. [h] Excludes high-styrene latex. [i] Beginning with 1985, includes neoprene; historical data are restated. Also includes butyl; polyisoprene; chlorosulfonated polyethylene; polyisobutylene; and acrylo, fluoro, and silicone elastomers.

Source : Society of the Plastics Industry, Fiber Economics Bureau, Rubber Manufacturers Association. Reprinted with permission from *Chem. Eng. News* 1991, 69(14),21. Copyright 1991 American Chemical Society.

TOP POLYMERS

The general conclusion that 1979 was an excellent year, 1980 and 1982 were bad years, and most of the 1980s were good years until 1989 is further exemplified in polymer production (Table 1.18, pages 27-28 and Figure 1.6). The five year period of 1985-90 was healthy. In 1989 all three main types of polymers were down in production. In 1990 plastics bounced back with a big 7.3% increase, but fibers and rubber decreased badly.

You should be somewhat familiar with general production totals. More plastics are made each year, about 48 billion lb, than fibers or rubber. The largest volume plastic is polyethylene, with combined low- and high-density types amounting to about 19 billion lb. The largest fiber market is polyester at 3.2 billion lb; the largest synthetic rubber is styrene-butadiene at 1.9 billion lb.

TOP 100 U.S. COMPANIES

The ranking for the top 100 U.S. companies (Table 1.19, pages 30-32) is based on the amount of chemical sales in millions of dollars, not total sales for the entire company. Notice that all the companies have nonchemical activities, some to much more of an extent than others. This is especially true of the oil companies in the top 100 (there are 13), which have very low chemical sales as a percentage of total sales. Note also that a number of companies are of foreign origin but have substantial U.S. operations, including Shell, BASF, Ciba-Geigy, Rhone-Poulenc, and Bayer.

The year 1990 was a record for these companies. For 1990 sales of chemicals for the 100 companies rose 5% from 1989, reaching a total of $185 billion. This is approximately 70% of the government's estimate of shipments for Chemicals and Allied Products for 1990, which is $266 billion. Of the 100 companies, however, 36 recorded sales declines for 1990. Fig. 1.7, page 35, shows that the 50 largest companies provide 86% of the sales for the top 100 in 1990. This is higher than the 79% for the big firms in 1985. Their chemical sales range from $15 billion to $1 billion. The second 50 only provide 14% of the top 100 sales. Their range is from $1 billion to $120 million. In terms of company type, chemical firms (those with large chemical sales as a percentage of total sales) account for 41% of the sales, diversified companies 20%, oil companies 21%, and foreign owned companies 18%. The percentage of foreign-owned companies increased from 8 to 18% between 1985 and 1990. As in the past, Du Pont is number one in chemical sales followed by Dow Chemical, but the gap between the two seems to be narrowing.

TABLE 1.19 Top 100 U.S. Chemical Companies

Rank 1990	1989		Chemical sales 1990 ($ milion)	Change from 1989	Chemical sales as % of total sales	Industry classification
1	1	Du Pont	$15,571	2.1%	38.9%	Diversified
2	2	Dow Chemical	14,690	3.6	74.3	Basic chemicals
3	3	Exxon	11,153	5.6	9.6	Petroleum
4	4	Union Carbide	7,621	-0.4	100.0	Basic chemicals
5	5	Monsanto	5,711	-1.2	63.5	Basic chemicals
6	6	Hoechst Celanese	5,499	-2.8	93.5	Basic chemicals
7	8	General Electric	5,167	4.8	8.8	Diversified
8	7	Occidental Petroleum	5,040	-2.4	23.2	Petroleum
9	9	BASF	4,366	-2.1	81.1	Basic chemicals
10	10	Amoco	4,087	-4.4	12.9	Petroleum
11	11	Mobil	4,084	1.1	7.7	Petroleum
12	12	Shell Oil	3,718	-3.0	15.2	Petroleum
13	13	Eastman Kodak	3,588	1.9	19.0	Photo equipment
14	15	W. R. Grace	3,570	10.8	52.9	Specialty chemicals
15	14	Chevron	3,325	9.1	8.0	Petroleum
16	17	Arco Chemical	2,830	6.3	100.0	Basic chemicals
17	18	Rohm & Haas	2,824	6.1	100.0	Basic chemicals
18	16	Allied-Signal	2,786	-6.9	22.6	Diversified
19	19	Air Products	2,614	5.3	90.3	Basic chemicals
20	21	Bayer USA	2,380	3.6	40.0	Diversified chemicals
21	32	Rhône-Poulenc	2,278	42.4	100.0	Specialty chemicals
22	22	American Cyanamld	2,255	0.6	49.3	Basic chemicals
23	23	Ashland Oil	2,245	0.7	25.0	Petroleum
24	30	Hercules	2,193	3.6	68.5	Basic chemicals
25	20	Phillips Petroleum	2,120	-13.4	15.6	Petroleum
26	42	Ciba-Geigy	1,976	10.0	52.0	Specialty chemicals
27	24	Quantum Chemical	1,959	-6.9	73.8	Basic chemicals
28	25	B. F. Goodrlch	1,940	-1.9	79.7	Basic chemicals
29	31	Akzo	1,893	1.0	100.0	Basic chemicals
30	26	Lyondell Petrochemical	1,889	19.6	29.1	Petrolcum products
31	29	Alcoa	1,842	5.7	17.2	Nonferrous metals
32	27	Texaco	1,723	-5.2	4.1	Petroleum
33	33	Dow Corning	1,718	9.1	100.0	Specialty chemicals
34	34	National Starch	1,670	6.4	100.0	Specialty chemicals
35	35	FMC	1,624	5.4	43.6	Machinery

TABLE 1.19 (continued)

Rank 1990	1989		Chemical sales 1990 ($ milion)	Change from 1989	Chemical sales as % of total sales	Industry classification
36	36	Ethyl	1,591	4.7	100.0	Basic chemicals
37	28	±Atochem	1,540	-13.4	100.0	Basic chemicals
38	56	Degussa	1,435	61.9	75.3	Specialty chemicals
39	45	Lubrizol	1,335	18.8	91.9	Specialty chemicals
40	42	Huntsman Chemical	1,300	8.3	100.0	Basic chemicals
41	40	Tenneco	1,298	6.0	8.9	Diversified
42	38	Olln	1,269	-1.2	49.0	Basic chemicals
43	39	Unocal	1,236	-3.1	10.5	Petroleum
44	48	Naico Chemical	1,212	13.2	100.0	Specialty chemicals
45	44	PPG Industries	1,150	-2.1	19.1	Glass products
46	52	Cabot	1,107	16.3	66.2	Basic chemicals
47	41	IMC Fertilizer	1,106	-9.5	100.0	Agrochemicals
48	61	Great Lakes Chemical	1,066	34.6	100.0	Specialty chemicals
49	67	Imcera	1,025	54.7	71.9	Specialty chemicals
50	54	Morton International	1,021	9.2	62.3	Specialty chemicals
51	—	Relchhold Chemicals	$1000	na	100.0%	Specialty chemicals
52	50	Solvay America	997	0%	100.0	Basic chemicals
53	51	Aristech	974	-2.2	100.0	Basic chemicals
54	55	Farmland Industries	965	9.2	28.6	Agricultural supplies
55	57	International Flavors	963	10.7	100.0	Specialty chemicals
56	53	Terra Chemicals	961	1.3	100.0	Agrochemicals
57	47	Georgia Gulf	932	-15.6	100.0	Basic chemicals
58	49	NL Industries	907	-9.4	100.0	Basic chemicals
59	58	CF Industries	903	5.5	100.0	Agrochemicals
60	46	Freeport-McMoRan	825	-25.3	52.2	Agrochemicals
61	66	American Petrofina	813	22.8	21.1	Petroleum
62	60	Uniroyal Chemical	808	0.4	100.0	Specialty chemicals
63	59	Engelhard	807	-0.7	27.4	Diversified
64	65	H. B. Fuller	792	5.2	100.0	Specialty chemicals
65	64	Witco	769	0.4	47.1	Specialty chemicals
66	62	Vista Chemicals	719	-7.7	100.0	Basic chemicals
67	69	Kerr-McGee	642	8.4	17.4	Petroleum
68	72	Betz Laboratories	597	15.5	100.0	Specialty chemicals
69	79	Loctite	555	17.2	100.0	Specialty chemicals
70	76	Merck	551	11.9	7.2	Drugs
71	74	Pfizer	548	7.5	8.5	Drugs
72	77	Ferro	513	5.5	45.6	Specialty materials
73	70	Sterling Chemical	506	-12.9	100.0	Basic chemicals
74	75	Hoffmann-La Roche	500	0	20.8	Drugs
75	63	Nova	489	-27.9	100.0	Basic chemicals

TABLE 1.19 (continued)

Rank 1990 1989		Chemical sales 1990 ($ milion)	Change from 1989	Chemical sales as % of total sales	Industry classification
76 73	Sun Co.	486	-4.6	3.7	Petroleum
77 68	GAF	481	-19.8	52.2	Specialty chemicals
78 —	Hüls	465	3.6	100.0	Specialty chemicals
79 83	Union Camp	460	12.8	16.2	Diversified
80 84	Sigma-Aldrich	440	22.6	83.2	Specialty chemicals
81 81	Dexter	438	-0.1	48.3	Specialty chemicals
82 78	Borden Chemicals	421	-9.7	100.0	Basic chemicals
83 82	Vulcan Materials	409	-4.9	37.0	Nonmetallic minerals
84 85	Stepan	390	12.5	100.0	Specialty chemicals
85 87	Crompton & Knowles	311	10.7	79.7	Specialty chemicals
86 86	First Mississippi	295	-4.1	59.3	Agrochemicals
87 90	Calgon Carbon	285	12.4	100.0	Basic chemicals
88 89	Petrolite	283	2.9	94.2	Specialty chemicals
89 88	Chemed	275	-0.9	45.9	Specialty chemicals
90 92	Mississippi Chemical	253	21.1	100.0	Farm cooperative
91 90	Georgia-Pacific	247	-2.4	2.0	Wood products
92 94	Quaker Chemical	201	10.9	100.0	Specialty chemicals
93 —	Sequa Corp.	201	42.4	10.8	Diversified
94 93	Sherex	177	-5.9	100.0	Specialty chemicals
95 98	Lawter International	150	10.3	100.0	Specialty chemicals
96 95	LeaRonal	141	-10.5	100.0	Specialty chemicals
97 99	Cambrex Corp.	134	2.0	100.0	Specialty chemicals
98 97	Eagle-Picher	123	-19.4	17.5	Diversified
99 —	Aceto	122	8.1	100.0	Specialty chemicals
100—	Sybron	120	na	100.0	Specialty chemicals

Source: Reprinted with permission from *Chem. Eng. News* **1991**, 69(18), 12-13. Copyright 1991 American Chemical Society.

TABLE 1.20 Top 50 World Chemical Companies

Rank 1990	Rank 1989		Chemical net sales ($ millions)	Chemical sales as % of total
1	1	BASF (Germany)	$18,520	64.4%
2	3	Hoechst (Germany)	17,804	64.3
3	2	ICI (U.K.)	17,515	76.0
4	5	Bayer (Germany)	16,312	63.5
5	4	Du Pont (U.S.)	15,571	38.9
6	6	Dow Chemical (U.S.)	14,690	74.3
7	8	EniChem (Italy)	12,571	100.0
8	7	Shell (U.K., Netherlands)	12,188	11.5
9	9	Exxon (U.S.)	11,153	9.6
10	10	Rhône-Poulenc (France)	10,122	70.0
11	16	Atochem (France)	9,690	30.0
12	13	Ciba-Geigy (Switzerland)	8,150	57.5
13	11	Union Carbide (U.S.)	7,621	100.0
14	15	Asahi Chemical (Japan)	6,699	100.0
15	24	Hüls (Germany)	6,300	100.0
16	19	Akzo (Netherlands)	5,832	61.2
17	17	Monsanto (U.S.)	5,711	63.5
18	25	DSM (Netherlands)	5,580	96.3
19	18	British Petroleum (U.K.)	5,448	6.7
20	20	Mitsubishi Kasei (Japan)	5,321	100.0
21	14	Solvay (Belgium)	5,292	69.3
22	23	General Electric (U.S.)	5,167	8.8
23	21	Occidental Petroleum (U.S.)	5,040	23.2
24	26	Sumitomo Chemical (Japan)	4,950	100.0
25	22	Norsk Hydro (Norway)	4,924	50.6
26	31	Seklsui Chemical (Japan)	4,144	100.0
27	27	Amoco (U.S.)	4,087	12.9
28	30	Mobil (U.S.)	4,084	7.7
29	32	Toray Industries (Japan)	4,040	100.0
30	33	Showa Denko (Japan)	3,934	100.0
31	28	Takeda Chemical (Japan)	3,833	100.0
32	34	Saudi Basic Industries (Saudi Arabia)	3,626	100.0
33	36	Eastman Kodak (U.S.)	3,588	19.0
34	40	W. R. Grace (U.S.)	3.570	52.9
35	37	Dainippon Ink & Chemicals (Japan)	3,489	100.0

TABLE 1.20 Top 50 World Chemical Companies

Rank 1990 1989		Chemical net sales ($ millions)	Chemical sales as % of total
36 39	Roche (Switzerland)	3,438	49.4
37 38	Chevron (U.S.)	3,325	8.0
38 —	Petrofina (Belgium)	3,259	18.9
39 43	Unilever (U.K., Netherlands)	3,089	7.8
40 42	Mitsul Toatsu Cheilcals (Japan)	3,039	100.0
41 44	Mitsubishi Petrochemical (Japan)	3,000	100.0
42 12	Degussa (Germany)	2,849	33.1
43 45	Arco Chemlcal (U.S.)	2,830	100.0
44 46	Rohm & Haas (U.S.)	2,824	100.0
45 41	Alllied-Signal (U.S.)	2,786	22.6
46 47	Air Products (U.S.)	2.614	90.3
47 49	Kemira (Finland)	2,497	83.8
48 —	Sandoz (Switzerland)	2,465	27.7
49 —	Courtaulds (U.K.)	2,288	67.0
50 —	Teljin (Japan)	2,279	100.0

Source: Reprinted with permission from *Chem. Eng. News 1991,* 69(31), 10. Copyright American Chemical Society.

TOP 50 WORLD CHEMICAL COMPANIES

Table 1.20, page. 33-34, is a list of worldwide companies ranked by chemical sales. The list includes 12 countries. Three of the top 4 are the large German companies BASF, Hoechst, and Bayer. DuPont, Dow Chemical, and Exxon from the U.S. are in the top 10. The U.S. dominates the list with a total of 16 companies. Japan has 11, Germany 5, and the United Kingdom 5.

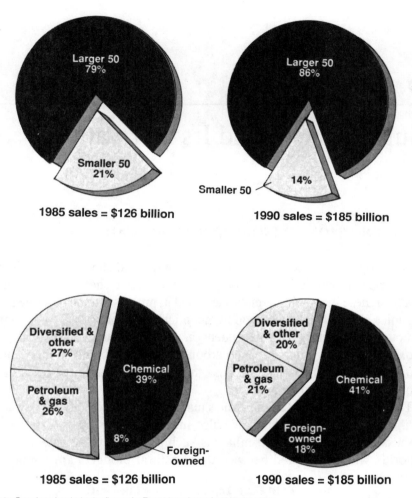

Figure 1.7 Types of companies in chemical sales. (*Source*: Reprinted with permission from *Chem. Eng. News* **1991**, 69(25), 11. Copyright 1991 American Chemical Society.)

2

Sulfuric Acid and Its Derivatives

INTRODUCTION TO INORGANIC CHEMICALS

It is appropriate that we begin our study of industrial chemicals with important inorganic compounds and then progress into organic chemicals and polymers. Many of these inorganic chemicals are used in processes to be described later for organics. Usually 21 of the top 50 chemicals are considered to be inorganic, although the exact figure is dependent on what you count. For instance, carbon dioxide, sodium carbonate, and carbon black are counted as inorganic even though they contain carbon, because their chemistry and uses resemble other inorganics more than organics.

Table 2.1 lists the top 21 inorganics made in the United States. They are listed in the order to be discussed. We also include various other materials in our discussion. Some important minerals such as sulfur, phosphate, and sodium chloride will be covered because these natural products are important raw

TABLE 2.1 Top 21 Inorganic Chemicals

Sulfuric acid derivatives	Sulfuric acid Phosphoric acid Aluminum sulfate	Limestone derivatives	Lime Sodium carbonate Calcium chloride Sodium silicate
Industrial gases	Nitrogen Oxygen Carbon dioxide	Sodium chloride derivatives	Sodium hydroxide Chlorine Hydrochloric acid
Inorganic nitrogen compounds	Ammonia Nitric Acid Ammonium nitrate Ammonium sulfate	Miscellaneous	Titanium dioxide Potash Carbon black Sodium sulfate

materials for inorganic chemical production. They are not strictly speaking chemicals because they are not made synthetically by a chemical reaction, although they are purified with some interesting chemistry taking place. Hydrogen will also be considered because it is used in the manufacture of ammonia and is coproduced with carbon dioxide in the steam-reforming of hydrocarbons.

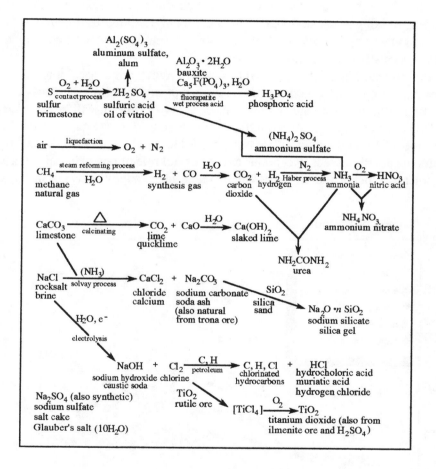

Important inorganic chemicals and their manufacture from earth, air, fire, and water. Air yields oxygen and nitrogen. Earth yields sulfur, bauxite, fluorapatite, natural gas, rock salt, sand, petroleum, rutile, ilmenite, some salt cake, and some trona ore. Fire is used in many of these processes. Water is used in many of these processes.

Figure 2.1 Manufacture of important inorganic chemicals. (*Source:* P. J. Chenier, *J. Chem. Educ. 1983, 60, 141.*)

Finally, urea is covered with inorganic nitrogen compounds because it is made from two "inorganics," ammonia and carbon dioxide.

The order of treatment of these chemicals is difficult to decide. Should it be alphabetical, according to the amount produced, according to important uses, etc.? We have chosen here an order that is dependent on raw material, which is summarized in Fig. 2.1. The most important, largest volume, basic (sometimes called heavy) chemicals from each important raw material are discussed first, followed by some of the derivatives for this chemical which also appear in the top 50. Although the uses of each chemical will be summarized, much of this discussion will be deferred until later chapters on selected specific technologies. Minor derivatives will not be considered.

Referring to Fig. 2.1, we proceed from left to right by first discussing sulfur's conversion into sulfuric acid, followed by some of sulfuric acid's derivatives, for example, aluminum sulfate and phosphoric acid. At times it will be necessary to delay covering a derivative until the other important starting material is described. Thus ammoniun sulfate is mentioned later, after both sulfuric acid and ammonia are discussed. Exceptions to the general rule of raw materials→basic chemicals→chemical derivatives will be made where appropriate. For instance,

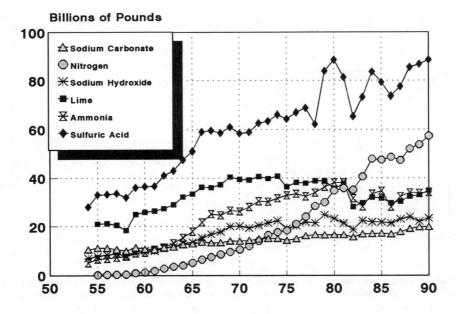

Figure 2.2 U.S. production of selected inorganic chemicals. *(Source:* L & M and *C & E News.* Reprinted by permission of John Wiley & Sons, Inc.)

the four industrial gases will be covered together even though nitrogen and oxygen have different sources as compared to carbon dioxide and hydrogen. After considering the inorganic nitrogen chemicals derived from ammonia we will continue with chemicals derived from limestone, and finally those made from sodium chloride. Note that all these chemicals are eventually made trom the original four basic "elements" of the Ionian Greeks dating from 500 B.C.: earth, air, fire, and water. Admittedly the "earth" element is now known to be quite complex.

Fig. 2.2 gives the U.S. production in billions of lb of one inorganic chemical from each of the main raw materials given in Fig. 2.1. This gives us some feel for the relative importance of these chemicals. Sulfuric acid, being the number one ranked chemical, has always had a large production compared to all other chemicals, even going back to the 1950s. Nitrogen has had a tremendous increase in production compared to most other chemicals, especially in the 1970s and '80s. It is now ranked number two mainly because of its increased use in enhanced oil recovery. Sodium hydroxide and ammonia have shown slow steady increases through the years. Lime has decreased in the 1970s and '80s with the suffering steel market.

The topics covered for each chemical will vary with their importance. The student should attempt to become familar at least with the reaction used in the chemical's manufacture and each chemical's important uses. Details of the large-scale manufacturing process and economic trends for selected chemicals will also be summarized. History of manufacture, characteristics of raw materials, and environmental or toxicological problems will be mentioned occasionally.

Before we begin this systematic discussion of important chemicals and chemical products, you will note that, although the chemistry is most important, the discussions will include some engineering and marketing concepts. Many readers using this book are probably primarily chemists. It is a good idea to keep in mind that chemists, to be successful in industry, must be able to understand and relate to nonchemists. Chemists must work with engineers and marketing specialists who may have a limited or no background in chemistry. For communication to be possible, chemists must know and appreciate the questions and problems confronting these people in their jobs. In the following sections we have attempted to include enough of these concepts so as to provide the chemist with a working knowledge of these disciplines. One obvious example of an important difference between a chemical reaction in the laboratory and a large-scale industrial process is that many industrial reactions are run via continuous rather than batch processes. White gives a good description of the differences. The batch technique resembles a laboratory scale, including loading the flask, doing the reaction, transferring the product, purifying the product, analyzing the product, and cleaning the equipment. Although many large-scale processes are also batch, with large stainless steel vessels and many safety features, there are disadvantages to this approach. In a continuous process the feed materials are continuously added to the reactor and the product is continuously withdrawn from the vessel. Advantages are eliminating "dead time" between batches, making product at higher rates, controlling the process more easily, and forming a more uniform product.

Keep in mind that many people from a variety of disciplines must be involved in making a process work and developing a successful product. The life cycle of most products includes basic research, applied research, development, scaleup, quality control, cost and profit evaluations, market research, market development, sales, and technical service to make a product grow and mature. Every person involved in this project must know something about the rest of the cycle, in addition to contributing a specific expertise to the cycle.

SULFURIC ACID, H_2SO_4; OIL OF VITRIOL

References

Kent, pp. 130-142
L & M, pp. 786-795
Austin, pp. 320-345
T, pp. 183-200
CP, 9-30-85, 8-22-88, 9-19-88, 9-9-91, 9-16-91
White, pp. 22-25

Raw Material

It is appropriate that we begin our discussion with what is by far the largest volume chemical produced in the United States: sulfuric acid. It is normally manufactured at about twice the amount of any other chemical and is a leading economic indicator of the strength of many industrialized nations. Since about 80% of all sulfuric acid is made by the contact process which involves oxidation of sulfur, we will examine this raw material in detail. The average per capita consumption of sulfur in the United States is a staggering 135 lb/yr.

Elemental sulfur (brimstone) can be obtained by mining with the Frasch method or by oxidation of hydrogen sulfide in the Claus process. Although the percentage of sulfur obtained by mining has decreased recently (76% in 1973, 54% in 1980, and 26% in 1991), the Frasch process is still important. Large deposits of sulfur along the Gulf Coast are released by heating the mineral with hot air and water under pressure (163°C, 250 psi) to make the yellow sulfur molten (mp 119°C) so that it is forced to the surface from a depth of 500-2500 ft. Alternatively, the Claus oxidation is performed on hydrogen sulfide obtained from "sour" natural gas wells or petroleum refineries. The hydrogen sulfide, being acidic, is readily separated from the gas or oil by extraction with potassium carbonate or ethanolamine, acidifying, and heating to release the gas. The hydrogen sulfide is then oxidized with air at 1000°C over a bauxite or alumina catalyst. The reactions taking place are given here. The Claus process is increasing in popularity and accounted for 24% of sulfur in 1973, 46% in 1980, and 74% in 1991.

$$H_2S + K_2CO_3 \longrightarrow H_2CO_3 + K_2S$$
$$HO-CH_2-CH_2-NH_2 + H_2S \longrightarrow HO-CH_2-CH_2-NH_3^+ + HS^-$$
$$H_2S + \tfrac{3}{2}O_2 \longrightarrow SO_2 + H_2O$$
$$SO_2 + 2H_2S \longrightarrow 3S + 2H_2O$$

overall:
$$3H_2S + \tfrac{3}{2}O_2 \longrightarrow 3S + 3H_2O$$
or
$$2H_2S + O_2 \longrightarrow 2S + 2H_2O$$

All but a small amount of sulfur is then used to manufacture sulfuric acid. Sulfur is one of the few materials whose quantity is often expressed in "long tons" (2240 lb) which are different from short tons (2000 lb) or metric tons (2204.6 lb). There is no advantage to this unit. It has simply been used in the industry for years and has resisted change without good reason.

Manufacture

Sulfuric acid has been known for centuries. It was first mentioned in the tenth century; its preparation was first described in the fifteenth century by burning sulfur with potassium nitrate. In 1746 Roebuck in England introduced the "lead chamber process," the name being derived from the type of lead enclosure where the acid was condensed. This process involves oxidation of sulfur to sulfur dioxide by oxygen, further oxidation of sulfur dioxide to sulfur trioxide with nitrogen dioxide, and, finally, hydrolysis of sulfur trioxide. The chemistry is more complex than that shown because a mixture of nitrogen oxides is used (from oxidation of ammonia). Modifications of the process by Gay-Lussac in 1827 and Glover in 1859 to include towers to recover excess nitrogen oxides and to increase the final acid concentration from 65% ("chamber acid") to 78%

("tower acid") made it very economical for many years, until the "contact process" displaced it in the 1940s. There have been no new lead chamber plants built since 1956.

$$S + O_2 \longrightarrow SO_2$$
$$NO + \tfrac{1}{2}O_2 \longrightarrow NO_2$$
$$SO_2 + NO_2 \longrightarrow SO_3 + NO$$
$$SO_3 + H_2O \longrightarrow H_2SO_4$$

overall: $\quad S + \tfrac{3}{2}O_2 + H_2O \longrightarrow H_2SO_4$

The contact process was invented by Phillips in England in 1831 but was not used commercially until many years later. Today 99% of all sulfuric acid is manufactured by this method. It was developed mainly because of the demand for stronger acid. All new contact plants use interpass absorption, also known as double absorption or double catalysis. This process is described in detail in Fig. 2.3.

Reactions:

$$S + O_2 \longrightarrow SO_2$$
$$SO_2 + \tfrac{1}{2}O_2 \rightleftharpoons SO_3$$
$$SO_3 + H_2O \longrightarrow H_2SO_4$$

overall: $\quad S + \tfrac{3}{2}O_2 + H_2O \longrightarrow H_2SO_4$

Description. Sulfur and oxygen are burned to SO_2, (Fig. 2.4, page 44, about 10% SO_2 by volume) at 1000°C and then cooled to 420°C. The SO_2 and O_2 enter the converter (Fig. 2.5), which contains four different chambers of V_2O_5 catalyst. About 60-65% SO_2 is converted to SO_3 in the first layer with a 2-4 sec contact time. It is an exothermic reaction so the gas leaves at 600°C. It is cooled to 400°C and enters the second layer of catalyst. After the third layer about 95-96% of the SO_2 is converted into SO_3, near the limit of conversion unless SO_3 is removed. The mixture is fed to the initial absorption tower, where SO_3 is hydrated to H_2SO_4 with a 0.5-1% rise in acid strength in the tower. The mixture is then reheated to 420°C and enters the fourth layer of catalyst which gives overall a 99.7% conversion of SO_2 to SO_3. It is cooled and then fed to the final absorption tower (Fig. 2.6) and hydrated to H_2SO_4. The final H_2SO_4 concentration is 98-99% (1-2% H_2O). A small amount of this is recycled by adding some water and recirculating into the towers to pick up more SO_3, but most of it goes to product storage.

The V_2O_5 catalyst has been the catalyst of choice since the 1920s. It is absorbed on an inert silicate support. It is not subject to poisoning and has about a 20-year lifetime.

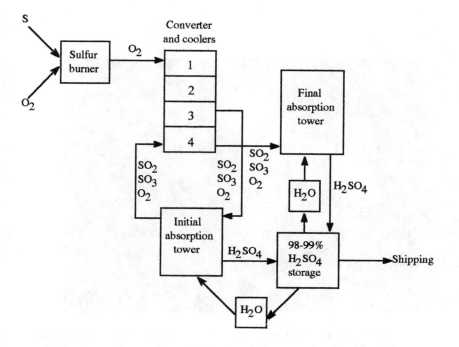

Figure 2.3 Contact process for sulfuric acid manufacture.

As we will see, many industrial processes are successes because the right catalyst was found. Around 70% of all industrial chemical conversions involve a catalyst. Sometimes the catalysis is not understood. In this case it is known that the V_2O_5 catalysis is promoted by the presence of small amounts of alkali metal surfates, usually Na_2SO_4, which react in the presence of SO_3 to give $S_2O_7^=$ in an initial step. This is the source of the oxide ion, $O^=$, which then reduces V^{+5} to V^{+4}. In turn the V^{+4} is reoxidized to V^{+5} by oxygen.

$$\text{initiation} \qquad SO_4^= + SO_3 \longrightarrow S_2O_7^= \text{ (containing } O^=)$$
$$(1) \qquad 2V^{+5} + O^= + SO_2 \longrightarrow SO_3 + 2V^{+4}$$
$$(2) \qquad 2V^{+4} + {}^1\!/_2O_2 \longrightarrow 2V^{+5} + O^=$$
$$\text{overall, (1) + (2)} \qquad SO_2 + {}^1\!/_2O_2 \longrightarrow SO_3$$

This exothermic process enables heat recovery in many places: after the sulfur burner, after the converter pass, and after the absorption towers. The waste heat can be used to generate steam for heating. A plant operating at 10% SO_2 feed and at a conversion rate of 99.7% SO_2 to SO_3 has a stack gas of 350 ppm of SO_2. The equilibrium conversion (theoretical best) is 100 ppm of SO_2. Regulations require that no more than 4 lb of SO_2 come out of the stack for each ton of H_2SO_4 made. This is not an appreciable source of acid rain, primarily caused by

Figure 2.4 A sulfur burner where sulfur and oxygen are burned at high temperatures to make sulfur dioxide. This plant makes both sulfuric acid (300 tons/day) and 20% oleum (400 tons/day). The plant uses some sulfur, but also recycles used sulfuric acid from a refinery, both of which are burned to sulfur dioxide in the same sulfur burner. (Courtesy of Du Pont, LaPorte, TX.)

electrical generating plants burning coal containing sulfur. In fact, the total sulfur emitted from coal-burning power station stacks is more than the total sulfur feed used in sulfuric acid plants. Nevertheless, efforts are continuing to reduce sulfur emissions from acid plants. A low-temperature process is being studied which would make lower SO_2 emission possible.

Although sulfur is the common starting raw material, other sources of SO_2 can be used, including iron, copper, lead, nickel, and zinc sulfides. Hydrogen sulfide, a by-product of natural gas, can be burned to SO_2. Some countries use gypsum, $CaSO_4$, which is cheap and plentiful but needs high temperatures to be converted to SO_2, O_2, and H_2O and the SO_2 recycled to make more H_2SO_4. About 5% of all H_2SO_4 is recycled.

Properties

Anhydrous, 100% sulfuric acid, is a colorless, odorless, heavy, oily liquid, bp 338°C, where it decomposes by losing SO_3 to give 98.3% H_2SO_4. It is soluble in all ratios with water. This dissolution in water is very exothermic. It is

Figure 2.5 Sulfur converters transform sulfur dioxide and oxygen into sulfur trioxide over a vanadium pentoxide catalyst. Note the recycling tubes at various levels that are external to the reactor itself. This is a smaller unit 6.5 ft in diameter and 25 ft high but it converts 20 tons of sulfur into 60 tons sulfuric acid per day. (Courtesy of Georgia-Pacific Corporation, Bellingham, WA.)

corrosive to the skin and is a strong oxidizing and dehydrating agent. Common concentrations and names are battery acid, 33.5% H_2SO_4; chamber or fertilizer acid, 62.18%; tower or Glover acid, 77.67%; and reagent, 98%.

Oleum is also manufactured. This is excess SO_3 dissolved in H_2SO_4, for example, 20% oleum is 20% SO_3 in 80% H_2SO_4 (no H_2O). If water were added to 20% oleum so that the SO_3 and H_2O made H_2SO_4, then 104.5 lb of H_2SO_4 could be made from 100 lb of 20% oleum. This is sometimes called "104.5% H_2SO_4." Other common oleum concentrations are 40% oleum (109% H_2SO_4) and 65% oleum (114.6% H_2SO_4).

Figure 2.6 Adsorption towers convert sulfur trioxide into sulfuric acid by addition of water. This tower is 6 ft x 24 ft. (Courtesy of Georgia-Pacific Corporation, Bellingham, WA.)

Sulfuric acid comes in different grades: technical, which is colored and contains impurities but which can be used to make fertilizer, steel, and bulk chemicals; certified pure (CP); and U.S. Pharmacopeia (USP), the last two used for making batteries, rayon, dyes, and drugs. Rubber or lead-lined containers can be used for dilute acid; iron, steel, or glass can be used for concentrated acid. Shipments require a white DOT label.

Economics

Fig. 2.7 gives the production of sulfur and sulfuric acid from the 1950s to the present. Notice the similarities in the curves for both, since one is made primarily

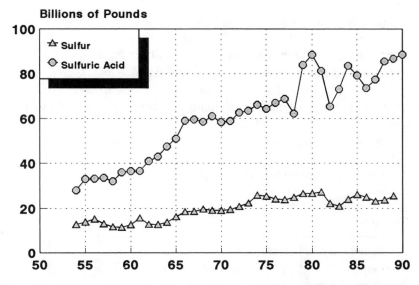

Figure 2.7 U.S. production of sulfur and sulfuric acid. *(Source:* L & M, *C & E News,* CEH. Reprinted by permission of John Wiley & Sons, Inc.)

from the other. With some exceptions the general pattern is a slow steady increase. Note the slump in the early 1980s, indicative of the chemical industry's and all of manufacturing's general slowdown in those years, as well as in 1986. The difference in the ratio of sulfur to sulfuric acid through the years is a reflection on other uses of sulfur (agricultural chemicals, petroleum refining, etc.) or other sources of raw material besides sulfuric acid (metal pyrites, recycling of used sulfuric acid, etc.).

Fig. 2.8 gives the average price trends for these two chemicals. Notice the sharp rise in the 1970s for both chemicals. We will see this phenomenon for many chemicals, especially in 1974-1975, when the Arab oil embargo occurred. Throughout the 1970s many years of double-digit inflation, in part caused by the oil embargo, produced a steep rise in prices of many chemical products, more so for organic chemicals derived from oil, but even spilling over to inorganics because of increasing energy costs in production.

The commercial value of a chemical is another method of measuring the importance of a chemical. It is estimated by multiplying the price by the amount produced, giving an indication of the total money value of the chemical manufactured in the U.S. each year. The more important chemicals and polymers have well over $1 billion/yr commercial value. For example, for 1990 the average price of sulfuric acid was $86/ton or 4.3¢/lb and the amount produced was 88.6 billion lb or 44.3 million tons. Some chemicals are also routinely quoted as ¢/lb. To convert $/ton to ¢/lb we multiply by 0.05:

$$\frac{\$86 \text{ dollars}}{\text{ton}} \times \frac{1 \text{ ton}}{2000 \text{ lb}} \times \frac{100 \text{ cents}}{\$ \text{ dollars}} = \frac{4.3 \text{ cents}}{\text{lb}}$$

To convert from million tons to billion lb we multiply by 2:

$$44.3 \text{ million tons} \times \frac{2000 \text{ lb}}{\text{ton}} \times \frac{1 \text{ billion}}{1000 \text{ million}} = 88.6 \text{ billion lb}$$

Either of these units can be used to calculate a commercial value of $3.8 billion for sulfuric acid:

$$\frac{\$86}{\text{ton}} \times 44.3 \text{ million tons} = \$3810 \text{ million} = \$3.8 \text{ billion}$$

$$\frac{\$0.043}{\text{lb}} \times 88.6 \text{ billion lb} = \$3.8 \text{ billion}$$

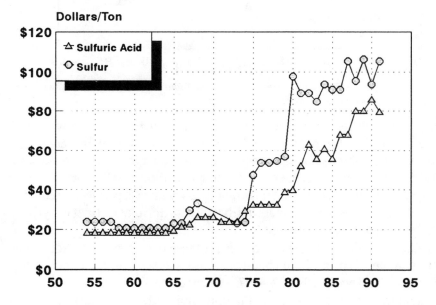

Figure 2.8 U.S. prices of sulfur and sulfuric acid. *(Source:* L & M and CMR. Reprinted by permission of John Wiley & Sons, Inc.)

A good indicator of the economic strength of a chemical is its high percentage of capacity being used. If production is 70-90% of capacity, it usually means that the product is in appropriate demand. Table 2.1 shows the total nameplate capacity of sulfuric acid plants in the U.S. for selected years and production as a percent of capacity. Nameplate capacity means what the plant could routinely produce, though at times some plants can actually make more than this amount if necessary. Most sulfuric acid plants manufacture between 200-2400 tons/day. There are about 90 plants in the U.S. representing nearly 60 companies.

TABLE 2.1 U.S. Sulfuric Acid Capacity

Year	Capacity, billion lb	Production as % of Capacity
1981	104	79
1985	95	82
1990	92	96

Source: CP.

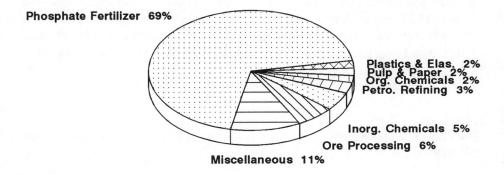

Figure 2.9 Uses of sulfuric acid. *(Source:* CP.)

Fig. 2.9 shows the uses of sulfuric acid. The largest use by far is in the manufacture of phosphate fertilizers, as we will see in the next section. It is the fastest growing use as well, being only 36% of sulfuric acid in 1957, 58% in 1975, and 69% in 1991.

PHOSPHORIC ACID, H₃PO₄; ORTHOPHOSPHORIC ACID

References

Kent, pp. 236-250
L & M, pp. 628-639
CP, 7-18-83, 8-28-89
White, pp. 10-17

Manufacture

By far the most important derivative of sulfuric acid is phosphoric acid. It has been unknowingly used as fertilizer for hundreds of years. The wet process method of manufacture was important until 1920, when furnace acid began increasing in popularity. The wet process, however, has made a comeback because of plant design improvements; 60% of phosphoric acid was made by this method in 1954, 88% in 1974, and 91% in 1989. The furnace process is used only to make concentrated acid (75-85%) and pure product. It is very expensive because of the 2000°C temperature required. In the furnace process phosphate rock is heated with sand and coke to give elemental phosphorus, which is then oxidized and hydrated to phosphoric acid. A simplified chemical reaction is:

$$2Ca_3(PO_4)_2 + 6SiO_2 + 10C \longrightarrow P_4 + 10CO + 6CaSiO_3$$
$$P_4 + 5O_2 + 6H_2O \longrightarrow 4H_3PO_4$$

Since almost all phosphoric acid is now made by the wet process, we will discuss this more fully.

Reaction:

$$Ca_3(PO_4)_2 + 3H_2SO_4 \longrightarrow 2H_3PO_4 + 3CaSO_4$$

or $CaF_2 \cdot Ca_3(PO_4)_2 + 10H_2SO_4 + 20H_2O \longrightarrow 10(CaSO_4 \cdot 2H_2O) + 2HF$
$$+ 6H_3PO_4$$

or $Ca_5F(PO_4)_3 + 5H_2SO_4 + 10H_2O \longrightarrow 5(CaSO_4 \cdot 2H_2O) + HF$
$$+ 3H_3PO_4$$

These three equations represent the wet process method in varying degrees of simplicity and depend on the phosphate source used. There is usually a high percentage of fluorine in the phosphate, in which case the mineral is called fluorapetite. It is mined in Florida, Texas, North Carolina, Idaho, and Montana. The United States has 30% of known phosphate reserves.

Description. Fig. 2.10 outlines the wet process. The phosphate rock is ground and mixed with dilute H_3PO_4 in a mill. It is transferred to a reactor and H_2SO_4 is added. The reactors are heated to 75-80°C for 4-8 hr. Air-cooling carries the HF and SiF_4 side products to an adsorber which transforms them into H_2SiF_6.

Filtration of the solid $CaSO_4 \cdot 2H_2O$ (gypsum) gives a dilute H_3PO_4 solution (28-35% P_2O_5 content). Evaporation of water to 54% P_2O_5 content is optional.

The H_2SiF_6 is formed in the process by the following reactions. SiO_2 is present in most phosphate rock.

$$4HF + SiO_2 \longrightarrow SiF_4 + 2H_2O$$
$$2HF + SiF_4 \longrightarrow H_2SiF_6$$
$$3SiF_4 + 2H_2O \longrightarrow 2H_2SiF_6 + SiO_2$$

There are two useful side products. The H_2SiF_6 is shipped as a 20-25% aqueous solution for fluoridation of drinking water. Fluorosilicate salts find use in ceramics, pesticides, wood preservatives, and concrete hardeners. Uranium, which occurs in many phosphate rocks in the range of 0.005-0.03% of U_3O_8, can be extracted from the dilute phosphoric acid after the filtration step, but this is not a primary source of the radioactive substance. The extraction plants are expensive and can only be justified when uranium prices are high.

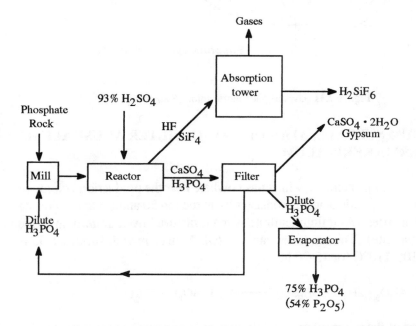

Figure 2.10 Wet process for phosphoric acid. *(Source:* L & M . Reprinted with permission by John Wiley & Sons, Inc.)

Properties

One hundred percent H_3PO_4 is a colorless solid, mp 42°C. The usual laboratory concentration is 85% H_3PO_4, since a crystalline hydrate separates at 88% concentration. Fig. 2.11 shows the percentages for phosphoric acid. Diammonium phosphate (DAP), superphosphoric acid, monoammonium phosphate (MAP), and triple superphosphate (TSP) are all fertilizer applications and total 82% of all phosphoric acid use.

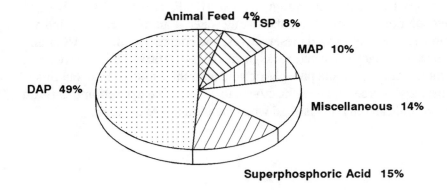

Figure 2.11 Uses of phosphoric acid. (*Source:* CP.)

ALUMINUM SULFATE, $Al_2(SO_4)_3 \cdot 18H_2O$; FILTER ALUM, ALUM, OR PAPERMAKER'S ALUM

This lower-ranking chemical, which has nowhere near the production volume of sulfuric and phosphoric acids, is consistently in the top 50 and is very important to some industries. Aluminum sulfate is manufactured from aluminum oxide (alumina, bauxite). The crude ore can be used. A mixture with sulfuric acid is heated at 105-110°C for 15-20 hr.

$$Al_2O_3 \cdot 2H_2O + 3H_2SO_4 \longrightarrow Al_2(SO_4)_3 + 5H_2O$$

Filtration of the water solution is followed by evaporation of the water to give the product, which is processed into a white powder. Alum has two prime uses. It is bought by the pulp and paper industry for coagulating and coating pulp fibers into a hard paper surface by reacting with small amounts of sodium carboxylates (soap) present. Aluminum salts of carboxylic acids are very gelatinous. In water purification it serves as a coagulant, pH conditioner, and phosphate and bacteria remover. It reacts with alkali to give an aluminum hydroxide floc which drags down such impurities in the water. For this reason it also helps the taste of water.

$$\text{Pulp and paper} \qquad 65\%$$
$$\text{Water purification} \qquad 32\%$$

$$6RCO_2^- Na^+ + Al_2(SO_4)_3 \longrightarrow 2(RCO_2^-)_3Al^{+3} + 3Na_2SO_4$$

$$Al_2(SO_4)_3 + 6NaOH \longrightarrow 2Al(OH)_3 + 3Na_2SO_4$$

SODIUM TRIPOLYPHOSPHATE (STPP); SODIUM TRIPHOSPHATE

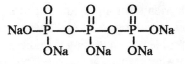

$$Na_5P_3O_{10}$$

Sodium tripolyphosphate has not been in the top 50 chemicals since 1988, but it was and still is an important derivative of sulfuric acid so we will mention it here. There are many sodium phosphates produced. STPP is 81% of all sodium phosphates. It is the second largest selling pure phosphorus-containing chemical, behind phosphoric acid. The reactions can be separated into three distinct steps, but really STPP is made directly in one step by mixing phosphoric acid and sodium carbonate (soda ash) in the proper amounts to give a 1:2 ratio of monosodium and disodium phosphates and then heating to effect dehydration at 300-500°C.

$$2H_3PO_4 + Na_2CO_3 \longrightarrow 2NaH_2PO_4 + H_2O + CO_2$$
$$\text{Monosodium phosphate}$$

$$4H_3PO_4 + 4Na_2CO_3 \longrightarrow 4Na_2HPO_4 + 4H_2O + 4CO_2$$
$$\text{Disodium phosphate}$$

$$NaH_2PO_4 + 2Na_2HPO_4 \longrightarrow Na_5P_3O_{10} + 2H_2O$$

Sodium tripolyphosphate is used almost solely in one type of product: detergents. Some detergents contain up to 50% by weight STPP. It has the unique property of complexing or sequestering dipositive ions such as Ca^{+2} and Mg^{+2} which are present in water. Since 1947 when it was placed in its first synthetic detergent, Tide®, it has revolutionized the industry. It is commonly referred to as a builder and has the following functions: (1) sequesters ions, (2) prevents redeposition of dirt, (3) buffers the aqueous medium at pH = 9-10, (4) kills some bacteria, and (5) controls corrosion and deposits in the lines of automatic washers.

Home laundry detergents	45%
Industrial and institutional detergents	24
Dishwashing detergents	16
Food additive	4

Phosphates are prime nutrients for algae and for this reason contribute to greening, eutrophication, and fast aging of lakes. There is some debate on how much of the total phosphate problem is caused by the detergent industry. We will discuss this further when these substances are studied in detail later in the book. In any event the recent economic history of STPP is interesting. Production increased in the '50s and '60s, but dropped dramatically in 1971-73 when phosphate in detergents was limited or outlawed by many states. A residual market is projected for a few reasons: (1) exports are picking up, (2) cold-water detergents (energy efficient) require more phosphate to do the job, (3) dishwashing and institutional detergents are not being restricted, and (4) inadequacies of various phosphate substitutes are well-known.

Year	STPP Production (billions of lb)
1955	1.10
1960	1.40
1965	1.82
1970	2.44
1975	1.55
1980	1.40
1985	1.21
1989	1.10
1993 est.	0.94

3

Industrial Gases

INTRODUCTION

Three inorganic gases, nitrogen, oxygen, and carbon dioxide, appear in the top 50 chemicals. A fourth gas, hydrogen, would also be included if it were not for the large amounts of captive use of hydrogen to manufacture ammonia. It is convenient to discuss all four at this time in our study of inorganic chemicals. Two of them, nitrogen and hydrogen, are used to produce ammonia, which in turn has many important derivatives that will be discussed in the next chapter. Not all four major gases are manufactured by the same method. Nitrogen and oxygen, obtained by the liquefaction of air, will be discussed first. Next, carbon dioxide and hydrogen, made by the process of steam-reforming of hydrocarbons, will be considered.

NITROGEN

References

Kent, pp. 607-618
Austin, pp. 115-124
L & M, pp. 579-588
T, pp. 273-288
KC, 7-13-87

Manufacture

The large-scale availability of nitrogen, oxygen, and argon from liquefaction of air began about 1939-1940. A 90% recovery is now feasible for these three major components in air. Nitrogen makes up 78% of all air, oxygen 21%, and argon 0.9%. Two major processes are used, differing only in the way in which the expansion of air occurs. The Linde-Frankl cycle is based on the classic Joule-Thompson effect of a gas, which means that there is a tremendous cooling effect of a gas when it is rapidly expanded, even though no external work is done on the system. Alternatively, the Claude process employs an expansion engine doing useful work on the gas. The temperature is reduced because of the removal of energy. This process is more efficient than relying on the Joule-Thompson effect.

Description. Figure 3.1 outlines the liquefaction of air. Air is filtered to remove particulates and then compressed to 77 psi. An oxidation chamber converts traces of hydrocarbons into carbon dioxide and water. The air is then passed through a water separator which gets some of the water out. A heat exchanger cools the sample down to very low temperatures, causing solid water and carbon dioxide to be separated from the main components.

Most of the nitrogen-oxygen mixture, now at -168°C and 72 psi, enters the bottom of a fractionating column (Figures 3.2 and 3.3). An expansion valve at this point causes further cooling. The more volatile nitrogen rises to the top of the column as a gas since nitrogen (bp = -196°C, 77°K) has a lower boiling point than oxygen (bp = -183°C, 90°K), and the column at 83°K is able to separate the two. The oxygen stays at the bottom of the column as a liquid because it is less volatile.

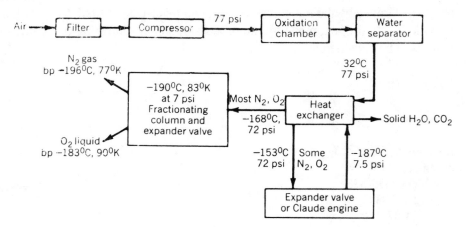

Figure 3.1 Liquefaction of air. *(Source:* L & M.)

Figure 3.2 This cryogenic distillation column is 95 ft high and contains two sections at different pressures. Oxygen is removed from the bottom and nitrogen from the top. The plant has a capacity of 200 tons/day total pure gases and liquids processed. (Courtesy of Air Products and Chemicals, Inc., Allentown, PA.)

A small amount of nitrogen-oxygen mixture after being recooled in the heat exchanger is shunted to the main expander valve (operated by the Joule-Thompson effect or by a Claude engine). This extremely cold gas is recycled into the heat exchanger to keep the system cold. Some argon remains in the oxygen fraction and this mixture can be sold as 90-95% oxygen. If purer oxygen is required, a more elaborate fractionating column with a greater number of plates gives an oxygen-argon separation. Oxygen can be obtained in 99.5% purity in this fashion. Not only argon, but other rare gases, neon, krypton, and xenon, can also be obtained in separations. Helium is *not* obtained from liquefaction of air. It occurs in much greater concentrations (2%) in natural gas wells and is isolated in the petroleum refinery.

Figure 3.3 Air separation plant. The dominant features are the nitrogen and oxygen storage tanks on the left and the cryogenic distillation column on the right. (Courtesy of Air Products and Chemicals, Inc., Allentown, PA.)

By far the largest use of nitrogen is in ammonia synthesis. However, this use is not included here because it is "captive," that is, the same company immediately reuses the gas internally to make another product, in this case ammonia. This nitrogen is not isolated, sold, or inventoried. Only "merchant" use is included in Fig. 3.4. The fastest growing use of nitrogen is in enhanced oil recovery (EOR), where it maintains pressure in oil fields so that a vacuum is not formed underground when natural gas and oil are pumped out. It is competing with carbon dioxide in this application.

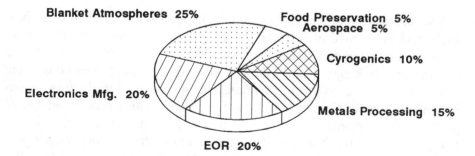

Figure 3.4 Merchant uses of nitrogen. (*Source*: KC.)

OXYGEN

References

KC, 7-13-87

Manufacture

The manufacture of oxygen is described along with that of nitrogen. Both are formed from the liquefaction of air. Oxygen gas is colorless, odorless, and tasteless, but it is slightly blue in the liquid state. Up to 99.995% purity is available commercially. It is commonly used from seamless steel cylinders under 2,000 psi pressure. A 1.5 cu ft cylinder holds 15 lb of oxygen, equivalent to 244 cu ft at standard temperature and pressure.

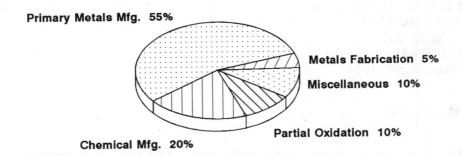

Figure 3.5 Uses of oxygen. (*Source*: KC.)

Fig. 3.5 gives the uses of oxygen. The steel industry prefers to use pure oxygen rather than air in processing iron. The oxygen reacts with elemental carbon to form carbon monoxide, which is processed with iron oxide so that carbon is incorporated into the iron metal, making it much lower melting and more pliable. This material is called fusible pig iron. Common pig irons contain 4.3% carbon and melt at 1130°C, whereas pure iron has a melting point of 1539°C. The following equations summarize some of this chemistry.

$$2C + O_2 \longrightarrow 2CO$$
$$Fe_2O_3 + 3CO \longrightarrow 2Fe + 3CO_2$$
$$2CO \longrightarrow C(\text{in Fe}) + CO_2$$

Steel is a mixture of several physical forms of iron and iron carbides. Properties are controlled by the amount of carbon and other elements present, such as manganese, cobalt, and nickel. Since the steel industry uses more than half of all oxygen, the production of oxygen is very dependent on this one use.

In other oxygen applications, metal fabrication involves cutting and welding with an oxygen-acetylene torch. Chemical manufacture includes the formation of ethylene oxide, acrylic acid, propylene oxide, and vinyl acetate. Miscellaneous uses include sewage treatment, aeration, pulp and paper bleaching, and missile fuel.

HYDROGEN

References

Kent, pp. 621-629
L & M, pp. 215-223,468-481
Austin, pp. 106-115
T, pp. 295-299
KC, 7-29-85
Wiseman, pp. 141-148

Manufacture

Hydrogen does not actually appear in the top 50. One reason is that most of it is captive and immediately reused to make ammonia, hydrogen chloride, and methanol—three other chemicals with high rankings. Since it is a feedstock for these chemicals it is even more important than these three and we will study its manufacture in detail. Hydrogen is our first example of a "petrochemical" even though it is not organic. Its primary manufacturing process is by steam-reforming of natural gas or hydrocarbons. Approximately 80% of the hydrogen used for ammonia manufacture comes from this process.

Reactions. A variety of low molecular weight hydrocarbons can be used as feedstock in the steam-reforming process. Equations are given for both methane (natural gas) and propane. The reaction occurs in two separate steps: reforming and shift conversion.

$$\text{\textit{Methane}}$$

Reforming	$CH_4 + H_2O \longrightarrow CO + 3H_2$	
Shift conversion	$CO + H_2O \longrightarrow CO_2 + H_2$	

$$\text{\textit{Propane}}$$

Reforming	$C_3H_8 + 3H_2O \longrightarrow 3CO_2 + 7H_2$	
Shift conversion	$3CO + 3H_2O \longrightarrow 3CO_2 + 3H_2$	

The reforming step makes a hydrogen: carbon monoxide mixture which is one of the most important materials known in the chemical industry. It is called *synthesis gas* and is used to produce a variety of other chemicals. The old method of making synthesis gas was from coke, but this gave a lower percentage of hydrogen in the mixture, which was called water gas or blue gas.

$$C + H_2O \longrightarrow CO + H_2$$

Higher H_2:CO ratios are now needed, and thus the newer hydrocarbon feedstocks are used. Coal gives a 1:1 ratio of H_2:CO, oil a 2:1 ratio, gasoline 2.4:1, and methane 4:1.

Note that in the second step, the shift conversion process (also known as the carbon monoxide or water gas shift reaction), more hydrogen is formed along with the other product, carbon dioxide. A variety of methods is used to make carbon dioxide, but this process is the leading method.

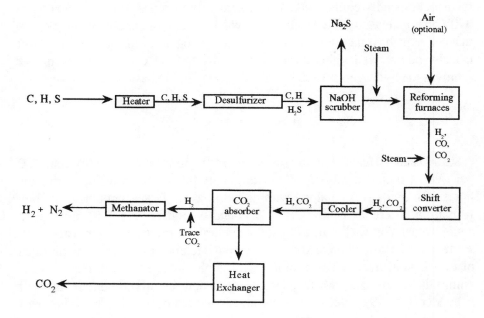

Figure 3.6 Steam-reforming of hydrocarbons. (*Source:* L & M. Reprinted by permission of John Wiley & Sons, Inc.)

Figure 3.7 The primary reformer for methane conversion to carbon monoxide and hydrogen. (Courtesy of Monsanto Agricultural Products Company, Luling, LA.)

Description. Fig. 3.6 diagrams the steam-reforming process. The hydrocarbon feedstock, usually contaminated with some organosulfur traces, is heated to 370°C before entering the desulfurizer, which contains a metallic oxide catalyst that converts the organosulfur compounds to hydrogen sulfide. Elemental sulfur can also be removed with activated carbon absorption. A caustic soda scrubber removes the hydrogen sulfide by salt formation in the basic aqueous solution.

$$H_2S + 2NaOH \longrightarrow Na_2S + 2H_2O$$

Steam is added and the mixture is heated in the furnace at 760-980°C and 600 psi over a nickel catalyst. When larger hydrocarbons are the feedstock, potassium oxide is used along with nickel to avoid larger amounts of carbon formation. There are primary (Fig. 3.7) and secondary (Fig. 3.8) furnaces in some plants. Air can be added to the secondary reformers. Oxygen reacts with some of the hydrocarbon feedstock to keep the temperature high. The nitrogen of the air is utilized when it, along with the hydrogen formed, reacts in the ammonia synthesizer. More steam is added and the mixture enters the shift converter (Fig. 3.9), where iron or chromic oxide catalysts at 425°C further react the gas to hydrogen and carbon dioxide. Some shift converters have high and low temperature sections, the high temperature section converting most of the CO to CO_2 relatively fast, the low temperature section completing the process and taking advantage of a more favorable equilibrium toward CO_2 at low temperatures in this exothermic reaction. Cooling to 38°C is followed by carbon dioxide absorption with monoethanolamine. The carbon dioxide is desorbed by heating the monoethanolamine and reversing this reaction. The carbon dioxide

Figure 3.8 A secondary reformer converts the last of the methane. (Courtesy of Monsanto Agricultural Products Company, Luling. LA.)

Figure 3.9 The shift converter reacts carbon monoxide and water to give carbon dioxide and more hydrogen. (Courtesy of Monsanto Agricultural Products Company, Luling. LA.)

is an important by-product. Alternatively, hot carbonate solutions can replace the monoethanolamine.

$$HO-CH_2-CH_2-NH_2 \ + \ H_2O \ + \ CO_2 \ \rightleftharpoons \ HO-CH_2-CH_2-NH_3^+ \\ + \ HCO_3^-$$

A methanator converts the last traces of carbon dioxide to methane, a less interfering contaminant in hydrogen used for ammonia manufacture.

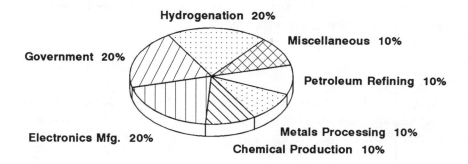

Figure 3.10 Merchant uses of hydrogen. (*Source*: KC.)

Uses

It is very difficult to define a percentage use chart for hydrogen because so much is captive, nearly 99%. Approximately one half is used in ammonia production. Another one third is employed in the metallurgical industries to reduce the oxides of metals to the free metals. Smaller amounts are used for methanol production, hydrogenations in petroleum refineries, pure hydrogen chloride manufacture, and as liquid hydrogen. The merchant sale of hydrogen, a minor amount compared to its captive use, is given in Fig.3.10.

CARBON DIOXIDE

References

KC, 7-13-87

Manufacture

Over 90% of all carbon dioxide is made by steam-reforming of hydrocarbons, and much of the time natural gas is the feedstock. It is an important by-product of hydrogen and ammonia manufacture.

$$CH_4 + 2H_2O \longrightarrow 4H_2 + CO_2$$

A small amount (1%) of carbon dioxide is still made from fermentation of grain. Ethyl alcohol is the main product.

$$C_6H_{12}O_6 \xrightarrow{\text{yeast}} 2C_2H_5OH + 2CO_2$$

Another 1% is recovered as a by-product of ethylene oxide manufacture from ethylene and oxygen. When the oxidation goes too far some carbon dioxide is formed.

$$CH_2{=}CH_2 + \tfrac{1}{2}O_2 \xrightarrow{\text{Ag}} \overset{\displaystyle O}{\overset{\displaystyle /\backslash}{CH_2{-}CH_2}} (+ CO_2 + H_2O)$$

Other small amounts are obtained from coke burning, the calcination of lime, and in the manufacture of sodium phosphates from soda ash and phosphoric acid.

$$C\,(\text{coke}) + O_2 \longrightarrow CO_2$$
$$CaCO_3 \xrightarrow{\Delta} CaO + CO_2$$
$$Na_2CO_3 + H_3PO_4 \longrightarrow Na_2HPO_4 + CO_2 + H_2O$$

Carbon dioxide is a gas at room temperature. Below -78°C it is a solid and is commonly referred to as "dry ice". At that temperature it sublimes and changes directly from a solid to a vapor. Because of this unique property, as well as its noncombustible nature, it is a common refrigerant and inert blanket. Fig. 3.11 shows the important merchant uses of carbon dioxide. Refrigeration using dry ice is especially important in the food industry. Enhanced oil recovery is a newer use of carbon dioxide and it will compete with nitrogen for future expansion in this market.

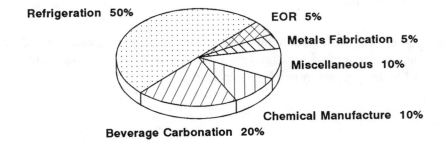

Refrigeration 50% EOR 5% Metals Fabrication 5% Miscellaneous 10% Chemical Manufacture 10% Beverage Carbonation 20%

Figure 3.11 Merchant uses of carbon dioxide. (*Source:* KC.)

Other important industrial uses include chemical manufacture (especially soda ash), inert atmospheres, pressure regulation, low-temperature grinding, and fire extinguishers. The growing use is in pH control of wastewater. It therefore competes with nitrogen in many applications involving cooling, noncombustible blanketing, and pressure regulation.

A new use of carbon dioxide is as a fumigant for stored grain. Its use may soar depending on the outcome of the toxicity question of ethylene dibromide (EDB), which was previously used in this application. The use chart reflects only the merchant market for carbon dioxide which is only 5% of consumption. The rest is captive and is employed in petroleum operations.

ECONOMICS OF INDUSTRIAL GASES

U.S. production of industrial gases is given in Fig. 3.12. Hydrogen is not included because so much of its production is captive, so as to make its production profile meaningless. Nor is the amount of nitrogen used to make ammonia included. Even without this captive nitrogen, notice the much steeper nitrogen production curve, especially in the late '70s and '80s. In 1980 nitrogen was ranked fifth in chemical production. It is now a strong second. Nitrogen

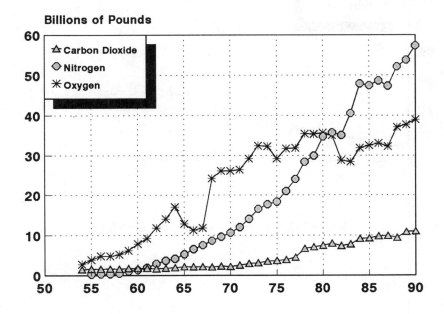

Figure 3.12 U.S. production of gases. (*Source*: L & M, *C & E News*. Reprinted by permission of John Wiley & Sons, Inc.)

and carbon dioxide were two of the fastest growing chemicals in the 1970s (12% per year) and 1980s (6% per year), especially because of their EOR uses. Oxygen on the other hand has had a more linear growth, no doubt due to the suffering steel industry. Prices for nitrogen, oxygen, and hydrogen are given in Fig. 3.13. Carbon dioxide varies considerably in price depending on its form, dry ice or gaseous. Note that hydrogen is much more expensive, coming from expensive hydrocarbons, compared to nitrogen and oxygen, which share equivalent prices because of being manufactured together. The key in gas pricing is the all important shipping charges, which are not included here. They are very expensive since a heavy container must be used to withstand the high pressures of even light weights of gases. The commercial value of the gases are about the same for nitrogen and oxygen, $600 million. Merchant hydrogen and carbon dioxide are near $200 million.

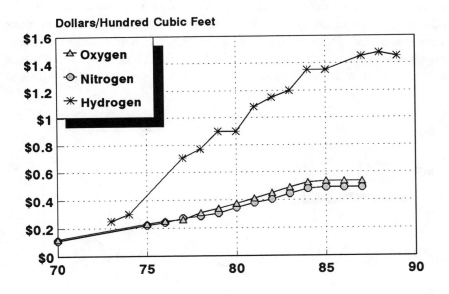

Figure 3.13 U.S. prices of gases. (*Source*: CEH.)

4

Inorganic Nitrogen Compounds

INTRODUCTION

Next, we consider ammonia and its derivatives in the top 50 chemicals. We have completed a study of the number one inorganic chemical sulfuric acid and its derivatives and have also studied industrial gases from which ammonia is made. Ammonia is in the top 10 chemicals and some important ammonia derivatives are listed in the top 50: ammonium nitrate, nitric acid, urea, and ammonium sulfate. Most ammonia eventually ends up in fertilizers of one type or another. The manufacturing chemistry for these chemicals is summarized here:

$$N_2 + 3H_2 \rightleftharpoons 2NH_3$$
$$\text{ammonia}$$

$$NH_3 + 2O_2 \longrightarrow HNO_3 + H_2O$$
$$\text{nitric acid}$$

$$NH_3 + HNO_3 \longrightarrow NH_4NO_3$$
$$\text{ammonium nitrate}$$

$$2NH_3 + CO_2 \longrightarrow NH_2CONH_2 + H_2O$$
$$\text{urea}$$

$$2NH_3 + H_2SO_4 \longrightarrow (NH_4)_2SO_4$$
$$\text{ammonium sulfate}$$

It might be argued that ammonia and its derivatives are all petrochemicals since the hydrogen is derived from methane or natural gas. Many ammonia plants are near oil refineries. Urea even contains carbon and is considered an organic

69

chemical. But because all these nitrogen derivatives have been traditionally thought of as being inorganic, we will consider them at this time.

AMMONIA, NH₃

References

Kent, pp. 143-164
L & M, pp. 83-92
T, pp. 201-221
KC, 2-17-86
CP, 10-3-88, 10-7-91
Wiseman, pp. 141-148

Manufacture

The process for ammonia manufacture will vary somewhat with the source of hydrogen, but 90% of ammonia plants generate the hydrogen by steam-reforming of natural gas. This has been the primary source of hydrogen since the early 1930s. Steam-reforming has already been discussed in the previous chapter and the process will not be repeated here, even though most of the design of an ammonia plant is concerned with the generation of hydrogen by steam reforming.

If the hydrogen is made by steam-reforming, air is introduced at the secondary reformer stage. This provides nitrogen for the ammonia reaction. The oxygen of the air reacts with the hydrocarbon feedstock in combustion and helps to elevate the temperature of the reformer. Otherwise nitrogen can be added from liquefaction of air. In either case a hydrogen-nitrogen mixture is furnished for ammonia manufacture.

Reaction

$$N_2 + 3H_2 \xrightarrow{\text{Fe}} 2NH_3$$

85-90% yield

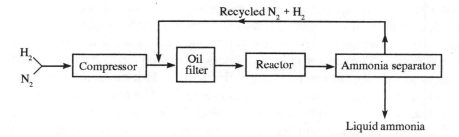

Figure 4.1 Haber process for making ammonia. (*Source:* L & M. Reprinted with permission by John Wiley & Sons, Inc.)

Figure 4.2 In the reactor for ammonia sythesis, nitrogen and hydrogen gases are used. (Courtesy of Monsanto Agricultural Products Company, Luling, LA.)

Description: Fig. 4.1 outlines the Haber process to make ammonia. The reaction of nitrogen and hydrogen gases was first studied by Haber with Nernst filtered to remove traces of oil, joined to recycled gases, and is fed to the reactor at 400-600°C. The reactor (Fig. 4.2) contains an iron oxide catalyst which reduces to a porous iron metal in the N_2:H_2 mixture. Exit gases are cooled out to -10 to -20°C and part of the ammonia liquefies. The remaining gases are recycled. The conversion to ammonia is 20-22%; the overall yield is 85-90%.

This is our first distinction between conversion and yield, and it is important to know the difference. Conversion is the amount of product made per pass in a given reaction and can at times be small. With recycling of a raw material, however, the final percent yield, the overall transformation of reactants into products, can be very high.

Properties

Anhydrous ammonia is a colorless gas with a pungent odor, bp -33°C. It can be liquefied at 25°C under 175 psi. The gas is usually shipped as a liquid under pressure. It is very soluble in water. The water solution can be called ammonia water, aqua ammonia, ammonium hydroxide, or sometimes just "ammonia" misleadingly.

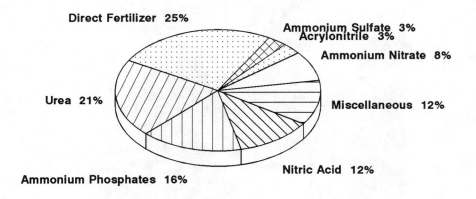

Figure 4.3 Uses of ammonia. (*Source*: CP.)

Uses

Fig. 4.3 gives the use profile for ammonia. It can be applied directly for fertilizer or made into other nitrogen-containing compounds used for fertilizer or, to a lesser extent, explosives, plastics, and fibers. Overall approximately 80% of ammonia has an end use as fertilizer. Explosives made from ammonia are ammonium nitrate and, via nitric acid, the nitroglycerin used in dynamite. Plastics include, via urea, urea-formaldehyde and melamine-formaldehyde resins. Some ammonia ends up in fibers since it is used to make hexamethylenediamine, adipic acid, or caprolactam, all nylon precursors.

Economics

Fig. 2.2, page 38, includes a production profile for ammonia. Ammonia production is tied to agriculture. Although the agricultural industry is quite variable, overall ammonia production is more constant. The great demand for fertilizers since World War II has spurred production. Nonfertilizer use is leveling or dropping. A modest increase of 2-3% per year is forecast. Production is always a high percentage of capacity in this country and is sometimes very near 100%. An average plant will manufacture 900-1500 tons/day. There are over 50 manufacturers in this country. Many of them are in the oil-rich Gulf region and are connected with oil refineries. Much ammonia is captive and usedon site or internally within the company. Ammonia's price is very dependent on that of natural gas. Since it is one of the largest volume petrochemicals, its price has varied similarly to that of organic chemicals with a large increase in the 1970s. Its commercial value is high. At $148/ton and a production of 33.9 billion lb its worth is calculated as $2.5 billion.

NITRIC ACID, HNO₃

References

Kent, pp. 164-171
L & M, pp. 563-570
Thompson, pp. 221-228
Austin, pp. 313-318
Chang, pp. 54-58

Manufacture

For many years nitric acid was made by the reaction of sulfuric acid and saltpeter, but this method is no longer used. Direct oxidation of ammonia is now the only process.

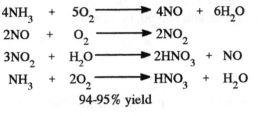

Reactions:

$$4NH_3 + 5O_2 \longrightarrow 4NO + 6H_2O$$
$$2NO + O_2 \longrightarrow 2NO_2$$
$$3NO_2 + H_2O \longrightarrow 2HNO_3 + NO$$

overall:
$$NH_3 + 2O_2 \longrightarrow HNO_3 + H_2O$$

94-95% yield

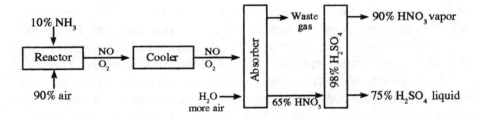

Figure 4.4 Manufacture of nitric acid by ammnoia oxidation. (*Source:* L & M. Reprinted with the permission of John Wiley & Sons, Inc.)

Description: Fig. 4.4 describes nitric acid manufacturing. A high-pressure process is most often used. It involves lower capital costs, increased acid strength obtained, increased rate of reaction, and a smaller tower volume required. The reactor contains a rhodium-platinium catalyst (2-10% rhodium) as wire gauzes in layers of 10-30 sheets at 750-920°C, 100 psi , and a contact time of 3 x 10⁻⁴ sec. After cooling, it enters the absorption tower with water and more air to oxidize the nitric oxide and hydrate it to 61-65% nitric acid in water. Waste gases contain nitric oxide or nitrogen dioxide. These are reduced with hydrogen or methane to ammonia or nitrogen gas. Traces of nitrogen oxides can be expelled. Concentration of the nitric acid in a silicon-iron or stoneware tower containing 98% sulfuric acid will give 90% nitric acid off the top and 70-75% sulfuric acid as the bottoms. This last step is necessary because simple distillation of nitric acid is not applicable; it forms an azeotrope with water at 68% acid. An alternative drying agent is magnesium nitrate, which can concentrate the acid to 100% HNO_3.

Properties

Pure 100% nitric acid is a colorless, highly corrosive liquid and a very powerful oxidizing agent, bp 86°C. It gradually yellows because of decomposition to nitrogen dioxide. Solutions containing more than 80% nitric acid are called "fuming nitric acids." The azeotrope is 68% HNO_3, 15 M, and has a bp of 110°C.

Uses

Nitric acid has a 65:25 fertilizer: explosive ratio. Ammonium nitrate makes up nearly all of these two uses. The other 10% is made into miscellaneous compounds: adipic acid, nitroglycerin, nitrocellulose, ammonium picrate, trinitrotoluene, nitrobenzene, silver nitrate, and various isocyanates.

AMMONIUM NITRATE, NH₄NO₃

References

Kent, pp. 171-173
L & M, pp. 97-102
T, pp. 228-231
Chang, pp. 63-65

Manufacture

$$NH_3 + HNO_3 \longrightarrow NH_4NO_3$$
99% yield

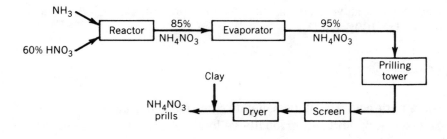

Figure 4.5 Ammonium nitrate manufacture. (*Source:* L & M. Reprinted by permission of John Wiley & Sons, Inc.)

Description. Fig. 4.5 outlines the manufacture of ammonium nitrate. Although the basic reaction is the same, there are many modifications in ammonium nitrate manufacture and product form. Crystals, granules, and prills are made with the same chemistry but different engineering. The prilling technique is described here. In a stainless steel reactor the heat of neutralization boils the mixture, concentrating it to 85% nitrate. Vacuum evaporation at 125-140°C further concentrates the solution to 95%. The last water of this hygroscopic material is very difficult to remove. The hot solution is pumped to the top of a spray or prilling tower 60-70 m high. It is discharged through a spray head and solidifies as it falls in the air to form small spherical pellets, prills, of 2mm diameter. The prills are screened, further dried, and dusted with clay to minimize sticking.

There are two alternative processes to prilling. In vacuum crystallization a rotating crystallizer forms crystals in a good size for fertilizer. These are centrifuged from the water and dried. In the Stengel process water is removed by heating. The molten mass solidifies with cooling into a solid sheet, which is ground into granular form mechanically.

Properties

Ammonium nitrate is a white hygroscopic solid, mp 169.6°C, is relatively unstable, and forms explosive mixtures with combustible materials or when contaminated with certain organic compounds. It is very soluble in water, 55% at 0°C.

Uses

Ammonium nitrate has an 82: 18 fertilizer: explosive ratio. The chief use of ammonium nitrate until after World War II was as an explosive. Although it still accounts for more than 75% of all explosives, its major use is now as a fertilizer because of its high nitrogen content, 33.5% N.

UREA $H_2N-\overset{\overset{\textstyle O}{\|}}{C}-NH_2$

References

Kent, pp. 173-179
L & M, pp. 854-861
W & RI, pp. 114-116
CP, 10-16-85, 9-26-88, 9-23-91

Manufacture

Although urea is an organic compound, it is best discussed with other ammonia-derived synthetic nitrogen compounds, especially in view of its importance to the fertilizer industry like the other compounds in this section.

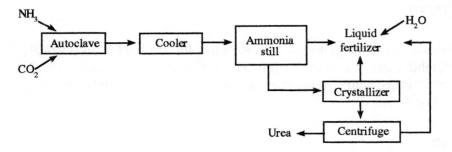

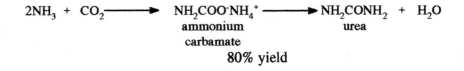

Figure 4.6 Urea synthesis. (*Source:* L & M. Reprinted with the permission of John Wiley & Sons, Inc.)

Reaction:

$$2NH_3 + CO_2 \longrightarrow \underset{\substack{\text{ammonium}\\\text{carbamate}}}{NH_2COO^-NH_4^+} \longrightarrow \underset{\text{urea}}{NH_2CONH_2} + H_2O$$

80% yield

Description. As shown in Fig. 4.6, a 3:1 molar ratio of ammonia and carbon dioxide (excess ammonia) are heated in the autoclave for 2 hr at 190°C and 1500-3000 psi. The mixture formed is approximately 35% urea, 8% ammonium carbamate, 10% water, and 47% ammonia. It is cooled to 150°C and the ammonia is distilled at 60°C. The residue from the ammonia still enters the crystallizer at 15°C. More ammonia is removed by vacuum. The resulting slurry is centrifuged. All excess nitrogenous materials are combined and processed into liquid fertilizer, which contains a mixture of all these materials.

Properties

Urea is a white solid, somewhat hygroscopic, mp 135°C, solubility 108 g/100g water at 20°C.

Uses

Urea is used in solid fertilizer (53%), liquid fertilizer (31%), and miscellaneous applications such as animal feed, urea, formaldehyde resins, melamine and adhesives. Presently, the most popular nitrogen fertilizer is a urea-ammonium nitrate solution. Urea-formaldehyde resins have large use as a plywood adhesive. Melamine-formaldehyde resins are used as dinnerware and for extra-hard surfaces (Formica®). The melamine is synthesized by condensation of urea molecules:

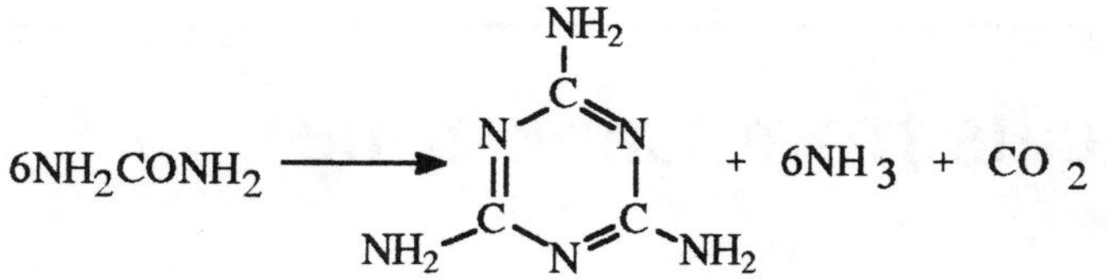

$$6NH_2CONH_2 \longrightarrow \text{melamine} + 6NH_3 + CO_2$$

melamine

AMMONIUM SULFATE, (NH4)2SO4

Manufacture

Although not nearly so important as the other nitrogen compounds, ammonium sulfate is still in the top 50 and is important in the fertilizer industry. Most of it is synthesized by the direct reaction of ammonia and sulfuric acid. Water is removed by evaporation and the product is crystallized to large, white uniform crystals, mp 513°C dec.

$$2NH_3 + H_2SO_4 \longrightarrow (NH_4)_2SO_4$$

The ammonium sulfate is used to the extent of 97% in fertilizer. Other uses include water treatment, fermentation processes, fireproofing agents, and leather tanning.

5

Chemicals from Limestone

INTRODUCTION

The next major raw material for which we discuss the derived chemicals is calcium carbonate, common limestone (Fig. 5.1). It is the source of some carbon dioxide, but, more importantly, it is used to make lime (calcium oxide) and slaked lime (calcium hydroxide). Limestone, together with salt and ammonia, are the ingredients for the Solvay manufacture of sodium carbonate, soda ash. Soda ash is also mined directly from trona ore. The Solvay process manufactures calcium chloride as an important by-product. Soda ash in turn is combined with sand to produce sodium silicates to complete the chemicals in the top 50 that are derived from limestone. Since lime is the highest ranking derivative of limestone in terms of total amount produced, we discuss it first. Refer to Fig. 2.1, page 37, for a diagram of limestone derivatives.

LIME, CaO

References

Austin, pp. 181-185
J. Chem. Educ. **1983**, *60*, 60-63
KC, 6-24-85

Before going further, let us clarify the various common names of limestone and lime. The following is a summary of the nomenclature and the chemicals. Industrial chemists quite often use the common names for these substances rather than the chemically descriptive names.

Figure 5.1 A limestone quarry at Jamesville, New York. This open pit furnished 30-40 rail cars per day for the soda ash plant in Solvay, New York, 14 mi away. (Courtesy of Allied-Signal, Inc., Solvay, NY.)

$CaCO_3$:	Limestone, calcite, calcium carbonate, marble chips, chalk
CaO:	Lime, quicklime, unslaked lime, calcium oxide
$Ca(OH)_2$:	Slaked lime, hydrated lime, calcium hydroxide. A saturated solution in water is called limewater. A suspension in water is called milk of lime.

Manufacture

Lime is one of the oldest materials known to humankind. It was used by Romans, Greeks, and Egyptians for the production of cement and was employed in agriculture as well. One of the first things done by American settlers was to set up a lime kiln for the "calcining" or heating of limestone.

Reaction

$$CaCO_3 \xrightarrow{\Delta} CaO + CO_2$$
$$CaO + H_2O \longrightarrow Ca(OH)_2$$

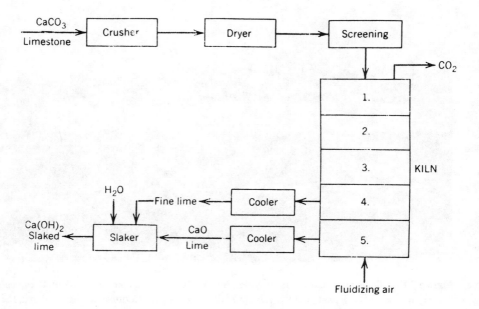

Figure 5.2 Lime manufacture.

Common temperatures used in converting limestone into lime are 1200-1300°C. For this reason lime is a very energy-intensive product. It takes the energy from a third of a ton of coal to produce 1 ton of lime.

Description. Fig 5.2 outlines lime production. The limestone is crushed and screened to a size of approximately 4-8 in. There are different heating techniques and kiln styles. The one diagramed is a vertical Dorrco Fluo Solids system. The limestone enters the top. Air entering the bottom "fluidizes" the solids to get better circulation and reaction. Approximately 98% decarbonation is typical. When a kiln is used in conjunction with the Solvay process and the manufacture of soda ash, coke can be fired in the kiln (Fig. 5.3) along with limestone to give the larger percentages of carbon dioxide needed for efficient soda ash production by the reaction $C + O_2 \rightarrow CO_2$. If a purer lime product is desired, the fine lime can be taken from area 4. A less pure product is obtained from the bottom kiln section. Another kind of kiln is the rotating, nearly horizontal type. These can be as much as 12 ft in diameter and 450 ft. long. Limestone enters one end. It is heated, rotated, and slowly moves at a slight decline to the other end of the kiln, where lime is obtained.

Figure 5.3 Lime kilns in a Solvay plant furnised both carbon dioxide and slaked lime. These were 75 ft high with an i.d. of 16 ft and an o.d. of 20 ft because of 4 ft of insulation for the 2300°F temperature. Anthracite coal was used in the kilns along with limestone. It took 36 hr for the limestone to go from top to bottom. As much as 850 tons of rock could be mixed in each kiln continuously. (Courtesy of Allied-Signal, Inc., Solvay, NY.)

Figure 5.4 A rotary slaker 6 ft in diameter aand 60 ft long. (Courtesy of Allied-Signal, Inc., Solvay, NY.)

Properties

For most applications slaked lime is sold (Fig. 5.4). The hydration of lime is very exothermic and could ignite paper or wood containers of the unslaked material. Slaked lime is slightly soluble in water to give a weakly basic solution.

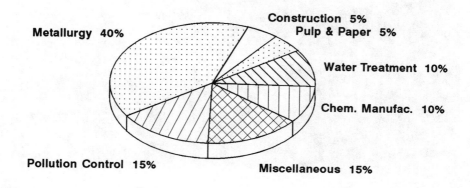

Figure 5.5 Uses of lime. (*Source*: KC.)

Fig 5.5 divides the uses of lime. Lime is used as a basic flux in the manufacture of steel. Silicon dioxide is a common impurity in iron ore which cannot be melted unless it combines with another substance first to convert it to a more fluid lava called slag. Silicon dioxide is a Lewis acid and therefore it reacts with the Lewis base lime. The molten silicate slag is less dense than the molten iron and collects at the top of the reactor, where it can be drawn off. Over 100 lb of lime must be used to manufacture a ton of steel.

$$CaO + SiO_2 \longrightarrow CaSiO_3$$

The uses of lime in chemical manufacture are too numerous to discuss since over 150 important chemicals are made with this basic material. In fact, only five other raw materials are used more frequently than lime for chemical manufacture: salt, coal, sulfur, air, and water. The most important chemical derivative of lime is soda ash, although the synthetic product has been a small percentage of all soda ash in recent years. Lime is used in water treatment to remove calcium and bicarbonate ions.

$$Ca(OH)_2 + Ca^{2+} + 2HCO_3^- \longrightarrow 2CaCO_3 + 2H_2O$$

A growing use of lime is in pollution control, where lime scrubbers placed in combustion stacks remove sulfur dioxide present in combustion gases from the burning of high sulfur coal.

$$SO_2 + H_2O \longrightarrow H_2SO_3$$
$$Ca(OH)_2 + H_2SO_3 \longrightarrow CaSO_3 + 2H_2O$$

Lime is employed in the kraft pulping process to be discussed in detail later. Most of it is recycled. Without this recycling the pulp and paper industry would be the largest lime user. The main reaction of lime in the kraft process is for the purpose of regenerating caustic soda (sodium hydroxide).

$$Na_2CO_3 + CaO + H_2O \longrightarrow CaCO_3 + 2NaOH$$

The caustic soda is then used in the digestion of wood. The lime is regenerated from the limestone by heating in a lime kiln. A large part of Portland cement is lime-based. Sand, alumina, and iron ore are mixed and heated with limestone to 1500°C. Average percentages of the final materials in the cement and their structures are given here.

21%	$2CaO \cdot SiO_2$		Dicalcium silicate
52	$3CaO \cdot SiO_2$		Tricalcium silicate
11	$3CaO \cdot Al_2O_3$	Tri	Tetracalcium aluminate
9	$4CaO \cdot Al_2O_3 \cdot Fe_2O_3$		Tetracalcium aluminoferrite
3	MgO		Magnesium oxide

The percentage of dicalcium silicate, sometimes abbreviated as C_2S in the industry, determines the final strength of the cement. The amount of tricalcium silicate, C_3S, is related to the early strength (7-8 days) required of the cement. Tricalcium aluminate, C_3A, relates to the set in the cement. Tetracalcium aluminoferrite, C_4AF, reduces the heat necessary in manufacture.

Economics

The production history of lime is given with other chemicals in Fig. 2.2, page 38. Production dropped more for lime than most other chemicals in the 1980s, 2.4% per year. Lime production is very dependent on the steel industry, which in turn fluctuates directly with automobile and housing demand. Lime, being an energy intensive chemical because of the high temperatures required to make it from limestone, fluctuates more with energy prices than most other inorganic chemicals. From 1970-1975 the price rose from \$12/ton to \$28/ton, mainly because the oil embargo increased energy costs. Presently it sells for \$49/ton or about 2.5¢/lb and the commercial value of its 34.8 billion lb is \$0.9 billion.

SODA ASH, Na₂CO₃; SODIUM CARBONATE

References

Kent, pp. 213-217
L. & M, pp. 706-715
T, pp. 124-133
CP, 9-23-85, 9-5-88, 8-5-91

Manufacture

The LeBlanc process for the manufacture of soda ash was discovered in 1773 and was used universally for many years in Europe. Salt cake (sodium sulfate) reacts with limestone to give soda ash and a troublesome side product gypsum (calcium sulfate). The process is no longer used.

$$Na_2SO_4 + CaCO_3 \xrightarrow{\Delta} Na_2CO_3 + CaSO_4$$

In 1864 Ernest Solvay, a Belgian chemist, invented his ammonia-soda process. A few years later the soda ash price was reduced one third. The Solvay process had completely replaced the LeBlanc method by 1915. The Solvay method is still very popular worldwide. However, in this country large deposits of natural trona ore were found in the 1940s in Green River, Wyoming. In the last few years there has been a tremendous conversion from synthetic to natural soda ash. The first and last Solvay plant in the United States (a large Allied Chemical plant in Solvay, NY) closed in 1986. Trona ore is found about 500 m below the surface. It is called sodium sesquicarbonate and is mostly $2Na_2CO_3 \cdot NaHCO_3 \cdot 2H_2O$ (45% Na_2CO_3, 36% $NaHCO_3$, 15% water + impurities). Heating this ore gives soda ash. The conversion from the Solvay process to natural soda ash has been called one of the most successful chemical industry transformations of the late 1970s and early 1980s. The ratio of production for selected years certainly proves this point.

	1948	1974	1981	1985	1986
Solvay	94	46	9	6	0
Natural	6	54	91	94	100

A new solution mining method that is claimed to be 25% cheaper may take over in the 1990s. Despite the fact that no new Solvay plants have been started since 1934 in this country, it is still an important method worldwide. There is some fascinating chemistry in this involved process and we will discuss it in detail.

Solvay Reactions

$$\left[\begin{array}{l} CaCO_3 \longrightarrow CaO + CO_2 \\ CaO + H_2O \longrightarrow Ca(OH)_2 \end{array}\right] \text{ source of } CO_2$$

$$2NH_3 + 2H_2O \longrightarrow 2NH_4OH \text{ source of } NH_4OH$$

$$\left[\begin{array}{l} 2NH_4OH + 2CO_2 \longrightarrow 2NH_4HCO_3 \\ 2NH_4HCO_3 + 2NaCl \longrightarrow 2NaHCO_3 + 2NH_4Cl \\ 2NaHCO_3 \longrightarrow Na_2CO_3 + CO_2 + H_2O \end{array}\right] \begin{array}{l} \text{main Solvay} \\ \text{reactions} \end{array}$$

$$2NH_4Cl + Ca(OH)_2 \longrightarrow 2NH_3 + CaCl_2 + 2H_2O \quad \text{recycle of } NH_3$$

overall reaction: $CaCO_3 + 2NaCl \longrightarrow Na_2CO_3 + CaCl_2$

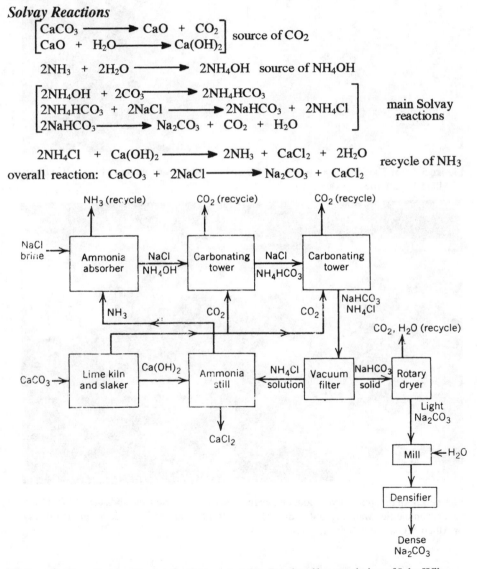

Figure 5.6 Manufacture of soda ash (*Source:* L & M. Reprinted by permission of John Wiley & Sons, Inc.)

Figure 5.7 Solid sodium bicarbonate was caught on rotating vacuum drum filters. (Courtesy of Allied-Signal, Inc., Solvay, NY)

Figure 5.8 In the rotary dryers sodium bicarbonate was converted into soda ash at 175°C and the carbon dioxide was recycled. This dryer was 100 ft long and 8 ft in diameter. (Courtesy of Allied-Signal, Inc., Solvay, NY.)

Figure 5.9 An ammonia still for recovery of the ammonia in the Solvay process. It was many stories high. (Courtesy of Allied -Signal, Inc., Solvay, NY.)

Description. A detailed description of salt mining will be postponed until the next chapter, but it is important to note that soda ash is made from both limestone and salt, the two major raw materials. As outlined in Fig. 5.6, the brine (salt solution) is mixed with ammonia in a large ammonia absorber. A lime kiln, using technology similar to that discussed earlier, serves as the source of carbon dioxide, which is mixed with the salt and ammonia in carbonation towers to form ammonium bicarbonate and finally sodium bicarbonate and ammonium chloride. Filtration (Fig. 5.7) separates the less soluble sodium bicarbonate from the ammonium chloride in solution.

The sodium bicarbonate is heated to 175°C in rotary dryers (Fig. 5.8) to give light soda ash. The carbon dioxide is recycled. Light soda ash is less dense than the natural material because holes are left in the crystals of sodium bicarbonate as the carbon dioxide is liberated. Dense soda ash, used by the glass industry, is manufactured from light ash by adding water and drying.

The ammonium chloride solution goes to an ammonia still (Fig. 5.9) where the ammonia is recovered and recycled. The remaining calcium chloride solution is an important by-product of this process, although in large amounts it is difficult to sell and causes a disposal problem.

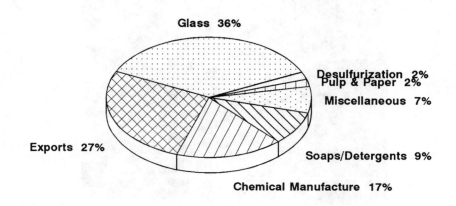

Figure 5.10 Uses of soda ash. (*Source*: CP.)

Uses

Fig. 5.10 outlines the uses of soda ash. Glass is the biggest industry using soda ash. The 36% used by this industry is divided into 60% bottles and containers, 29% flat glass, 7% fiberglass, and 4% other. Although exports of a chemical are usually included under miscellaneous, we list it separately here at 27% because it is a large amount and shows that, for some chemicals such as soda ash, a large percentage of the U.S. production is sometimes exported.

The glass industry is very complex and would take some time to discuss at length. There are about 500 different kinds of glass. However, 90% of all glass made is soda-lime-silica glass, which incorporates ingredients to be heated to give an approximate weight ratio of 70-74% SiO_2, 10-13% CaO, and 13-16% Na_2O. These glasses can be used for windows, containers, and many transparent fixtures. The sand must be nearly pure quartz, a crystalline form of silicon dioxide. These deposits often determine the location of glass factories. Sodium oxide is principally supplied from dense soda ash, but other sources of the oxide include sodium bicarbonate, sodium sulfate, and sodium nitrate. Some nitrate is generally used because it will oxidize iron impurities and avoid coloration of the glass. Limestone is the source of lime. When these substances are heated the following reactions occur.

$$\text{Na}_2\text{CO}_3 + a\,\text{SiO}_2 \longrightarrow \text{Na}_2\text{O} \cdot a\,\text{SiO}_2 + \text{CO}_2$$
$$\text{CaCO}_3 + b\,\text{SiO}_2 \longrightarrow \text{CaCO} \cdot b\,\text{SiO}_2 + \text{CO}_2$$

For common window glass the mole ratio may be 2 mol Na$_2$O, 1 mol CaO, and 5 mol SiO$_2$. Glass is essentially an amorphous, multicomponent solid mixture. Specific CaO—SiO$_2$ or Na$_2$O—SiO$_2$ compounds do not exist. The addition of borax increases the glass resistance to acids and thermal shock. This is called Pyrex® glass.

In many other uses soda ash competes directly with caustic soda as an alkali. The chemical of choice is then dependent on price and availability of the two.

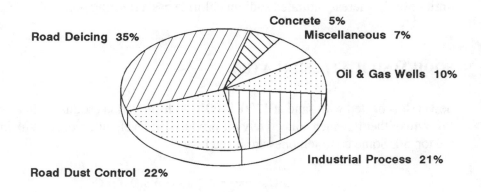

Figure 5.11 Uses of calcium chloride. (*Source*: CP.)

CALCIUM CHLORIDE, $CaCl_2$

Calcium chloride is obtained from natural brines (especially in Michigan). A typical brine contains 14% NaCl, 9% $CaCl_2$, and 3% $MgCl_2$. Evaporation precipitates the sodium chloride. The magnesium chloride is removed by adding slaked lime to precipitate magnesium hydroxide.

$$MgCl_2 + Ca(OH)_2 \longrightarrow Mg(OH)_2 + CaCl_2$$

The uses of calcium chloride are given in Fig. 5.11. A large amount of calcium chloride is used on roads for dust control in the summer and deicing in the winter. It is less corrosive to concrete than is sodium chloride. A debate on which is worse environmentally on local plant life because of high salt concentrations remains to be resolved. The home ice-melt market for calcium chloride has grown 7-8% per year recently. Local governments are also using more calcium chloride. Another recent competitor in the market is calcium magnesium acetate, made by reaction of high-magnesium content lime with acetic acid. If a more inexpensive route to acetic acid can be found, this salt could prove to be a noncorrosive alternative to the chlorides. Calcium chloride is used for some industrial refrigeration applications. Saturated calcium chloride does not freeze until - 50°C, whereas saturated sodium chloride has a freezing point of -20°C.

SODIUM SILICATE; SILICA GEL

Soda ash is heated with sand at 1200-1400°C to form various sodium silicates (over 40 of them), which collectively are produced at levels sufficient to rank in the top 50. Some common ones are listed here.

$$Na_2CO_3 + n\,SiO_2 \longrightarrow Na_2O \cdot n\,SiO_2 + CO_2$$

		Ratio of SiO_2/Na_2O
Sodium tetrasilicate	$Na_2Si_4O_9$	4
Sodium metasilicate	Na_2SiO_3	1
Sodium sesquisilicate	$Na_3HSiO_4 \cdot 5H_2O$	0.67
Sodium orthosilicate	Na_4SiO_4	0.50

As a fine *silica gel* with a large surface area they are used for catalysis and column chromatography. They are also popular as partial phosphate replacements in soaps and detergents. Fig. 5.12 gives the uses of sodium silicate.

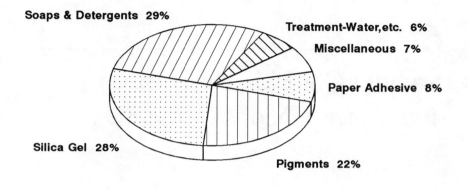

Figure 5.12 Uses of sodium silicate. (*Source*: CP.)

6

Sodium Chloride Derivatives and Miscellaneous Inorganics

SODIUM CHLORIDE, NaCl

References

Kent, pp. 212-213
L & M, pp. 722-730
Austin, pp. 213-215

This very important chemical, which is known by many names such as salt, common salt, rock salt, grainer salt, and brine solution, is not included in the top 50 because it is really a naturally occurring mineral. We sometimes forget this because, although it is a mineral, it occurs as a very pure chemical and is readily isolated. If it were included in the top 50 list, it would be near number 1 with sulfuric acid, since some 80 billion lb are processed each year in the United States. Salt mining must be nearly as old as humankind. It has been used as an object of worship, as a medium of exchange, and as a political weapon with a distribution dependent on high taxes. It is the oldest inorganic chemical industry. Sodium hydroxide, chlorine, hydrochloric acid, and some titanium dioxide, soda ash, and sodium sulfate are all top 50 chemicals that are made from salt or salt derivatives.

Isolation

There are three important methods of salt isolation and purification: brine solution, rock salt mining, and the open pan or grainer process. The percentages of these methods have not changed dramatically in the last few years and are 46% brine, 36% rock salt, and 18% grainer salt.

Brine. In this method water is pumped into the salt deposit and the saturated salt solution is removed containing 26% salt, 73.5% water, and 0.5% impurities. Hydrogen sulfide is removed by aeration and oxidation with chlorine. Ca^{+2}, Mg^{+2}, and Fe^{+3} are precipitated as the carbonates using soda ash. These are removed in a settling tank. The brine solution can be sold directly or it can be evaporated to give salt of 99.8% purity.

Rock Salt. Deep mines averaging 1000 ft are used to take the solid material directly from the deposit. Salt obtained by this method is 98.5-99.4% pure. Leading states producing rock salt and their percentages are Louisiana (30%), Texas (21%), Ohio (13%), New York (13%), and Michigan (10%).

Open Pan or Grainer Salt. Hot brine solution is held in an open pan approximately 4-6 m wide, 45-60 m long, and 60 cm deep at 96° C. Flat, pure sodium chloride crystals form on the surface and fall to the bottom. The crystals are raked to a centrifuge, separated from the brine, and dried. A purity of 99.98% is obtained. Grainer salt dissolves more readily and is preferred in some applications, such as the butter and cheese industries. It is more expensive because of energy use for the hot brine.

Uses and Economics

Fig. 6.1 outlines the uses of salt. Half is consumed in the important electrolysis of brine to form two top 50 chemicals, sodium hydroxide and chlorine. One fifth is used on highways for deicing and competes with calcium chloride in this application. The food industry and animal feeds make up a third important use. Salt is surpassed only by phosphate rock in total production of all minerals, as shown in Fig. 6.16, page 111. Salt has decreased somewhat in the last few years. It is still one of the cheapest chemicals at 3 ¢/lb in 1991. Its current production is 77 billion lb, giving a commercial value of $2.3 billion.

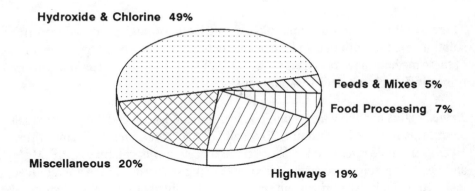

Figure 6.1 Uses of Sodium Chloride. (*Source:* Kent.)

CAUSTIC SODA, NaOH; SODIUM HYDROXIDE, CAUSTIC

References

Kent, pp. 222-231
L & M, pp. 737-745
Austin, pp. 231-239
T, pp. 106-120
CP, 4-18-83, 5-5-86, 6-19-89

Manufacture

For many years since its discovery in 1853 the "lime causticization" method of manufacturing caustic soda was used which involves reaction of slaked lime and soda ash.

$$Na_2CO_3 + Ca(OH)_2 \longrightarrow 2NaOH + CaCO_3$$

In 1892 the electrolysis of brine was discovered as a method for making both sodium hydroxide and chlorine. This rapidly grew in importance and since the 1960s it has been the only method of manufacture. Among electrolytic industries it is the second largest consumer of electricity, aluminum manufacture being the largest.

Year	% of NaOH by Electrolysis
1935	44
1940	50
1954	85
1962	100

Reaction

$$2NaCl + 2H_2O \xrightarrow{e-} 2NaOH + H_2 + Cl_2$$

95-97% current efficiency

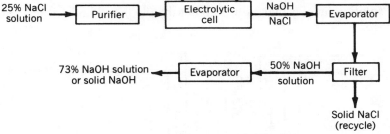

Figure 6.2 Electrolysis of brine. (*Source:* L & M. Reprinted by permission of John Wiley & Sons, Inc.)

Description. Fig. 6.2 shows the basics of an electrolysis plant. The brine that is used must be purified for this electrolytic process. Calcium, sulfate, and magnesium ions are removed by precipitation reactions.

$$Na_2CO_3 + CaCl_2 \longrightarrow CaCO_3 + 2NaCl$$
$$BaCl_2 + Na_2SO_4 \longrightarrow BaSO_4 + 2NaCl$$
$$2NaOH + MgCl_2 \longrightarrow Mg(OH)_2 + 2NaCl$$

Different types of electrolytic cells are employed and will be discussed in the next section. Details of chlorine gas purification will be covered later. The hydrogen that is generated can be used as fuel or can be combined with some of the chlorine to give high-purity hydrogen chloride gas. There are cheaper methods for making hydrochloric acid. The hydrogen can also be used in a neighboring ammonia manufacturing plant. Evaporation and filtration of the basic solution after electrolysis gives solid salt, which can be recycled, and an industrially popular 50% caustic solution. Further evaporation gives the solid caustic product.

Types of Electrolytic Cells

Two important types of cells are employed, the diaphragm cell (now 78% of all production) and the mercury cell (19%).

Diaphragm Cell. A simple diaphragm cell and the reactions occurring at the anode and cathode are summarized in Fig. 6.3.

The diaphragm prevents the diffusion of sodium hydroxide toward the anode. This wall allows for the slow passage of solution and the free passage of sodium ions. It is made of asbestos fibers supported on an iron screen (Fig. 6.4). The anode solution level is maintained higher than in the cathode compartment to

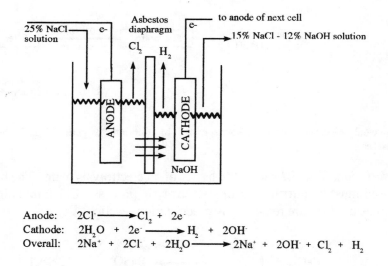

Anode: $2Cl^- \longrightarrow Cl_2 + 2e^-$
Cathode: $2H_2O + 2e^- \longrightarrow H_2 + 2OH^-$
Overall: $2Na^+ + 2Cl^- + 2H_2O \longrightarrow 2Na^+ + 2OH^- + Cl_2 + H_2$

Figure 6.3 Simplified diaphragm cell.

retard back migration. If sodium hydroxide built up near the anode it would react with chlorine to give sodium hypochlorite as a side product.

$$Cl_2 + 2NaOH \longrightarrow NaOCl + NaCl + H_2O$$

The anodes formerly were made of graphite but had to be replaced approximately every 250 days. New dimensionally stable anodes (DSA) are made of titanium with a coating of platinum, ruthenium, or iridium. They were developed around 1970 and have replaced graphite anodes in the United States. The cathodes are steel boxes with perforated steel plates. There are advantages of the diaphragm cell. They are much less polluting than the mercury cells. They are probably more economical to run. Most new installations are diaphragms.

Figure 6.4 A series of cathodes are fitted between the anodes. The cathodes are covered with asbestos, the membrane in most diaphragm cells. (Courtesy of Oxy Tech Systems, Tacoma, WA.)

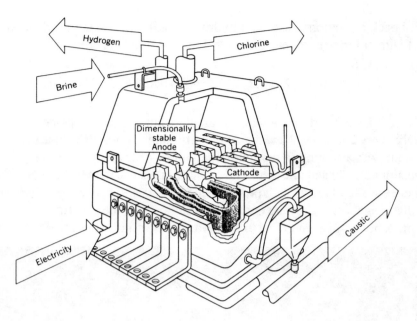

Figure 6.5 Typical diaphragm cell. (*Source*: Oxy Tech Systems, Tacoma, WA.)

Figure 6.6 One diaphragm cell. Brine enters through a small gray pipe on top. Chlorine exits through a large black rubber pipe on top; hydrogen exits through the large black rubber pipe on the front side leading up. Caustic leaves from the diagonal tube near the bottom of the cell and enters the black funnel shown. Each cell is linked to the next one electrically on the left and right sides. Compare this picture with the diagram shown in Fig. 6.5. Cells are approximately 6 ft cubes. (Courtesy of Oxy Tech Systems, Tacoma, Wa.)

They are definitely less expensive to operate in terms of electrical usage. Finally, the brine system is simpler and is a more economical operation. A more accurate design of a diaphragm cell is pictured in Fig. 6.5 and 6.6. Each cell is about 6 ft square and may contain 100 anodes and cathodes. A typical plant would have several circuits with approximately 90 cells in each circuit.

Mercury Cell. The mercury cell has no diaphragm but is made of two separate compartments as shown in Fig. 6.7. In the electrolyzing chamber the dimensionally stable anodes of ruthenium-titanium cause the chloride ion oxidation which is identical to that of a diaphragm cell. However, the cathode is simply a sodium amalgam flowing across the steel bottom of the cell at a slight angle from the horizontal. Notice that the cathode reaction, unlike the diaphragm cell cathode, involves the reduction of sodium ions to the metal.

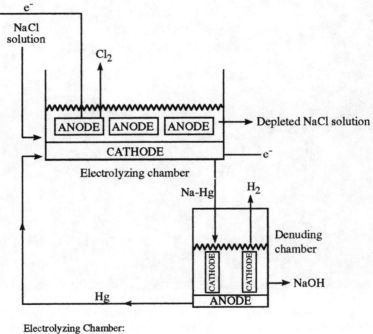

Electrolyzing Chamber:
Anode: $2\,Cl^- \longrightarrow Cl_2 + 2e^-$
Cathode: $2\,Na^+ + 2\,e^- \longrightarrow 2\,Na$
Denuding Chamber:
Anode: $2\,Na \longrightarrow 2\,Na^+ + 2\,e^-$
Cathode: $2\,H_2O + 2\,e^- \longrightarrow 2\,OH^- + H_2$
Overall: $2\,Na^+ + 2\,Cl^- + 2\,H_2O \longrightarrow 2\,Na^+ + 2\,OH^- + Cl_2 + H_2$

Figure 6.7 Simplified mercury cell.

This sodium amalgam enters a separate denuding chamber, sometimes called an amalgam decomposer, where the sodium metal reacts with water. Here the amalgam is the anode and the cathodes are graphite or iron. Hydrogen and caustic soda are the products here. The overall reaction, however, is identical to that of the diaphragm cell. The mercury is recycled into the electrolyzing chamber. Typical electrolyzing chambers measure 4 x 50 ft and are 1 ft high (Fig. 6.8). The decomposers, one for each cell, are 2 x 16 ft high cylinders. The main advantage of the mercury cell is the low contamination of sodium chloride in the final caustic soda. This caustic has only 230 ppm NaCl impurity as compared to the diaphragm cell's caustic at 1,000 ppm NaCl. Also, the sodium hydroxide solution does not require evaporation, and the chlorine is produced separately from the hydrogen and caustic, minimizing the hypochlorite side reaction and potential explosions.

Figure 6.8 These mercury cell electrolyzing chambers are 4 ft x 50 ft and 1 ft high. There are 32 cells in this room. A rubber cover resistant to chlorine is on top of each cell. The pipe leading from the top is for chlorine gas. This plant utilizes 400 tons/day of sodium chloride and manufactures approximately 220 tons/day of both chlorine and caustic. (Courtesy of Georgia-Pacific Corporation, Bellingham, WA.)

Newer Cell Membranes. Recently, a new type of diaphragm has been impacted on the industry for a few new plants. The new membranes are perfluorinated polymers with occasional sulfonate and/or carboxylate groups. They have the general structure outlined here. The anionic groups almost completely inhibit transport of hydroxide ions from the cathode, at the same time letting current flow in the form of sodium ions. The resulting caustic is purer and more concentrated while still avoiding the potential pollution of mercury cells. These cells have larger power requirements than asbestos diaphragm cells.

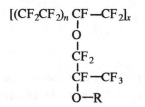

$$R = -CF_2CF_2CF_2CO_2^- \, Na^+ \quad \text{or} \quad -CF_2CF_2SO_3^-Na^+$$

Uses of Caustic Soda

Fig. 6.9 shows the important applications of sodium hydroxide. It has a very diverse use profile. It is an important reactant in organic and inorganic chemical manufacturing processes. It is used in many industrial sectors, including petroleum and pulp and paper, as the chief industrial alkali in most reactions.

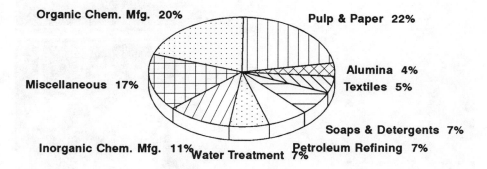

Figure 6.9 Uses of caustic soda. (*Source*: CP.)

CHLORINE

References

Kent, pp. 222-231
L & M, pp. 224-253
Austin, pp. 231-239
T, pp. 106-123
CP, 4-11-83, 4-28-86, 6-12-89

Manufacture

The electrolysis reaction and types of cells were described adequately under caustic soda. The chlorine gas, contaminated with water from the electrolytic cell, is cooled to 12-14°C to liquefy most of the water, then dried in a tower of

Figure 6.10 Two large caustic storage tanks hold a combined 4000 dry tons or 8000 wet tons of 50% caustic. The horizontal cylindrical chlorine storage tanks hold 300 tons each. (Courtesy of Georgia-Pacific Corporation. Bellingham, WA.)

sulfuric acid. The pure chlorine gas is compressed to 40 psi and condensed by cooling at -20 to -40°C to liquefy the gas. Storage vessels for both caustic and chlorine are shown in Fig. 6.10.

Properties

Chlorine is a toxic, greenish-yellow gas with a pungent, irritating odor, bp -35°C at atmospheric pressure. It is very corrosive when wet, and is soluble in water to the extent of 177 mL/100 g at 30°C.

Uses of Chlorine

Fig. 6.11 outlines the diverse uses of chlorine. Many important organic chemicals are made by chlorinations, some of which are in the top 50 list such as ethylene dichloride. Propylene oxide does not contain chlorine, but a chlorohydrin process is used to make it from propylene. This is followed by loss of hydrogen chloride. These chemicals, together with various other chlorinated organics, combine to require nearly two thirds of all chlorine for the organic chemicals sector.

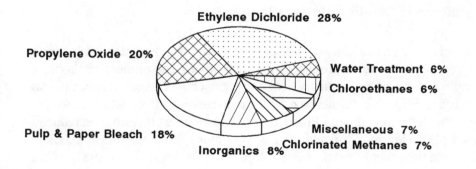

Figure 6.11 Uses of chlorine. (*Source*: CP.)

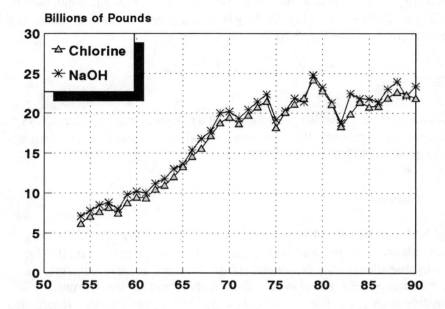

Figure 6.12 U.S. production of sodium hydroxide and chlorine. (*Source*: L & M and *C & E News*. Reprinted by permission of John Wiley & Sons, Inc.)

Economics of Caustic Soda and Chlorine

Fig. 6.12 very dramatically demonstrates the complete dependence of sodium hydroxide and chlorine production on each other. The production problem continually facing this industry is that two key chemicals are produced in an unvarying ratio while the demand for the two chemicals fluctuates. Lately, the chlorine demand is down. Production of caustic must therefore be reduced since a large inventory of chlorine gas costs money. The result is that caustic soda is in tight supply and the price is higher in some years (1974, 1982, 1990), shown in Fig. 6.13.

Chlorine is always produced at a slightly less amount than caustic. The invariable ratio is 1:1.13, which makes the balance of demand difficult. In most recent years chlorine has been in excess supply, primarily because of (1) the bad market slump of poly (vinyl chloride) plastic, made from ethylene dichloride through vinyl chloride, in housing and automobiles and (2) newer safety regulations on chlorinated hydrocarbons in general. The outlook for the next

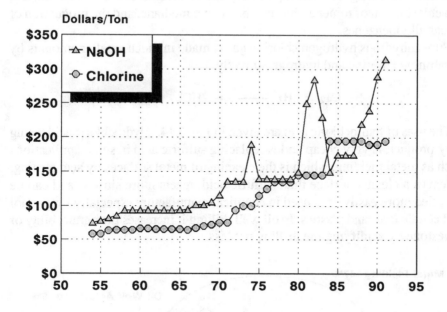

Figure 6.13 U.S. prices of sodium hydroxide and chlorine. (*Source*: L & M and CMR. Reprinted by permission of John Wiley & Sons, Inc.)

few years is a continuation of the stronger demand for caustic. What the industry really needs is an alternate method of economically making both compounds to more easily meet the diverging demands for the two chemicals.

HYDROCHLORIC ACID, HCl; MURIATIC ACID; AQUEOUS HYDROGEN CHLORIDE

Remember not to confuse hydrogen chloride, a colorless, poisonous gas with a pungent odor and a bp of -35°C, with hydrochloric acid, an aqueous solution of HCl typically with 24-36% HCl by weight. The principal manufacture of hydrochloric acid is as a by-product from the chlorination of many organic compounds. We will be discussing these processes under organic chemicals, but a single example here would be the chlorination of benzene. Thus hydrochloric acid is a derivative of chlorine. About 90% of it is made by various

$$\langle\bigcirc\rangle + Cl_2 \longrightarrow \langle\bigcirc\rangle\text{—Cl} + HCl$$

reactions including the cracking of ethylene dichloride and tetrachloroethane, the chlorination of toluene, fluorocarbons, and methane, and the production of linear alkylbenzenes.

Pure anhydrous hydrogen chloride gas is made in much smaller amounts by combining chlorine and hydrogen directly.

$$Cl_2 + H_2 \longrightarrow 2HCl$$

The uses of hydrochloric acid are given in Fig. 6.14. Hydrochloric acid, being a by-product, is very cheap and is replacing sulfuric acid in some applications such as metal pickling, which is the cleaning of metal surfaces by acid etching. It leaves a cleaner surface than sulfuric acid, reacts more slowly, and can be recycled more easily. It is used in chemical manufacture especially for phenol and certain dyes and plastics. In oil well drilling it increases the permeability of limestone by acidifying the drilling process.

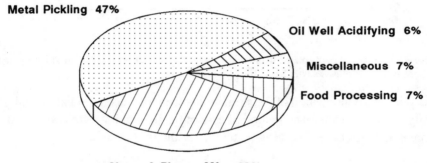

Figure 6.14 Uses of hydrochloric acid.

TITANIUM DIOXIDE

Presently there are two main processes for manufacturing this important white pigment. The main one involves reaction of rutile ore (about 95% TiO_2) with chlorine to give titanium tetrachloride. For this reason we have chosen to group this key chemical under chlorine and sodium chloride. The titanium tetrachloride is a liquid and can be purified by distillation, bp 136°C. It is then oxidized to pure titanium dioxide and the chlorine is regenerated. Approximately 90% of all titanium dioxide is made by this process.

$$3TiO_2 + 4C + 6Cl_2 \xrightarrow{900°C} 3TiCl_4 + 2CO + 2CO_2$$
$$TiCl_4 + O_2 \xrightarrow{1200-1400°C} TiO_2 + 2Cl_2$$

The other 10% of the product is made by taking ilmenite ore (45-60% TiO_2) and treating it with sulfuric acid for digestion and filtration. Hydrolysis of the sulfate and final heating gives pure titanium dioxide.

$$FeO \cdot TiO_2 + 2H_2SO_4 \longrightarrow FeSO_4 + TiOSO_4 + 2H_2O$$
$$TiOSO_4 + 2H_2O \longrightarrow TiO_2 \cdot H_2O + H_2SO_4$$
$$TiO_2 \cdot H_2O \xrightarrow{\Delta} TiO_2 + H_2O$$

The iron sulfate crystallizes out from the $TiOSO_4$ solution and can be recycled to make more sulfuric acid.

The use profile of titanium dioxide is given in Fig. 6.15. Titanium dioxide has been the best selling white pigment since 1939. We will discuss why this is so later when we study coatings as a unit.

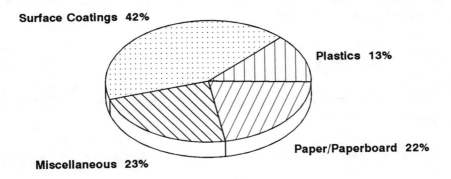

Figure 6.15 Uses of titanium dioxide. (*Source*: KC.)

MISCELLANEOUS INORGANIC CHEMICALS

There are three inorganic chemicals in the top 50 that we have not yet covered: sodium sulfate, potash, and carbon black, which are difficult to classify under a previous category.

Sodium Sulfate, Na₂SO₄; Salt Cake; Glauber's Salt (The Decahydrate)

Salt cake is obtained from a variety of sources. The Mannheim process previoulsy was the main source, involving the reaction of sodium chloride and sulfuric acid at very high temperatures (800-900°C). It no longer compares in energy costs to other methods and accounts for a very small percentage.

$$2NaCl + H_2SO_4 \longrightarrow Na_2SO_4 + 2HCl$$

Approximately one half (48%) of all salt cake is now obtained directly from natural salt sources in Searles Lake, California, or in in Texas and Utah. Brines with 7-11% Na_2SO_4 are used and pumped through a salt deposit to lower the solubility of the sodium sulfate so that upon cooling Glauber's salt $(Na_2SO_4 \cdot 10H_2O)$ will crystallize and can be separated. Heating then forms the anhydrous salt cake.

Around 10% of all sodium sulfate is obtained as a by-product in the production of vicose rayon. Sulfuric acid and sodium hydroxide are used to degrade the cellulose to rayon in a fiber-spinning bath. More about this later, but the simple salk cake reaction is given here.

$$2NaOH + H_2SO_4 \longrightarrow Na_2SO_4 + 2H_2O$$

Sodium dichromate manufacture gives another 13% of sodium sulfate as a by-product.

$$2Na_2CrO_4 + H_2SO_4 + H_2O \longrightarrow Na_2Cr_2O_7 \cdot 2H_2O + Na_2SO_4$$

Finally, the Hargreaves method of manufacture accounts for 5% of all sodium sulfate.

$$4NaCl + 2SO_2 + 2H_2O + O_2 \longrightarrow 2Na_2SO_4 + 4HCl$$

The current use of sodium sulfate includes detergents (55%), kraft sulfate pulping (25%), and glass (3%). The percentage of salt cake used in the kraft pulping digestion process has been steadily falling (it was 74% in 1968) because of a trend away from this method of making paper products. At the same time the amount used in detergents as a phosphate substitute has been increasing (it was 16% in 1968).

Potash

The industrial term *potash* can be very misleading. It can refer to potassium carbonate (K_2CO_3), potassium hydroxide (KOH), potassium chloride (KCl), potassium sulfate (K_2SO_4), potassium nitrate (KNO_3), or collectively to all potassium salts and to the oxide K_2O. More correctly KOH is called caustic potash and KCl is called muriate of potash. Production is recorded in weight equivalents of K_2O since almost all potash is used as fertilizer and this industry quotes weight percentages of K_2O in its trade.

Large deposits of sylvinite (42.7% KCl, 56.6% NaCl) near Carlsbad, New Mexico, account for 85% of the potassium products produced in the United States. The potassium chloride can be separated by either fractional crystallization or flotation. Potassium chloride is also obtained from the brines of Searles Lake, California. All these sources give potash (97% potassium chloride) with a 60% K_2O equivalent for fertilizer use. A chemical-grade product can be obtained to a purity of 99.9% potassium chloride. Almost all potash produced is potassium chloride. Potash is used almost exclusively as fertilizer (97%) with only a small amount (3%) used in chemical manufacture.

A small amount of potassium sulfate is isolated from natural deposits. Potassium nitrate is made by two synthetic processes.

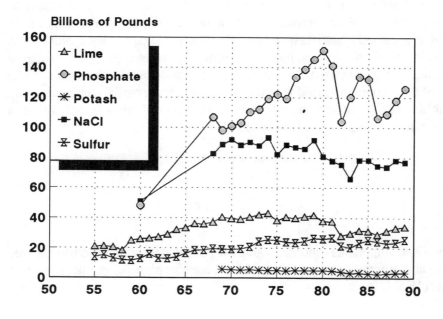

Figure 6.16 U.S. production of minerals. (*Source*: *C & E News*, "Facts and Figures.")

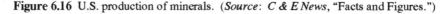

$$NaNO_3 + KCl \longrightarrow KNO_3 + NaCl$$
$$2KCl + 2HNO_3 + {}^1/_2O_2 \longrightarrow 2KNO_3 + Cl_2 + H_2O$$

Potassium hydroxide is made by electrolysis of potassium chloride solutions in cells that are exactly analogous to sodium hydroxide production.

$$2KCl + 2H_2O \xrightarrow{\ e^-\ } 2KOH + H_2 + Cl_2$$

A large amount of potash is imported, in contrast to other mineral production in the United States. In 1989, 3.4 billion lb of potash were processed in the U.S., but 9.4 billion lb were imported. This is a good time to compare some important minerals that we use to make various key inorganic chemicals, shown in Fig. 6.16. Phosphate rock leads in production and exporting. Salt is second in production, followed by lime, sulfur, and potash.

Carbon Black, C

This includes furnace black, colloidal black, thermal black, channel black, and acetylene black.

We will not debate whether carbon black is best treated as an inorganic or organic chemical. Both approaches have merit. Although certainly derived from petroleum, carbon black has uses similar to some inorganics; for example, it is the most widely used black pigment. Perhaps it is an appropriate bridge to complete our discussion of inorganics and to introduce petrochemicals. Over 90% of the amount produced is furnace black involving partial combustion or a combination of combustion and thermal cracking of hydrocarbons, and to a lesser extent natural gas, at 1200-1400°C.

$$-CH_2- \longrightarrow C + H_2$$
$$-CH_2- + 1{}^1/_2O_2 \longrightarrow CO_2 + H_2O$$
or
$$CH_4 \longrightarrow C + 2H_2$$
$$CH_4 + 2O_2 \longrightarrow 2CO_2 + H_2O$$

Carbon is an important reinforcing agent for various elastomers. It is used in tires (65%) and other elastomers (29%). It is a leading pigment in inks (3%) and paints (1%).

7

Petroleum Refining Processes

References

Kent, pp. 488-518
W & R I, pp. 35-54
D. Kolb and K. E. Kolb, "Petroleum Chemistry," *J. Chem. Educ..* **1979**, *56*, 465-469
H. Wittcoff, "Nonleaded Gasoline: Its Importance in the Chemical Industry," *J. Chem. Educ.* **1987**, *64*, 773-776
CEH, pp. 229. 3001 A-Y
Wiseman, pp. 13-42, 90-95
Szmant, pp. 33-35, 64-66

INTRODUCTION

Before beginning our study of pure organic chemicals, we need to obtain some background into the chemistry of petroleum, since it is from this source that nearly all the major organic chemicals are derived. Table 7.1 lists the seven important organic chemicals, all of which are obtained by petroleum-refining processes: ethylene, propylene, the butylenes, benzene, toluene, xylene, and methane. From these are made all 29 highest volume organic chemicals as shown in Table 7.1 (some have more than one source and are listed twice). It seems appropriate that we study petroleum and its major refining processes in detail before discussing these chemicals.

TABLE 7.1 Highest Volume Organic Chemicals Listed by Source

Ethylene	*Propylene*	*Benzene*
Ethylene dichloride	Acetone	Ethylbenzene
Ethylbenzene	Isopropanol	Styrene
Styrene	Propylene oxide	Cyclohexane
Vinyl chloride	Acrylonitrile	Phenol
Ethylene oxide	Cumene	Acetone
Ethylene glycol	Phenol	Adipic acid
Acetic acid		Cumene
Vinyl acetate	*C4 Fraction*	Caprolactam
Toluene	Butadiene	*Xylene*
	Acetic acid	
Benzene	Methyl *t-* butyl ether	Terephthalic acid
	Vinyl acetate	*p*-Xylene
		Dimethyl terephthalate
	Methane	
	Urea	
	Methanol	
	Formaldehyde	
	Acetic acid	
	Methyl *t-* butyl ether	
	Dimethyl terephthalate	
	Vinyl acetate	

Petroleum refining is not a part of our usual definition of the chemical industry, which includes Chemicals and Allied Products (SIC 28). However, the chemical process industries include those sectors of manufacturing as shown in Fig. 7.1: Paper and Allied Products (SIC 26), Chemicals and Allied Products (SIC 28), Petroleum and Coal Products (SIC 29), Rubber and Miscellaneous Plastics (SIC 30), and Stone, Clay, and Glass Products (SIC 32). All of these are important to the chemical industry. Petroleum refining is the largest part of Petroleum and Coal Products, which is about the same size as Chemicals and Allied Products. It provides the raw materials for a large portion of the chemical industry and employs many chemists. SIC 29 was down to $125 billion of shipments in 1986 after a record $224 billion in 1981. The slump is due mainly to decreased prices rather than production. In 1988 it still did not increase very much and was at $131 billion.

By far the major product of this industry is the gasoline fraction from petroleum. Fig. 7.2 demonstrates this, since U.S. shipments of gasoline were down in 1986-1988 as well. Other products such as jet fuel, kerosene, and fuel oils contribute substantially less to the total value of petroleum.

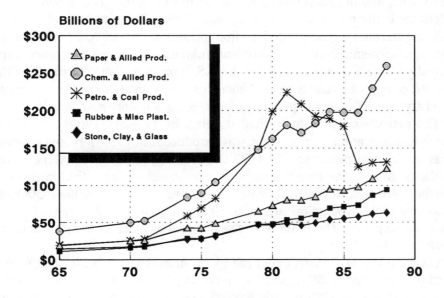

Figure 7.1 U.S. shipments in the chemical process industries. (*Source*: AS.)

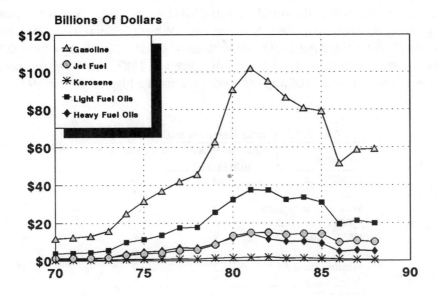

Figure 7.2 U.S. shipments of petroleum products. (*Source*: AS and CEH.)

Where are the chemicals derived from petroleum? The olefins--ethylene, propylene, and the butylenes--are derived from natural gas and petroleum. Methane is the major constituent in natural gas. The aromatics--benzene, toluene, and the xylenes--are derived from petroleum. About 90% by weight of the organic chemicals in the world comes from natural gas and petroleum. But actually only 3% of this crude oil in the U.S. is processed into chemicals, with the rest going as the various fuels. Although we are a small user of the petroleum industry, this 3% going to petrochemical feedstock is important to us!

The petrochemical industry had its birth in the early 1900s. In 1913 propylene, a byproduct of cracking, was introduced. In 1920 isopropyl alcohol was made from petroleum. In 1923 the first derivatives of ethylene were commercialized: ethylene chlorohydrin, ethylene glycol, and dichloroethane. By the 1940s petrochemicals were fully developed in the U.S. and the 1950s and '60s saw rapid production increases. The oil crisis of 1973 caused huge increases in prices. The 1980s were characterized by much slower growth rates than the '50s and '60s.

Oil is the largest segment of our energy raw materials use, being 46%, while coal use accounts for 28%, gas 19%, and hydroelectric and nuclear 7%. Table 7.2 summarizes the known world reserves of oil and the production by region. We immediately see that most countries, including the U.S., outside the Middle East region import oil in large amounts for their production and use. Two thirds of the known reserves in the world are in the Middle East. Fig. 7.3 demonstrates the growing dependence of the U.S. on imports. While our domestic production has grown some since the 1950s, imports have grown dramatically from 0.3 billion barrels of oil in 1955 to 1.7 billion barrels in 1987. Although we have decreased our percentage of imports somewhat from a high of 45% in 1977, it

TABLE 7.2 World Reserves and Production

Area	Known World Reserves, %	Oil Production, %
Middle East	64	23
U.S.	3	15
Western Hemisphere Other Than U.S.	14	13
USSR, China	9	28
Africa	6	9
Western Europe	3	7
Asia-Pacific	2	5

Source: CEH

was still 37% in 1987. The estimate for 1989 is unfortunately back up to a record 55%. A barrel (bbl) of crude oil is 42 gallons and 1 ton of crude oil is approximately 7.3 bbl. More meaningful figures of our tremendous use of oil can be quoted in terms of bbl used per day. We use approximately 18 million bbl/day of oil in this country. Worldwide production is about 56 million bbl/day. With only known reserves, this level of worldwide production could remain constant for only 43 years.

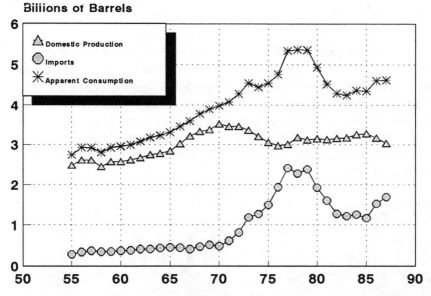

Figure 7.3 U.S. supply/demand for crude petroleum. (*Source*: CEH.)

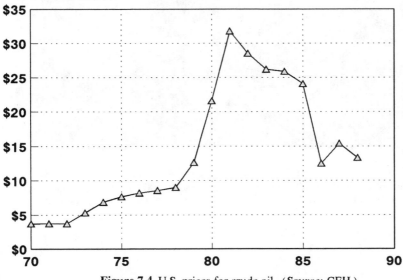

Figure 7.4 U.S. prices for crude oil. (*Source*: CEH.)

Finally we look at the price of oil. Fig. 7.4 shows the average U.S., domestic wholesale price for a barrel of oil. Note the very low prices in the early 1970s, the large increases in the late 1970s after the oil embargo, and the gradual leveling and final drop of prices in the late 1980s. An uncertain future lies ahead. The highest price of oil thus far has been a brief period at $40 /bbl in late 1990, immediately after Iraq entered Kuwait.

DISTILLATION

Several thousand compounds are present in petroleum. Few are separated as pure substances. Many of the uses of petroleum can be served by certain fractions from the distillation of crude oil. Typical distillation fractions and their uses are given in Table 7.3 and a distillation unit is shown in Fig. 7.5. The complexity of the molecules, molecular weight, and carbon number increase with the boiling point. The higher boiling fractions are usually distilled *in vacuo* at lower temperature than their atmospheric boiling points to avoid excessive decomposition to tars.

Figure 7.5 Large petroleum distillation columns like this one in the foreground can process over 400,000 barrels of crude oil per day into nearly 210,000-230,000 barrels of gasoline. That's enough to fill 678,000 13-gallon automobile tanks or 441,000 20-gallon automobile tanks. (Courtesy of Amoco Oil Co., Texas City, TX.)

TABLE 7.3 Fractions of Petroleum

Approximate bp (°C)	Name	Uses
<20°C	Gases	CH_4, C_2H_6, C_3H_8, C_4H_{10}—similar to natural gas and useful for fuel and chemicals. Much of it, however, is flared because of expensive recovery.
20-150°C	Light naphtha (straight run gasoline)	Predominantly C_4-C_{10} aliphatic and cycloaliphatic compounds. May contain some aromatics. Useful for both fuel and chemicals.
150-200°C	Heavy naphtha	
175-275°C	Kerosene	Contains C_9—C_{16} compounds useful for jet, tractor, and heating fuel.
200-400°C	Gas oil	Contains C_{15}—C_{25} compounds useful for diesel and heating fuel.
>350°C	Lubricating oil	Used for lubrication. May be catalytically cracked to lighter fractions.
>350°C	Heavy fuel oil	Boiler fuel. May be catalytically cracked to lighter fractions.
	Asphalt	Paving, coating, and structural uses.

Source: W & R I. Reprinted by permission of John Wiley & Sons, Inc.

Each fraction of distilled petroleum still contains a complex mixture of chemicals but they can be somewhat categorized. A certain sample of straight-run gasoline (light naphtha) might contain nearly 30 aliphatic, noncyclic hydrocarbons, nearly 20 cycloaliphatic hydrocarbons (mainly cyclopentanes and cyclohexanes) sometimes called napthenes, and 20 aromatic compounds. Although petroleum is basically made up of hydrocarbons, there are smaller amounts of other types of materials. For example carboxylic acids occur to the extent of about 0.1-3%. These can be isolated quite easily by base extraction and the mixture is known as naphthenic acid:

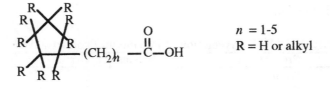

$n = 1-5$

$R = H$ or alkyl

A smaller amount of cyclohexyl derivatives are included in the mixture as well. A small percentage of petroleum is made up of compounds containing sulfur in one form or another. Examples of such compounds follow:

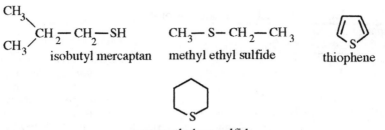

isobutyl mercaptan   methyl ethyl sulfide   thiophene

pentamethylene sulfide

When gas or heating oil is used as a source of energy and burned to CO_2 and H_2O, the sulfur ends up as SO_2 in the air. The SO_2 is a major air contaminant, especially in larger cities. With air moisture it can form H_2SO_4 and H_2SO_3. Much of the sulfur-containing compounds must be taken out of petroleum before it can be used for this purpose. The current maximum percentage allowable in gasoline is 0.10% S. Nitrogen-containing compounds are present in petroleum and form NO_2 upon combustion. The problem is not quite so bad, however, since most crude petroleum contains only 0.008% N, although even at this percentage it causes environmental concerns. Inorganic compounds are also present (sand, clay, and salt) but are more easily removed.

OCTANE NUMBER

The petroleum fraction that is the most important for the United States is gasoline. Let us take a closer look at some of the important aspects of gasoline. U.S. production was approximately 2.5 billion bbl in 1987. This is about 51% of the 5.3 billion bbl of refinery products and is way ahead of all other products, as shown in Table 7.4.

TABLE 7.4 U.S. Production of Petroleum Products, 1987

Product	Billions of bbl/yr
Gasoline	2.50
Distillate Fuel Oil	1.00
Residual Fuel Oil	0.32
Jet Fuel	0.49
Petrochemical Feedstock	0.14
Other[a]	0.88
Total	5.33

[a]Other includes kerosene, gases, lubricants, wax, road oil, asphalt, and coke

One cannot talk about the chemistry of gasoline without understanding the octane number. When gasoline is burned in an internal combustion engine to CO_2 and H_2O, there is a tendency for many gasoline mixtures to burn unevenly. This is caused basically by ignition before the piston of the engine is in the proper position. Such nonconstant and unsmooth combustion creates a "knocking" noise in the engine. It has been found that certain hydrocarbons burn more smoothly than others in the gasoline mixture. In 1927 a scale was set up that attempted to define the "antiknock" properties of gasolines. At the time, 2,2,4-trimethylpentane (commonly called "isooctane") was the hydrocarbon that, when burned pure in an engine, gave the best antiknock properties (caused the least knocking). This compound was assigned the number 100, meaning it was the best hydrocarbon to use. The worst hydrocarbon they could find in gasoline that when burned pure gave the most knocking was *n*-heptane, assigned the number 0. When isooctane and heptane were mixed together they gave different amounts of knocking depending on their ratio: the higher the percentage of isooctane in the mixture, the lower was the amount of knocking. Then gasoline mixtures obtained from petroleum were burned for comparison. If a certain gasoline has the same amount of knocking as a 90% isooctane, 10% heptane (by volume) mixture, we now say that its octane number is 90. *Hence the octane number of a gasoline is the percent isooctane in an isooctane-heptane mixture that gives the same amount of knocking as the gasoline being measured.*

Thus a *high* octane number means a *low* amount of knocking. The development of very high compression engines, especially for jet airplanes, now makes it necessary to extend the octane number scale beyond 100 with the use of additives.

Now there are two octane scales, a research octane number (RON) and a motor octane number (MON). *RON* values reflect performance at 600 rpm, 125°F, and low speed. *MON* is a performance index of driving with 900 rpm, 300°F, and high speed. Before 1973 RON values were the ones usually quoted to the public, but since 1973 the octane values posted on station pumps have been RON—MON averages. The average value better relates to the actual performance of the gasoline in an automobile engine. Concurrently, with the introduction of this new average scale, refiners also lowered the octane quality of their gasolines by about two units. As a result, some motorists began noticing knocking noises in their engines, even though they thought they were using the same gasoline they had always used. The MON is about six units lower than the RON. The pump now gives the (R & M)/2 value. Regular is usually 87-89 and premium about 92 on this scale.

Certain rules have been developed for predicting the octane number of different types of gasoline, depending on the ratio of different types of hydrocarbons in the mixtures:

1. The octane number increases as the amount of branching or number of rings increases. Example: 2,2,4-trimethylpentane causes a higher octane number than *n*-octane; methylcyclohexane causes a higher octane number than *n*-heptane.

Octane number:

$$
\begin{array}{c}
\quad\quad CH_3 \quad\quad CH_3 \\
\quad\quad | \quad\quad\quad | \\
CH_3-C-CH_2-CH-CH_3 \quad > \quad CH_3-(CH_2)_6-CH_3 \\
\quad\quad | \\
\quad\quad CH_3
\end{array}
$$

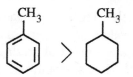

$$> CH_3-(CH_2)_5-CH_3$$

Some typical RON values are *n*-heptane, 0; *n*-octane, -19; ethylcyclohexane, 46; methylcyclohexane, 75.

2. The octane number increases as the number of multiple bonds increases. Example: Toluene causes a higher octane number than does methylcyclohexane.

RONs:

Ethylbenzene	107
Toluene	120
Xylenes	116-120

3. Summary:

Octane number: Aromatics, alkenes, and alkynes >
Cyclic alkanes and branched alkanes >
Straight-chain alkanes

If you recall that combustion is a free radical process, we can easily see why cyclic and branched alkanes burn more easily (and more smoothly) than straight-chain alkanes. The reason is that more stable free radicals are formed. This results in less knocking and a higher octane rating. Examples of free radical stability are the following:

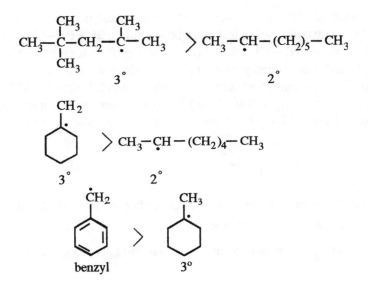

ADDITIVES

In 1922 two chemists working at General Motors, Midgley and Boyd, were looking at different substances that would aid combustion of gasoline and help the knocking problems of engines. In other words, they were seeking methods of increasing the octane rating of gasoline without altering the hydrocarbon makeup. They were also interested in cleaning up the exhaust of automobiles by eliminating pollutants such as unburned hydrocarbons and carbon monoxide through more complete combustion. By far the best substance that they found was tetraethyllead.

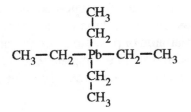

This relatively cheap material is made from a Pb—Na alloy and ethyl chloride:

$$4Pb\text{—}Na + 4Et\text{—}Cl \longrightarrow Pb(Et)_4 + 4NaCl + 3Pb$$
$$bp = 202^\circ$$

The material added to gasoline to increase octane is called "ethyl" fluid. A typical mixture contains the following: 63% $PbEt_4$, 26% Br—CH_2—CH_2—Br, 9% Cl—CH_2—CH_2—Cl, 2% dye (as a warning of its toxicity). About 1-6 ml of ethyl fluid is added per gallon of gasoline, depending on the octane number desired. The maximum lead content of gasoline was 0.5 g/gal in 1985, compared to 4 g/gal in the early 1970s. The EPA (Environmental Protection Agency) suggested 0.1 g in 1986. Tetraethyllead apparently burns to form lead dioxide.

$$Pb(Et)_4 \xrightarrow{\text{O}_2} PbO_2 + CO_2 + H_2O$$

Lead in this form complexes with hydrocarbons and aids in breaking carbon-carbon and carbon-hydrogen bonds.

$$PbO_2 + \text{hydrocarbons} \longrightarrow PbO_2 + CO_2 + H_2O$$

The lead oxide is not volatile and would accumulate in the engine if dibromoethane and dichloroethane were not added. These substances react with PbO_2 and form a volatile compound, $PbBr_2$ or $PbCl_2$, which is eliminated in the exhaust:

$$Br\text{—}CH_2\text{—}CH_2\text{—}Br + PbO_2 \xrightarrow{\text{O}_2} PbBr_2 + CO_2 + H_2O$$

In the environment the lead dihalide undergoes oxidation-reduction by sunlight to elemental lead and halogen, both of which are serious pollutants:

$$PbBr_2 \xrightarrow{h\nu} Pb^\circ + Br_2$$

Before 1970 there was very little unleaded gasoline on the market, but by 1974 all gas stations were offering it. In 1974, unleaded fuel had become a necessity for most new cars because of their catalytic converters placed in the exhaust system. These contain platinum or palladium compounds that act as a surface catalyst to burn the hydrocarbons more completely. But lead coats the platinum and palladium and deactivates the converters, so unleaded gas must be used.

Current federal regulations call for phasing out all leaded gasoline by 1995. Fig. 7.6 shows U.S. consumption of unleaded and leaded gasoline since 1975. The percentage of unleaded gasoline has increased steadily from 16% in 1975 to 76% in 1987.

Millions of Barrels/Day

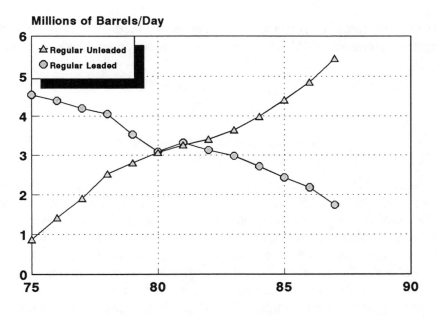

Figure 7.6 U.S. consumption of gasoline. (*Source*: CEH.)

This change to unleaded gasoline may or may not be a wise decision. Unleaded gasoline requires the much larger and more extensive use of modern refining processes such as cracking and reforming of straight-run or natural gasoline. These processes increase the percentage of aromatic, olefins, and branched hydrocarbons and thus increase the octane number. The resultant gasoline is more expensive. However, even more importantly, these new gasolines do *not* solve the pollution problem. They solve the lead pollution problem, but unleaded gasolines show larger emissions of other contaminants. Of particular importance to the environment is the increase of certain *hydrocarbon emissions,* especially *carbon monoxide.* Certain unburned aromatic hydrocarbons and alkenes absorb sunlight readily and cause *smog.* Other aromatic hydrocarbons, such as benzopyrene, have been identified as known *carcinogens.* The effect of having these types of contaminants in the atmosphere has not yet been thoroughly studied.

benzopyrene

For the past several years other additives have been tried. Methylcyclopentadienyl manganese tricarbonyl (MMT) has been used but is not presently added because

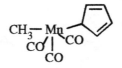

methylcyclopentadienyl manganese
tricarbonyl (MMT)

it has been found to be a potential health hazard. In 1977 EPA began allowing the use of *t*-butyl alcohol up to 7%. More recently 50:50 mixtures of *t*-butyl alcohol and methyl alcohol are being used, and ethyl alcohol has also been approved. The most attractive alternative to tetraethyllead is now methyl *t*-butyl ether (MTBE), although it is sometimes more expensive than tetraethyllead. MTBE has been approved at the 7% level since 1979. In 1984 MTBE broke into the top 50 chemicals for the first time and from 1985-90 its production grew 27% per year, the largest increase by any of the top 50 chemicals.Note that all these additives would be expected to be good free radical initiators. A weak bond (especially the carbon-metal bond) or a stable free radical formed after breaking a bond is the important feature of these additives.

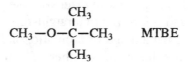

HYDROTREATING

Before other processes such as cracking and reforming are used to increase the octane rating, hydrotreating must occur. The distilled petroleum fractions are reacted with hydrogen at 400°C with a cobalt oxide/molybdenum oxide catalyst. The main reason for this reaction is to decrease the percentages of nitrogen- and sulfur-containing compounds, not only to lower pollution caused by these compounds when they were burned, but also to assure that no poisoning of catalysts in further refinery operations occurs. Sulfur compounds are notorious for this poisoning. Examples of hydrotreating reactions of molecules typically found in most oil feeds are the following:

$$\text{R-thiophene} + 4H_2 \longrightarrow CH_3-\underset{\underset{R}{|}}{C}H-CH_2-CH_3 + H_2S$$

$$R-CH_2-CH_2-CH_2-NH_2 + H_2 \longrightarrow R-CH_2-CH_2-CH_3 + NH_3$$

The hydrogen sulfide and ammonia can be removed by extraction with base and acid respectively.

As side reactions to this hydrotreating, some carbon-carbon double bonds are hydrogenated. Olefins are converted partially into alkanes and aromatics into cyclic alkanes. These reactions actually decrease the octane rating of the gasoline somewhat, but further refinery operations such as cracking and reforming will restore and increase the percentage of olefin and aromatic compounds. The temporary formation of more saturated compounds is necessary to get the sulfur and nitrogen percentages down.

$$R-CH_2-CH=CH_2 + H_2 \longrightarrow R-CH_2-CH_2-CH_3$$

$$R-\text{benzene} + 3H_2 \longrightarrow R-\text{cyclohexane}$$

CRACKING

There are other processes that are used to refine petroleum to make it more appealing to a specific use. These are important processes for the gasoline fraction because they increase the octane rating. Some are used to increase the percentage of crude oil which can be used for gasoline. These processes are also important in the production of the key organic chemicals shown in Table 7.1, so we should be familiar with them.

One such process is cracking (Fig. 7.7). In catalytic cracking, as the name implies, petroleum fractions of higher molecular weight than gasoline can be heated with a catalyst and cracked into smaller molecules that have a higher number of double bonds. This material can then be added to straight-run gasoline.

$$\text{gas oil or kerosene} \xrightarrow[\text{silica-alumina}]{450\text{-}550^\circ C} \text{cracked gasoline}$$
$$(C_{12} \text{ and higher}) \qquad\qquad\qquad (C_5\text{-}C_{10})$$

A simplified example would be the following:

$$C_8-CH_2-CH_2-CH_2-CH_3 \longrightarrow C_8-CH=CH_2 \; + \; CH_3-CH_3$$

Figure 7.7 A typical large catalytic cracking unit (on the right in this picture) can process 110,000 barrels/day. (Courtesy of Amoco Oil Co., Texas City, TX.)

Catalytic cracking usually involves carbocations, but the mechanism is uncertain.

Although important to the gasoline industry, catalytic cracking is not a major route to petrochemicals. Thermal cracking involves higher temperatures of 850-900°C in the absence of a catalyst. It gives much higher percentages of C_2, C_3, and C_4 olefins and relatively low yields of gasoline. It was superseded for gasoline production by catalytic cracking and was only revived with the demand for ethylene production in the chemical industry. Only 9% of total U.S. refinery cracking is thermal, but this is the only way in which olefins for the chemical industry are made. The lighter petroleum fractions such as naphtha are cracked thermally to give mixtures rich in ethylene, propylene, butadiene, and BTX (benzene, toluene, xylenes). Even ethane and propane are cracked. When ethane is "cracked" to ethylene it of course loses no carbons, but it does lose two hydrogens. More examples:

$$CH_3\text{---}CH_2\text{---}CH_3 \longrightarrow CH_2{=}CH_2 \ + \ CH_4$$

$$CH_3\text{---}CH_3 \longrightarrow CH_2{=}CH_2 \ + \ H_2$$

Thermal cracking is a free radical chain reaction. The mechanism follows.

$$n\text{-}C_{10}H_{22} \xrightarrow{\ \Delta\ } 2CH_3\text{---}CH_2\text{---}CH_2\text{---}CH_2\text{---}CH_2 \cdot \tag{1}$$

$$CH_3\text{---}CH_2\text{---}CH_2 \cdot \ \cdot CH_2\text{---}CH_2 \longrightarrow CH_2\text{---}CH_2\text{---}CH_2 \cdot \ + \ CH_2{=}CH_2 \tag{2}$$

(a ß scission)

$$CH_3 \cdot \ \cdot CH_2\text{---}CH_2 \cdot \longrightarrow CH_3 \cdot \ + \ CH_2{=}CH_2 \tag{3}$$

$$CH_3 \cdot \ + \ R\,H \longrightarrow CH_4 \ + \ R \cdot \tag{4}$$

then (2) - (4), etc.

$$\text{or} \ \ CH_3\text{---}CH\overset{H}{\text{---}}CH_2 \cdot \longrightarrow CH_3\text{---}CH{=}CH_2 \ + \ H \cdot \tag{5}$$

$$H \ + \ R\,H \longrightarrow H_2 \ + \ R \cdot \tag{6}$$

then (2), (5), (6) etc.

An alternative to step (1) for $n\text{-}C_{10}H_{22}$ involves breaking a C-H bond to give a 2° radical, which then can undergo its own ß-scission. Although ß-scissions of C-H bonds can also happen, C-C bonds are weaker so these are preferred.

$$n\text{-}\ C_{10}H_{22} \xrightarrow{\Delta} C_7H_{15}\text{---}CH_2\text{---}\overset{\bullet}{C}H\text{---}CH_3 \ + \ H\bullet \qquad (1)$$

$$C_7H_{15}\bullet\,CH_2\text{---}\overset{\bullet}{C}H\text{---}CH_3 \longrightarrow C_7H_{15}\bullet \ + \ CH_2\!\!=\!\!CH_2\text{---}CH_3 \qquad (2)$$

$$C_7H_{15}\bullet \longrightarrow C_5H_{11}\bullet \ + \ CH_2\!\!=\!\!CH_2 \qquad (3)$$

To maximize the amount of ethylene in the product, which is the idea in an olefin plant, the number of ß-scissions are maximized. Higher temperatures favor this, because ß-scissions have a high energy of activation. Also, since the ß-scission is a unimolecular process, whereas other possible reactions are bimolecular, a low concentration of hydrocarbon is preferred. Thus steam is used as a diluent. When thermal cracking is used, ethylene percentages can be as high as 76%; in catalytic cracking the percentage is less than 1%.

REFORMING

Catalytic reforming leaves the number of carbon atoms in the feedstock molecules unchanged but the mixture contains a higher number of double bonds and aromatic rings. Reforming has become the principal process for upgrading gasoline. High temperatures with typical catalysts of platinum or rhenium on alumina and short contact times are used. A typical example is the reforming of methylcyclohexane to toluene. It is done in the presence of hydrogen (hydroforming) to control the rate and extent of this dehydrogenation process. Straight-run gasoline can be reformed to as high as 40-50% aromatic hydrocarbons, of which 15-20% is toluene. Reformed petroleum is our main BTX source. Other examples follow on page 131:

Dehydrogenation

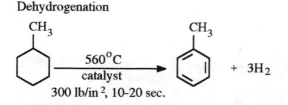

Dealkylation and Dehydrogention

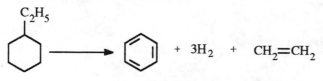

Rearrangement and Dehydrogenation

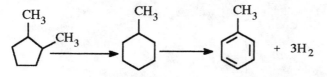

Cyclization and Dehydrogenation

$$CH_3-CH_2-CH_2-CH_2-CH_2-CH_3 \longrightarrow \text{⬡} + 4H_2$$

The "platforming" process, where reforming occurs with a platinum catalyst and the surface of this catalyst acts as a "platform" for the reaction, has been well named (Fig. 7.8).

Although the mechanism of the platinum catalysis is by no means completely understood, chemists do know a lot about how it works. It is an example of a dual catalyst: platinum metal on an alumina support. Platinum, a transition metal, is one of many metals known for its hydrogenation and dehydrogenation catalytic effects. Alumina is a good Lewis acid and as such easily isomerizes one carbocation to another through methyl shifts. Thus there is a hydrogenation function and an acidic function present in the catalyst, as diagramed in Fig. 7.9, page 132. Together simple aliphatics can be converted into aromatics. The mechanism for the conversion of hexane into benzene is given in Fig. 7.10, page 134. Basically it is a series of alternating dehydrogenations and carbocation rearrangements. Note that this conversion requires a 3° to 1° carbocation rearrangement to expand the ring size. Although this is unusual since 3° carbocations are more stable than 1° ions, we must remember that this occurs

Figure 7.8 An aerial view of a catalytic reforming processing plant. The reactors are the 21-ft spherical objects in the middle. These contain platinum and are in a series so that the octane is increased a little more in each reactor. (Courtesy of Amoco Oil Co., Texas City, TX.)

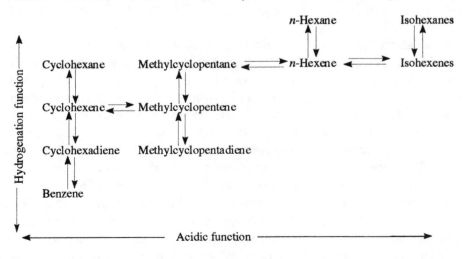

Figure 7.9 Functions of a dual catalyst. (*Source*: Reprinted with permission from *Chem. Eng. News* **1982**, 60(46), 12. Copyright 1982 American Chemical Society.)

catalytically. Complexation to the Lewis acid catalytic surface makes the 1°
carbocation stable enough to form, albeit as a reactive intermediate. The driving
force for this rearrangement is the resonance stabilization of the final aromatic
ring.

ALKYLATION AND POLYMERIZATION

Although cracking and reforming are by far the most important refinery
processes, especially for the production of petrochemicals, two other processes
deserve mention. In alkylation paraffins react with olefins in the presence of an

Reaction:

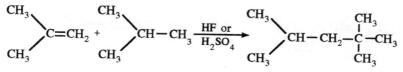

Mechanism:

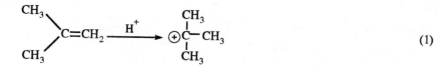

(1)

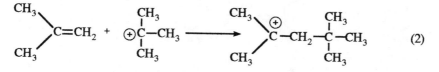

(2)

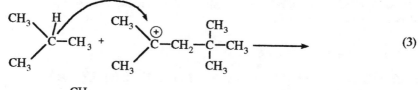

(3)

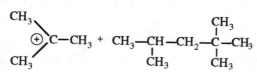

then (2), (3), (2), (3), etc.

acid catalyst to give highly branched alkanes. Isobutylene and isobutane can react to give 2,2,4-trimethylpentane ("isooctane") which can be added to straight-run gasoline to improve the octane. The mechanism is well understood as a carbocation chain process involving a hydride shift, shown on page 133.

In polymerization an olefin can react with another olefin to generate dimers, trimers and tetramers of the olefin. As a simple example, isobutylene reacts to give a highly branched C_8 olefin.

$$2CH_2-\underset{\underset{CH_3}{|}}{C}=CH_2 \xrightarrow{\Delta} CH_3-\underset{\underset{CH_3}{|}}{\overset{\overset{CH_3}{|}}{C}}-CH_2-\underset{\underset{CH_3}{|}}{C}=CH_2$$

In general, polymerization gives an average RON of 97, alkylation gives about 96.

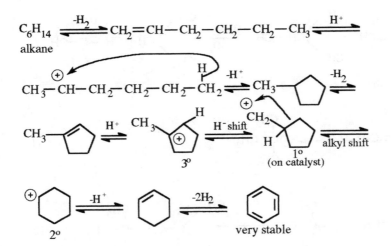

Figure 7.10 Mechanism of catalytic reforming. (*Source:* Wiseman. Reprinted with permission by Ellis Horwood Limited, Chichester.)

SEPARATION OF NATURAL GAS: METHANE PRODUCTION

You may have noticed that, of the seven basic organic building blocks given in Table 7.1, only six of them are considered "chemicals" and are included in the *C & E News Top 50*. Methane is certainly an important substance, but it is really not commercially made by a chemical reaction as are the other six, which we will study in more detail in the next chapter. Methane is naturally occurring and can be as high as 97% natural gas, the remainder being hydrogen, ethane, propane, butane, hydrogen sulfide, and heavier hydrocarbons. A typical mixture contains 85% methane, 9% ethane, 3% propane, 1% butanes, and 1% nitrogen. Most of the natural gas is used as fuel, but about 28% of the 23 trillion cu ft (TCF) per year in the United States is used by the chemical industry. If we estimate natural gas consumption in mass rather than volume, the 23 trillion cubic feet is approximately a trillion lb of methane. Of this, the one fourth used for chemical manufacture is about 250 billion lb of methane. So the methane used by the chemical industry does compare to other raw materials such as phosphate rock in amount consumed. A typical price is 7-9 ¢/lb. Uses of natural gas by all industry include fuel (72%), inorganic chemicals including ammonia (15%), organic chemicals (12%), and carbon black (1%). The ethane and propane are converted to ethylene and propylene. The methane is purified and used to make a number of chemicals.

A simplified schematic for natural gas separation is given here and consists of the following steps.

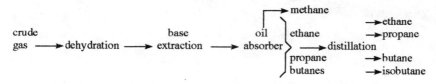

1. Dehydration by passing through diethylene glycol, in which water is very soluble.

$$HO-CH_2-CH_2-O-CH_2-CH_2-OH$$

2. Elimination of hydrogen sulfide and carbon dioxide with aqueous mono- or diethanolamine.

$$HO-CH_2-CH_2-\overset{\overset{\displaystyle H}{|}}{N}-CH_2-CH_2-OH \ + \ H_2S \longrightarrow$$

$$HO-CH_2-CH_2-\overset{\overset{\displaystyle H}{|}\oplus}{\underset{\displaystyle |}{\underset{\displaystyle H}{N}}}CH_2-CH_2-OH \ + \ HS^-$$

$$HO-CH_2-CH_2-\overset{\overset{\displaystyle H}{|}}{N}-CH_2-CH_2-OH \ + \ H_2CO_3 \longrightarrow$$

$$HO-CH_2-CH_2-\overset{\overset{\displaystyle H}{|}\oplus}{\underset{\displaystyle |}{\underset{\displaystyle H}{N}}}CH_2-CH_2-OH \ + \ HCO_3^-$$

3. Dissolution if the higher boiling gases in an oil absorber of hexane, leaving the methane separated.
4. Fractional distillation of the oil to recover the oil and to collect the ethane, propane, and isobutane separately.

Figure 7.11 Storage tanks for crude oil can be huge. These handle 750,000 barrels, about the size of one oil tanker and perhaps two days supply for the oil refinery, A floating, expandable top enables minimal pressure variation with temperature. (Courtesy of Amoco Oil Co., Texas City, TX.)

8

Basic Organic Chemicals

As we saw in Table 7.1 page 114, the major organic chemicals are all derived from seven basic ones: ethylene, propylene, the C_4 fraction, benzene, toluene, xylene, and methane. The production of methane, the major constituent in natural gas, has already been examined. We now consider in detail the manufacture, uses, and economic aspects of the other basic six organics. This will lead us into a discussion of the derivatives of each of them and their technology. We treat the seven basic ones first because, in addition to their importance, there is some similarity in their manufacture. For instance, ethylene, propylene, and the C_4 fraction are all made by steam (thermal) cracking of hydrocarbons. Just how significant are these seven basic organics? Nearly all organic chemicals and polymers are derived from them.

ETHYLENE (ETHENE) $CH_2{=}CH_2$

References

L & M, pp. 376-384
W & RI, pp. 55-58
Wiseman, pp. 30-42
CP, 3-18-91

Manufacture

Most ethylene and propylene is made by the *thermal cracking*, sometimes called *steam cracking*, of hydrocarbons at high temperatures with no catalyst. In contrast to the catalytic cracking used by the petroleum industry to obtain large amounts of gasoline, thermal cracking is used since it yields larger percentages of C_2, C_3, and C_4 olefins. In 1981 about 71% of the feedstock for this process was ethane and propane from natural gas. But naphtha and gas oil fractions from petroleum can be used, and recently, their use has increased dramatically with the high price and scarcer supply of natural gas. In 1986 only 55% ethane and propane were used, with 45% naphtha and gas oil serving as feedstock for U.S. ethylene production. Relative costs of running ethylene plants vary with the type of feed and are cheaper for natural gas feeds: ethane, 1.0; propane, 1.2; naphtha, 1.4; gas oil, 1.5. This must be weighed against the difference in prices of the feedstocks themselves. The shift to heavier feedstocks is predicted to continue in the 1990s.

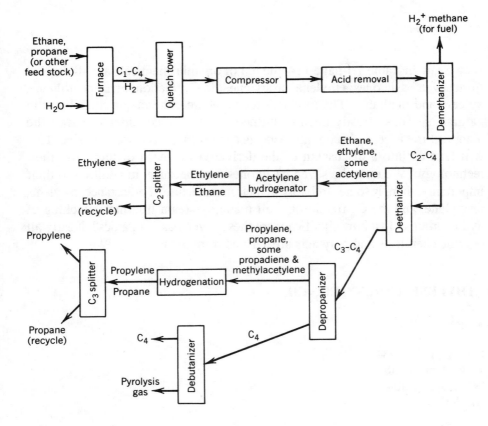

Figure 8.1 Manufacture of olefins by thermal cracking. (*Source:* L & M. Reprinted by permission of John Wiley & Sons, Inc.)

Reactions

$$CH_3\!-\!CH_3 \longrightarrow CH_2\!=\!CH_2 + H_2$$

$$2CH_3\!-\!CH_2\!-\!CH_3 \longrightarrow CH_3\!-\!CH\!=\!CH_2 + CH_2\!=\!CH_2 + H_2 + CH_4$$

Description . Steam cracking is pictured in Fig. 8.1. The furnace in which the cracking takes place is at 815-870°C (1600°F). Sometimes as many as 6 to 20 furnaces are in parallel to increase production (see Fig. 8.2). Steam is used as a dilutant to inhibit coking in the tubes and to increase the percentage of ethylene formed. The amount of steam changes with the molecular weight of the hydrocarbon and varies from 0.3 kg steam/kg ethane to 0.9 kg steam/kg gas oil. Contact time is 1 sec or less in the furnace. The exit gases are immediately cooled in the quench tower, then placed under 500 psi pressure by a compressor (Fig. 8.3). Monoethanolamine or caustic is used to remove hydrogen sulfide and carbon dioxide (see the natural gas discussion, Chapter 7).

The demethanizer, deethanizer, and debutanizer are fractionating columns that separate the lighter and heavier compounds from each other. Traces of

Figure 8.2 Distance view of two olefin plants. Note the furnace stacks and the large distillation columns. (Courtesy of Amoco Chemicals Corporation, Alvin, TX.)

triple bonds are removed by catalytic hydrogenation with a palladium catalyst in both the C_2 and C_3 stream. Cumulated double bonds are also hydrogenated in the C_3 fraction. These are more reactive in hydrogenation than ethylene or propylene. The C_2 and C_3 splitters (Fig. 8.4) are distillation columns that can be as high as 200 ft. A control room is shown in Fig. 8.5, page 143. The mechanism of cracking is discussed on page 127.

Figure 8.3 Exit gases from the furnace and quench tower of an olefin plant enter a compressor before distillation. Ice forms on the outside of the compressor even on warm days because of the cooling effect. (Courtesy of Amoco Chemicals Corporation, Alvin, TX.)

Lower molecular weight feedstocks, such as ethane and propane, give a high percentage of ethylene; higher molecular weight feedstocks, such as naphtha and gas oil, are used if propylene demand is up. The following table summarizes the typical yields of olefins obtained from various feeds:

	Feed			
Product	*Ethane*	*Propane*	*Naphtha*	*Gas Oil*
Ethylene	76	42	31	23
Propylene	3	16	16	14
C_4	2	5	9	9

Figure 8.4 Distillation columns used in a large olefin plant. The middle one is the C_2 splitter and the highest at 200 ft, separating ethane and ethylene. (Courtesy of Amoco Chemicals Corporation, Alvin, TX.)

Properties

Ethylene is a colorless, flammable gas with a faint, pleasant odor and a bp of -103.8°C. The flash point, the lowest temperature at which the vapors of a liquid decompose to a flammable gaseous mixture, is -136.1°C. The ignition temperature, the temperature at which a substance begins to burn, is 450°C. Ethylene is sold from 95% purity (technical) to 99.9% purity. It can be transported by pipeline or by tank car. Smaller amounts come in 100 lb cylinders. Much of it is used on site by the company to make other products.

Uses

Fig. 8.6, on page 144, shows the breakdown in uses of ethylene. Over half is polymerized directly to polyethylene, both high and low density, which are used in thousands of plastics applications. Major organic chemicals made from ethylene are ethylene oxide, which is in turn converted into ethylene glycol for antifreeze and polyester fibers; vinyl chloride, polymerized to poly(vinyl chloride), another important plastic; styrene, polymerized to polystyrene plastic and foam; and linear alcohols and olefins, whose important end-uses are in soaps and detergents and plastics. We will be covering these derivatives and specific uses in more detail in later chapters.

$$(CH_2-CH_2)_n \qquad CH_2-CH_2 \qquad CH_2=CH-Cl$$

polyethylene   ethylene oxide   vinyl chloride

$$\langle\bigcirc\rangle-CH=CH_2 \qquad CH_3-(CH_2-CH_2)_n-OH$$

styrene   linear alcohols

PROPYLENE (PROPENE) $CH_3-CH=CH_2$

References

L & M, pp. 376-384
W & RI, pp. 67-71
Wiseman, pp. 30-42
CP, 11-13-89

Manufacture

Propylene is manufactured by steam cracking of hydrocarbons as discussed under ethylene. The best feedstocks are propane, naphtha, or gas oil, depending on price and availability. About 50-75% of the propylene is consumed by the petroleum refining industry for alkylation and polymerization to oligomers which are added to gasoline. A smaller amount is made by steam cracking to give pure propylene for chemical manufacture.

$$2CH_3—CH_2—CH_3 \longrightarrow CH_3—CH=CH_2 + H_2 + CH_2=CH_2 + CH_4$$

Properties

Propylene is a colorless, flammable gas with a slightly sweet aroma, bp -47.7°C, flash point -107.8°C, ignition temperature 497.2°C. It is available in cylinders and tank cars and by pipeline.

Figure 8.5 Control room for a large olefin plant. Numerous graphs and switches allow control of temperature, pressure, and other variables throughout the plant. Computerization is the norm for these modern plants. (Courtesy of Amoco Chemicals Corporation, Alvin, TX.)

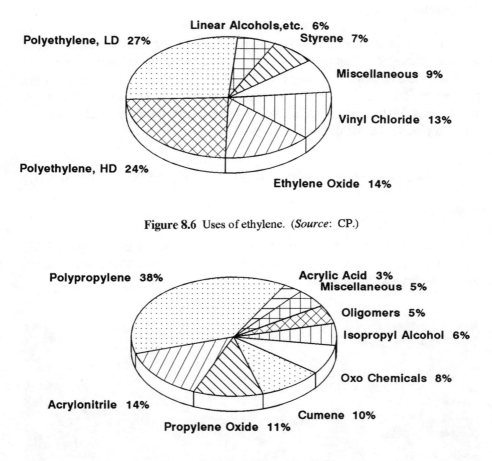

Figure 8.6 Uses of ethylene. (*Source*: CP.)

Figure 8.7 Merchant uses of propylene. (*Source*: CP.)

Uses

Fig. 8.7 outlines the merchant uses of propylene. The biggest use, polymerization to polypropylene, is growing since this polymer is competing in many plastics applications with high density polyethylene. Acrylonitrile is polymerized to plastics and fibers. Propylene oxide is used in polyurethane plastic and foam. Cumene is made from propylene and benzene. It is an important intermediate in the manufacture of two top 50 chemicals, phenol and acetone. Isopropyl alcohol is made from propylene. It is a common industrial solvent in coatings, chemical processes, pharmaceuticals, and household and personal products. Oxo chemicals are made by reacting propylene with synthesis gas (CO/H$_2$) to form C$_4$ alcohols. Small amounts of propylene are made into oligomers, where 3-5 propylene units are added to each other. These have importance in soaps and detergents, besides the very large captive use they have as mentioned above for petroleum refining.

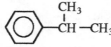

CH$_3$		
(CH—CH$_2$)$_n$	CH$_2$=CH—C≡N	CH$_3$—CH—CH$_2$
polypropylene	acrylonitrile	propylene oxide

CH$_3$		
⬡—CH—CH$_3$	CH$_3$—CH—CH$_3$	
	OH	
cumene	isopropyl alcohol	

THE C$_4$ STREAM

Besides ethylene and propylene, the steam-cracking of naphtha and catalytic cracking in the refinery produce appreciable amounts of C$_4$ compounds. This C$_4$ stream includes butane, isobutane, 1-butene (butylene), *cis-* and *trans-*2-butene, isobutene (isobutylene), and butadiene. The C$_4$ hydrocarbons can be used to alkylate gasoline. Of these, only butadiene appears in the top 50 chemicals as a separate pure chemical because of its large-scale use in the synthetic elastomer styrene-butadiene rubber (SBR). We will therefore consider only butadiene. The other C$_4$ hydrocarbons have specific uses but are not so important as butadiene.

BUTADIENE (1, 3-BUTADIENE) CH$_2$=CH-CH=CH$_2$

References

W & R I, pp. 81-82
L & M, pp. 164-172
CP, 4-11-88
Wiseman, pp 80-88

Manufacture

In the last ten years not enough butadiene could be made by steam-cracking alone. Thus about 70% is now made by dehydrogenation of butane or the butenes.

Reaction

$$n\text{-} C_4H_8 \xrightarrow{Fe_2O_3} CH_2{=}CH{-}CH{=}CH_2 + H_2$$

$$CH_3{-}CH_2{-}CH_2{-}CH_3 \xrightarrow{Fe_2O_3} CH_2{=}CH{-}CH{=}CH_2 + 2H_2$$

57-63% yield

Description. Fig. 8.8 diagrams the manufacture of butadiene. The crude C_4 fraction is extracted with acetone, furfural, or other solvents to remove alkanes such as *n*-butane, isobutane, and small amounts of pentanes, leaving only 1- and 2-butenes and isobutene. The isobutene is removed by extraction with sulfuric acid because it oligomerizes more easily, being able to form a tertiary carbocation.

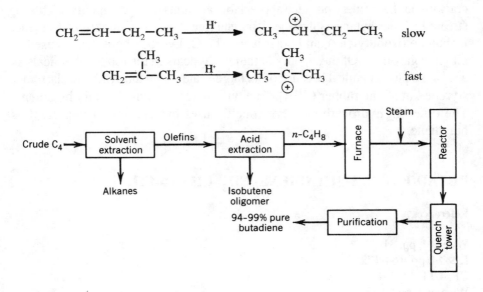

Figure 8.8 Butadiene manufacture from the C_4 stream by hydrogenation.

The straight-chain 1- and 2-butenes are preheated to 600°C in a furnace, mixed with steam as a diluent to minimize carbon formation, and passed through a 5-m diameter reactor with a bed of iron oxide pellets 90-120 cm deep. The reactor is at 620-675°C. The butenes take only 0.2 sec to pass through. An alternate catalyst is calcium nickel phosphate. The material is cooled and purified by fractional distillation and extraction with solvents such as furfural, acetonitrile, dimethylformamide (DMF), and N-methylpyrollidone (NMP). The conjugated π system of butadiene is attracted to these polar solvents more than the other C_4 compounds. *Extractive distillation* is used, where the C_4 compounds other than butadiene are distilled while the butadiene is complexed with the solvent. The solvent and butadiene pass from the bottom of the column and are then separated by distillation. Yields of 80% at conversions of 35% are common.

Properties

Butadiene is a colorless, odorless, flammable gas, with a bp of -4.7°C. As of 1984 butadiene has been on the "suspected human carcinogen" list with a time-weighted average threshold limit value of 10 ppm. This will be discussed more fully under benzene. Butadiene is expensive to store because it polymerizes easily and must be refrigerated.

Uses

In Fig. 8.9 we see that most butadiene is polymerized either by itself or with styrene or acrylonitrile. The most important synthetic elastomer is styrene-butadiene rubber (SBR). SBR, along with polybutadiene, has its biggest market

$$(CH_2-CH=CH-CH_2)_n$$
polybutadiene

$$CH_2=\overset{\overset{\displaystyle Cl}{|}}{C}-CH=CH_2$$
chloroprene

$$N\equiv C-CH_2-CH_2-CH_2-CH_2-C\equiv N$$
adiponitrile

$$H_2N-CH_2-CH_2-CH_2-CH_2-CH_2-CH_2-NH_2$$
hexamethylenediamine (HMDA)

in automobile tires. Specialty elastomers are polychloroprene and nitrile rubber, and an important plastic is acrylonitrile/butadiene/styrene (ABS) terpolymer. Butadiene is made into adiponitrile, which is converted into hexamethylenediamine (HMDA), one of the monomers for nylon.

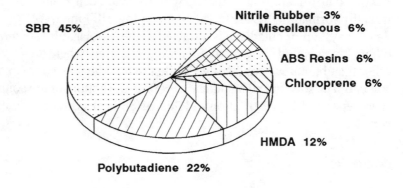

Figure 8.9 Uses of butadiene. (*Source:* CP.)

ECONOMIC ASPECTS OF THE OLEFINS

Figure 8.10 shows the U.S. production of ethylene, propylene, and butadiene over the years. Ethylene has shown a good, steady increase for many years since the 1950s, as it has replaced ethanol as the major C_2 raw material and is now used in nearly half of all organic polymers and chemicals produced by volume. Some would argue that it should be used in place of sulfuric acid as the main economic indicator of the chemical industry. Certainly for organic chemicals it has top billing. Propylene shows a similar but lower trend. Although the official production of propylene is usually about half that of ethylene, it is probably near ethylene if captive refinery-made material could be included accurately. Butadiene has definitely leveled off with the major slump in the automobile and tire industries during the 1980s, the only one of the three to drop in production during this decade. Three fourths of all butadiene ends up in tires. Butadiene is about one tenth of ethylene production.

Ethylene production as a percentage of capacity has become very tight in the last few years. The figures for three selected years are given on page 149. Note

Year	Capacity, Billion lb	Production as Percent of Capacity
1981	39	75
1984	37	84
1990	42	89

that production and capacity are now nearly even, because of increasing polyethylene demand for the most part. There has been a reluctance to open new ethylene plants because of present questions on this market future. There are 30 ethylene plants in the U.S. representing 20 companies. Many of the plants are owned by oil company subsidiaries and are located in Texas and Louisiana near the oil fields of the Gulf region. A typical large plant will manufacture 1.0-2.0 billion lb/yr of ethylene and 0.5-1.5 billion lb/yr of propylene. There are over 60 plants of propylene from 30 companies. The manufacture of propylene is a more diversified business.

Fig. 8.11 summarizes price trends for the olefins. This is perhaps one of the most startling examples of all price charts shown in this text. Note that prices

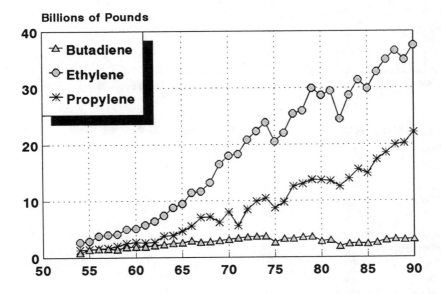

Figure 8.10 U.S. production of olefins. (*Source:* L & M and *C & E News*. Reprinted by permission of John Wiley & Sons, Inc.)

were amazingly steady or decreasing for many years through the 1950s, '60s, and early '70s. Then the oil embargo caused a record steep incline for most organics, and double-digit inflationary years until the early 1980s caused a jump in ethylene from 4¢/lb in 1973 to 28¢/lb in 1981, a 700% increase for an eight-year span. The ups and downs of the 1980s economy is also evident in the chart. Trends for propylene and butadiene follow a similar pattern.

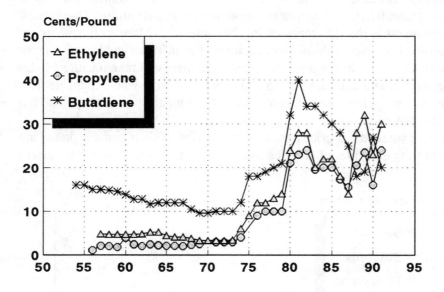

Figure 8.11 U.S. prices of olefins. (*Source*: L & M and CMR. Reprinted by permission of John Wiley & Sons, Inc.)

BENZENE (BENZOL)

References

W & R I, pp. 92-93
L & M, pp. 126-137
CP, 4-23-90

Manufacture: Catalytic Reforming

For many years benzene was made from coal tar even as late as 1949, when all of it was made by this old process. New processes began to take over in the 1950s, which were used for 50% of the benzene in 1959, for 94% in 1972, and for 96% in 1980. These new processes consist of catalytic reforming of naphtha and hydrodealkylation of toluene in a 70:30 use ratio or sometimes a 50:50 ratio depending on the relative prices of benzene and toluene.

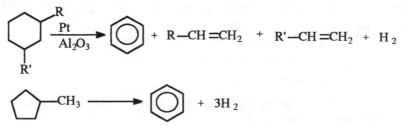

Diagram

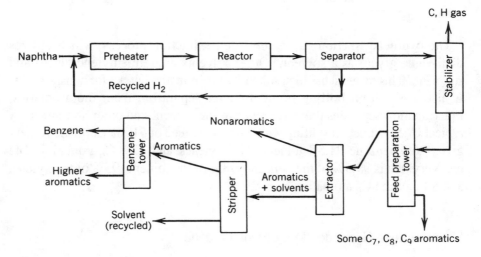

Figure 8.12 Manufacture of benzene, toluene, and xylenes by catalytic reforming. (*Source:* L & M. Reprinted by permission of John Wiley & Sons, Inc.)

Description. As seen in Fig. 8.12, the naphtha is preheated in a hydrogen atmosphere (to suppress coke formation) at 450-510°C and 250-800 psi. The reactor is filled with 3-6 mm pellets of platinum catalyst. In one pass 80% of the C_6 naphthenes form benzene and natural benzene (1-9% of the feedstock) remains unchanged. Since hydrogen is present as the recycle gas, this is often called *hydroforming*. When platinum is the catalyst it is called the *platforming process*. For the mechanism of catalytic reforming see page 130. The hydrogen is separated; the stabilizer removes light hydrocarbon gases. The feed preparation tower increases the benzene percentage via distillation and collection of the overhead cut.

HO—CH_2—CH_2—O—CH_2—CH_2—OH
diethylene glycol

HO—CH_2—CH_2—(O—CH_2—CH_2)$_2$—O—CH_2—CH_2—OH
tetraethylene glycol

SO$_2$ sulfolane

   The overhead fraction enters the Udex extraction process, which utilizes diethylene glycol as a solvent. Other solvents are tetraethylene glycol and sulfolane. This material has high solubility for aromatics but not for nonaromatics. It also has a high boiling point for later separation from the aromatics. Fractionation separates the benzene from the solvent and other aromatics. A typical Udex extraction starting with a reformed feed of 51.3% aromatic content gives 7.6% benzene, 21.5% toluene, 21% xylenes, and 1.2% C_9 aromatics. The recovery rate is 99.5% of the benzene, 98% of the toluene, 95% of the xylene, and 80% of the C_9 aromatics.

Manufacture: Hydrodealkylation of Toluene

More toluene is formed than is needed in the catalytic reforming of naphtha. Benzene is always in tight supply. When the price is right it is economical to hydrodealkylate (add hydrogen, lose the methyl) toluene to benzene. This is best done on pure toluene, where the yield can be as high as 98.5%. The reaction can be promoted thermally or catalytically. As much as 30-50% of all benzene is made this way.

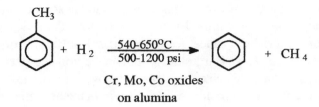

Properties

Benzene is a clear, colorless, flammable liquid with a pleasant characteristic odor, bp 80.1°C, flash point -11.1°C, and ignition temperature 538°C. Benzene has recently been found to be very toxic and is on the list of compounds that are "suspect of carcenogenic potential for man." It has a low threshold limit value or TLV. The time weighted average TLV (TWA) is the allowable exposure for an average 8 hr day or a 40 hr week. The short-term exposure limit TLV (STEL) is the maximum allowable exposure for any 15-min period. For benzene the TWA = 10 ppm. This allowable exposure is much lower than those for toluene and xylene, probably because these latter two compounds have benzylic positions that are easily oxidized in vivo to compounds that can be eliminated from the body.

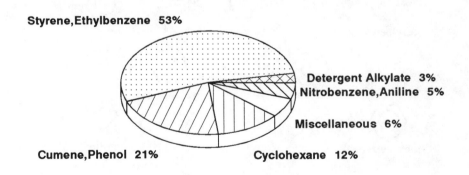

Figure 8.13 Uses of benzene. (*Source:* CP.)

As recently as 1989 the Environmental Protection Agency ordered a 90% reduction of industrial benzene emissions over the next several years at a cost of $1 billion. The new standard would leave more than 99% of the exposed population with risks of cancer less than one in 1 million, or one cancer case in the U.S. every 10 years. Hardest hit would be the iron and steel industry, where benzene emissions from coke by-product recovery plants are large. Chemical industry plants have already reduced their benzene emissions 98%. EPA estimates that the 390,000 or so gasoline service stations in the U.S. will all have to be fitted with devices to eliminate the escape vapors when fuel is put into underground storage tanks.

Uses

The important derivatives of benzene are shown in Fig. 8.13. Ethylbenzene is made from ethylene and benzene and then dehydrogenated to styrene, which is polymerized for various plastics applications. Cumene is manufactured from propylene and benzene and made into phenol and acetone. Cyclohexane, a starting material for some nylon, is made by hydrogenation of benzene. Nitration of benzene followed by reduction gives aniline, important in the manufacture of polyurethanes.

TOLUENE (TOLUOL)   CH₃—⟨O⟩

Reference

L & M, pp. 822-830

Manufacture

The platforming-Udex process for catalytic reforming of naphtha is also used for toluene. The feedstock should be rich in seven carbon naphthenes for higher toluene percentages. *n*-Heptane and dimethylhexane remain unchanged and contaminate the product. About 80-90% conversion of napthenes into toluene is usually realized. Shell has an extraction process using sulfolane as the solvent. It has higher solvent power, solvent circulation is reduced, and the equipment can be smaller. The toluene is purified by azeotropically distilling the nonaromatics with methyl ethyl ketone (MEK, 90%) and water (10%). The excess MEK is then distilled from the toluene.

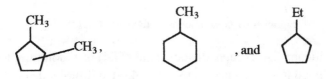

Properties

Toluene is a colorless, flammable liquid with a benzenelike odor, bp 110.8°C, flash point 4.4°C, ignition temperature 552°C, TLV (TWA) = 100 ppm, and TLV (STEL) = 150 ppm.

Uses

Fig. 8.14 shows the non-fuel uses of toluene. Some of the toluene goes into gasoline depending on its supply and price compared to other octane enhancers. Of the other uses of toluene about half is converted into benzene by hydrodealkylation, though this amount varies with the price difference between

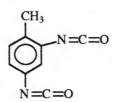

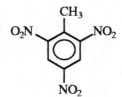

2,4- toluenediisocyanate (TDI)   2,4,6-trinitrotoluene (TNT)

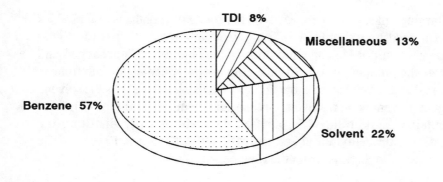

Figure 8.14 Uses of toluene. (*Source*: CEH.)

benzene and toluene. 2,4-Toluenediisocyanate (TDI) is a monomer for poly-urethanes. Included in miscellaneous uses is 2,4,6-trinitrotoluene (TNT) as an explosive.

XYLENES (XYLOLS)

References

L & M, pp. 874-881
CP, 8-7-89 (para)
CP, 7-31-89 (ortho)

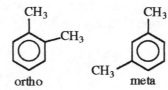

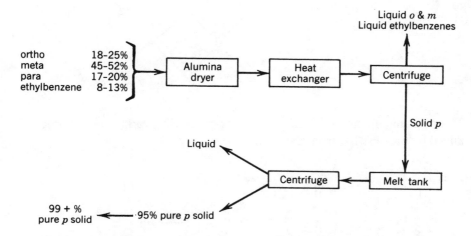

Figure 8.15 Separation of *p*-xylene by fractional distillation. (*Source:* L & M.)

Manufacture

The xylenes can be used as a mixture or separated into pure isomers, depending on the application. The mixture is obtained from catalytic reforming of naphtha and separated from benzene and toluene by distillation.

Separation of *p*-Xylene

As Fig. 8.15 depicts, the C_8 mixture is cooled to -70°C in the heat exchanger refrigerated by ethylene. Because of the difference in melting points (ortho, -25.0°C; meta, - 47.9°C; para, 13.2°C), the para isomer crystallizes preferentially. The other two remain liquid as a mixture. The solid para isomer is centrifuged and separated. A second cooling cycle needs only propane as coolant and 95% purity results. Complete separation is accomplished with an optional third cooling cycle.

Because of the large demand for *p*-xylene, another method is now being used by Amoco to increase the percentage of the para isomer in mixed xylenes. They are heated at 300°C with an acidic zeolite catalyst, which equilibrates the three xylenes to an *o, m, p* ratio of 10:72:18%. The para isomer is separated by fractional crystallization, whereas the *o, m* mixture is reisomerized with the catalyst to produce more para product. Theoretically, all the xylenes could be transformed into the desired para isomer. The zeolite catalyst has the following structure.

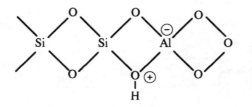

The rearrangements of the methyl groups occur via a carbocationic process induced by protonation from the zeolite.

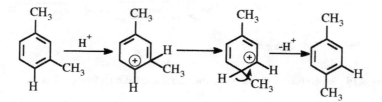

A third possibility of separating the para isomer has been used for about 20 years. This isomer can be selectively adsorbed on zeolites, then desorbed after the ortho and meta isomers have been separated.

Separation of *o*-Xylene by Fractional Distillation

The slightly different boiling point of the *o*-xylene is the basis for separation from the other two isomers through an elaborate column.

Isomer	Boiling Point (°C)
Ortho	144
Meta	139.1
Para	138.5

Properties

The xylenes are colorless, flammable liquids, flash point 17.2°C, ignition temperature 359°C, TLV (TWA) = 100 ppm, and TLV (STEL) = 150 ppm.

Uses

Mixed Xylenes

Pure para	39%
Pure ortho	18
Gasoline, benzene, solvent	37

Pure p-Xylene

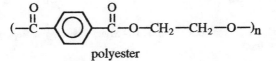

polyester

About half of polyester is made from terephthalic acid and half from dimethyl terephthalate. Either is reacted with ethylene glycol to give poly (ethylene terephthalate). Large amounts of this polyester are used in textile fibers, photographic film, and soft drink bottles. A small amount of p-xylene is exported.

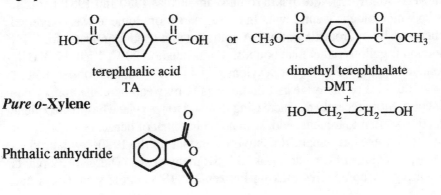

terephthalic acid
TA

dimethyl terephthalate
DMT

$+$

$HO-CH_2-CH_2-OH$

Pure o-Xylene

Phthalic anhydride

Phthalic anhydride is an intermediate in the synthesis of plasticizers, substances that make plastics more flexible. A common plasticizer is dioctyl phthalate.

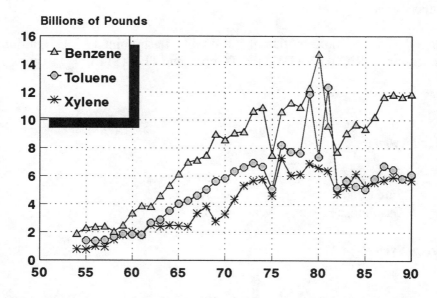

Figure 8.16 U.S. production of aromatics. (*Source*: L&M and *C & E News*. Reprinted by permission of John Wiley & Sons, Inc.)

ECONOMIC ASPECTS OF AROMATICS

In Fig. 8.16 the production of benzene, toluene, and xylene are summarized. Healthy gains over the years with a steady incline until 1975 occurred, when the erratic years followed. Note that in some years such as 1980 and 1981, benzene went up when toluene went down. In those years more toluene was converted into benzene by hydrodealkylation. A common unit used in industry for BTX production is gallons rather than pounds. For benzene at 20°C, 1 gal = 7.320 lb; for toluene, 1 gal = 7.210 lb; for p-xylene, 1 gal = 7.134 lb; and for o-xylene, 1 gal = 7.300 lb. For many years toluene was in between benzene and xylene production. More recently the increasing demand for p-xylene has made xylene production similar to toluene, with both about half that of benzene. Capacity of plants for benzene has remained relatively constant for the 1980s, with production as a percentage of capacity near 75% for many years. There are 37 plants representing 24 companies making benzene in 1990, mostly in Texas and Louisiana.

Year	Capacity, million gal	Production as % of Capacity
1981	2500	70
1987	2100	78
1990	2400	75

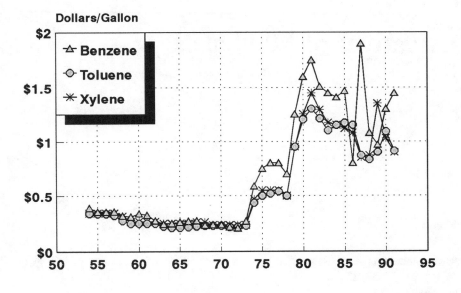

Figure 8.17 U.S. prices of aromatics. (*Source*: L & M and CMR. Reprinted by permission of John Wiley & Sons, Inc.)

Fig. 8.17 shows U.S. prices for the aromatics in dollars per gallon. As in the olefins we see very steady pricing to the mid '70s, then heavy increases through the late '70s and early '80s, followed by ups and downs, especially for benzene. For the hydrodealkylation of toluene to be profitable as a production method for benzene the price of toluene must be 50¢/gal lower than that for benzene. Sometimes this happens, sometimes not. *p*-Xylene is the more expensive isomer of the two commercial xylenes because of the crystallization process required. It is about 30¢/lb compared to the ortho isomer which is 17¢/lb. The commercial value of benzene is approximately $2.3 billion. Toluene has a value of $0.8 billion, and mixed xylenes about $0.7 billion. *p*-Xylene, the most important isomer, has a commercial value of $1.3 billion compared to *o*-xylene at $0.2 billion.

9

Derivatives of Ethylene

References

L & M, selected pages
W & R I, pp. 55-67, 231
Kent, pp. 923-942
Wiseman, pp. 43-63, 102-103, 151-152, 163-164
Szmant, pp. 188-264
White, pp. 62-79
Weissermel, p. 152

INTRODUCTION

Over 100 billion lb of chemicals and polymers per year are made from ethylene, by far the most important organic chemical. Over 40% of all organic chemicals by volume are derived from ethylene. Unfortunately, we cannot describe the interesting chemistry and uses for all the important derivatives of ethylene, some of which are listed in Fig. 9.1. We will limit our detailed discussion to those chemicals made from ethylene that appear in the top 50, which amount to 9 important organic chemicals. These are listed with their ranking in Table 9.1.

TABLE 9.1   Ethylene Derivatives in the Top 50

1990 Rank	Chemical
15	Ethylene dichloride
18	Vinyl chloride
19	Ethylbenzene
20	Styrene
27	Ethylene oxide
29	Ethylene glycol
33	Acetic acid
44	Vinyl acetate
—	Ethanol (previously in the top 50)

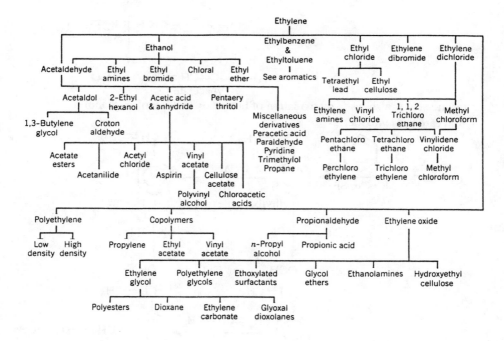

Figure 9.1 Major ethylene derivatives (*Source:* Kent.)

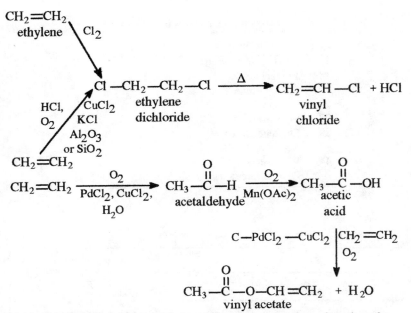

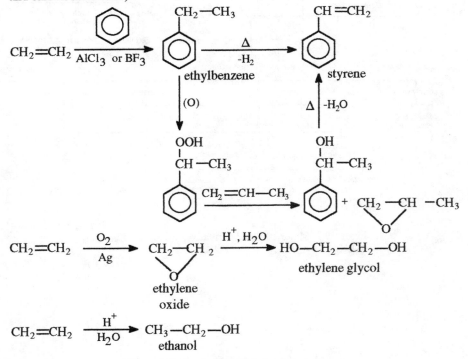

(acetic acid is also made by oxidation of butane or reaction of methanol and carbon monoxide)

Figure 9.2 Synthesis of ethylene derivatives.

Again, we are faced with the question of what order to treat these chemicals: by rank, alphabetically, and so on. Some of these chemicals can be grouped by manufacturing process, since one may be made from another, and both originally from ethylene, in a multistep synthetic sequence. We take advantage of this manufacturing relationship in our choice of order since it groups these chemicals together in one important feature that we wish to master: their chemistry of manufacture. The discussions of these chemicals will be different from that for inorganics. For inorganics the emphasis was on the simple reaction, the engineering aspects of the chemical's manufacture, uses, and economics. Because there are more organics than inorganics in the top 50, we must sacrifice some details in engineering and economics but still stress the chemistry of manufacture (including mechanism when known) and uses of the chemicals. Let us recall at this point that nearly half of all ethylene is polymerized to polyethylene. This process and the polymer will be discussed in a later chapter. Other than this the main large-scale industrial reactions of ethylene are summarized in this chapter. Figure 9.2 gives an outline of this chemistry with the order that we will use to consider these chemicals.

ETHYLENE DICHLORIDE (EDC), Cl-CH$_2$-CH$_2$-Cl

Reference

CP, 5-22-89

Ethylene dichloride is usually the highest-ranked derived organic chemical and is made in excess of 10 billion lb/yr. There are two major manufacturing methods for this chemical. The classical method for EDC manufacture is the electrophilic addition of chlorine to the double bond of ethylene. The yield is good (96-98%); it can be done in vapor or liquid phase at 40-50°C using ethylene dibromide as a solvent, and the product is easily purified by fractional distillation. The mechanism (page 166) is well understood and is a good example of the very general addition of an electrophile to a double bond. Here the intermediate is the bridged chloronium ion, since in this structure all atoms have a complete octet. A primary carbocation is less stable. Polarization of the chlorine-chlorine bond occurs as it approaches the π cloud of the double bond. Backside attack of chloride ion on the bridged ion completes the process.

Reaction:

$$CH_2{=}CH_2 + Cl_2 \xrightarrow{FeCl_3} Cl{-}CH_2{-}CH_2{-}Cl$$

Mechanism:

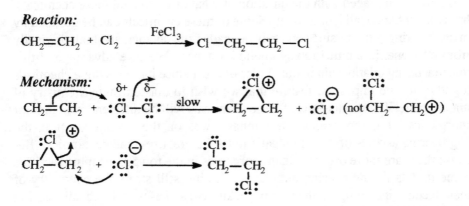

In contrast to this direct chlorination there is the oxychlorination of ethylene using hydrogen chloride and oxygen, the major method now used. Since the chlorine supply is sometimes short and it is difficult to balance the caustic soda and chlorine demand (both are made by the electrolysis of brine), hydrogen chloride provides a cheap alternate source for the chlorine atom. Most of the ethylene dichloride manufactured is converted into vinyl chloride by eliminating a mole of HCl, which can then be recycled and used to make more EDC by

$$CH_2{=}CH_2 + 2HCl + {}^{1}\!/_{2}O_2 \xrightarrow[Al_2O_3 \text{ or } SiO_2]{CuCl_2,\ KCl} Cl{-}CH_2{-}CH_2{-}Cl + H_2O$$

$$Cl{-}CH_2{-}CH_2{-}Cl \xrightarrow{\Delta} CH_2{=}CH{-}Cl + HCl$$
$$\text{(recycled)}$$

oxychlorination. EDC and vinyl chloride plants usually are physically linked.

What probably happens in the oxychlorination process is that chlorine is formed in situ. The reaction of hydrogen chloride and oxygen to give chlorine and water was discovered by Deacon in 1858. Once the chlorine is formed, it then adds to ethylene as in the direct chlorination mechanism. Cu^{+2} is the catalyst and helps to more rapidly react HCl and O_2 because of its ability to undergo reduction to Cu^{+1} (or Cu_2^{+2}) and reoxidation to Cu^{+2}. The KCl is used to reduce

$$2HCl + 2Cu^{+2} \longrightarrow Cl_2 + Cu_2^{+2} + 2H^+$$
$$Cu_2^{+2} + {}^{1}\!/_{2}O_2 + 2H^+ \longrightarrow 2Cu_2^{+2} + H_2O$$

the volatility of $CuCl_2$.

Ethylene dichloride is a colorless liquid with a bp of 84°C. As with many chlorinated hydrocarbons, it is quite toxic and has a TLV value of 10 (TWA).

Uses of Ethylene Dichloride

Vinyl chloride	90%	$CH_2=CH-Cl$
Perchloroethylene ⎫		$Cl_2 = CCl_2$
Methyl chloroform ⎬	3%	CH_3-CCl_3
Vinylidene chloride ⎬		$CH_2=CCl_2$
Ethylenamines ⎭		$NH_2-CH_2-CH_2-NH_2$

The vinyl chloride is polymerized to the important plastic poly (vinyl chloride) (PVC). Perchloroethylene (perc) is a dry cleaning agent (50%), metal degreaser (9%), and raw material for fluorochlorocarbon (F-113) manufacture (28%) for aerosols and refrigerants. Methyl chloroform is one of a few chlorinated compounds that has low toxicity. It is replacing trichloroethylene as a metal degreasing agent and competes with perc in this use. Vinylidene chloride is polymerized to a plastic (Saran ®). The ethylenediamines are used as chelating agents, the most important being ethylenediaminetetracetic acid (EDTA).

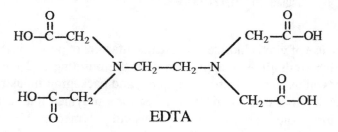

Despite these minor uses, the economics of EDC is linked to the demand for PVC plastic.

VINYL CHLORIDE, CH₂=CH-Cl; 1-CHLOROETHENE

Reference

CP, 5-29-89

Although there are two manufacturing methods for ethylene dichloride, all the vinyl chloride is made by a single process, thermal dehydrochlorination of EDC. This takes place at temperatures of 480-510°C under a pressure of 50 psi with a charcoal catalyst to give a 95% yield. Vinyl chloride is a gas at ambient pressure with a bp of -13°C. It is separated from ethylene dichoride by fractional distillation. Vinyl chloride readily polymerizes so it is stabilized with inhibitors to polymerization during storage. The mechanism of formation is a free-radical chain process as shown below. Although the conversion is low, 50-60%, recycling the EDC allows an overall 99% yield.

Reaction:

$$Cl-CH_2-CH_2-Cl \xrightarrow{\Delta} CH_2{=}CH-Cl + HCl$$

Mechanism

(1) $Cl-CH_2-CH_2-Cl \longrightarrow Cl\cdot + \cdot CH_2-CH_2-Cl$

(2) $Cl\cdot + Cl-CH_2-CH_2-Cl \longrightarrow HCl + Cl-\overset{\bullet}{C}H-CH_2-Cl$

(3) $Cl-\overset{\bullet}{C}H-CH_2\cdot\cdot Cl \longrightarrow Cl-CH{=}CH_2 + Cl\cdot$

then (2), (3), (2), (3), etc.

The single use of vinyl chloride is in the manufacture of poly (vinyl chloride) plastic, which finds diverse applications in the building and construction industry as well as in the electrical, apparel, and packaging industries. Poly (vinyl chloride) does degrade relatively fast for a polymer, but various heat, ozone, and ultraviolet stabilizers make it a useful polymer. A wide variety of desirable properties can be obtained by using various amounts of plasticizers, such that both rigid and plasticized PVC have large markets. PVC takes up 91% of all vinyl chloride with only 2% being used for chlorinated solvents.

After some tough years in the 1970s vinyl chloride had a good economic gain in the 1980s. The 1990 production of 10.65 billion lb is up nearly 11% over 1989. At a price of 21¢/lb that gives a total commercial value of $2.2 billion. One of the reasons vinyl chloride has had some bad years is the findings of toxicity. It causes liver cancer, and is on the list of chemicals "recognized to have carcinogenic potential." In 1973 its TLV was 200 ppm. This was reduced in 1974 to 50 ppm. As of 1980 the TLV (TWA) is 5. However, apparently this causes no health problems for poly (vinyl chloride) uses. Only the monomer is a health hazard. As a result, the economic situation looks good. Let us summarize here the possible carcinogens in the top 50. To date, only vinyl chloride is on the "worst" list and is a recognized carcinogen. Five other chemicals are now on the "suspect" list: acrylonitrile, benzene, 1,3-butadiene, ethylene oxide, and formaldehyde.

$$\overset{\displaystyle O}{\overset{\displaystyle \|}{}}$$

ACETIC ACID, CH$_3$-C-OH; ETHANOIC ACID, GLACIAL ACETIC ACID

(A 3-5% aqueous solution is called vinegar.)

Reference

CP, 4-10-89

If there is a prime example of an organic chemical that is in a state of flux and turnover in regards to the manufacturing method, it is probably acetic acid. There are now three industrial processes for making acetic acid. Domestic capacity in 1978 was almost equal among acetaldehyde oxidation, *n-* butane oxidation, and methanol carbonylation. In 1980 methanol carbonylation exceeded 40% of the capacity and will continue to increase in its share of capacity because of economic advantages. In 1989 methanol carbonylation was 70% of capacity, butane oxidation 15%, acetaldehyde oxidation 10%, and coal gasification 5%. Perhaps acetic acid should be covered in the derivatives of methane chapter, but it is appropriate to also cover it here since both it and vinyl acetate are still made from ethylene.

Ethylene is the exclusive organic raw material for making acetaldehyde, 70% of which is further oxidized to acetic acid or acetic anhydride. The Wacker process, named after a German company, for making acetaldehyde involves cupric chloride and a small amount of palladium chloride in aqueous solution as a catalyst. The mechanism is partially understood: (1) A π complex between ethylene and palladium chloride is formed and decomposes to acetaldehyde and palladium metal; (2) the palladium is reoxidized to palladium chloride by the cupric chloride; and (3) the cuprous chloride thus formed is reoxidized to the cupric state by oxygen fed to the system. The three equations that follow indicate the series of redox reactions that occur. When added together they give the overall reaction. The yield is 95%.

(1) $CH_2{=}CH_2 + PdCl_2 + H_2O \longrightarrow CH_3CHO + Pd^\circ + 2HCl$

(2) $Pd^\circ + 2CuCl_2 \longrightarrow PdCl_2 + 2CuCl$

(3) $2CuCl + \frac{1}{2}O_2 + 2HCl \longrightarrow 2CuCl_2 + H_2O$

overall: $CH_2{=}CH_2 + \frac{1}{2}O_2 \xrightarrow[PdCl_2]{CuCl_2} CH_3{-}\overset{\displaystyle O}{\overset{\displaystyle \|}{C}}{-}H$

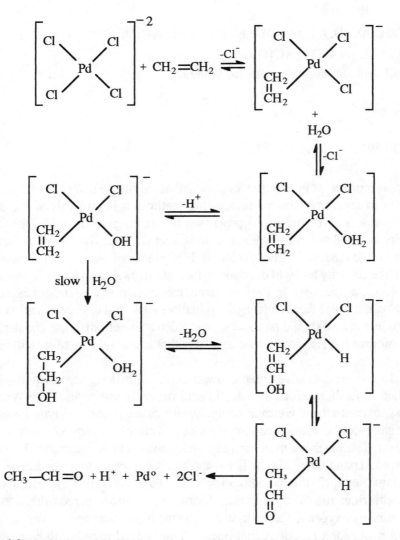

Figure 9.3 Mechanism of the Wacker reaction. (*Source:* White. Reprinted by permission of John Wiley & Sons, Inc.)

The details of the mechanism of the reaction of ethylene with $PdCl_2$ (equation (1) above) are also known and are shown in Fig. 9.3. The palladium ion complexes with ethylene and water molecules and the water adds across the π bond while still complexed to palladium. The palladium then serves as a hydrogen acceptor while the double bond reforms. Keto-enol tautomerism takes place, followed by release of an acetaldehyde molecule from the palladium.

When it was a major source for acetic acid, acetaldehyde was in the top 50 at about 1.5 billion lb. Now it is under a billion pounds but it is still used to manufacture acetic acid by further oxidation. Here a manganese or cobalt acetate catalyst is used with air as the oxidizing agent. Temperatures range from 55-80°C and 15-75 psi. The yield is 95%.

$$2CH_3-\overset{O}{\overset{\|}{C}}-H \ + O_2 \ \xrightarrow[\text{or } Co(OAc)_2]{Mn(OAc)_2} \ 2CH_3-\overset{O}{\overset{\|}{C}}-OH$$

Although the oxidation reaction is simple, the mechanism is quite complex and involves the formation of peracetic acid first.

(1) $\quad CH_3-\overset{O}{\overset{\|}{C}}-H \ + \ Mn^{+3} \ \longrightarrow \ CH_3-\overset{O}{\overset{\|}{C}}\cdot \ + \ Mn\cdot^{+2} \ + \ H^+$

(2) $\quad CH_3-\overset{O}{\overset{\|}{C}}\cdot \ + \ O_2 \ \longrightarrow \ CH_3-\overset{O}{\overset{\|}{C}}-O-O\cdot$

(3) $\quad CH_3-\overset{O}{\overset{\|}{C}}-O-O\cdot \ + \ CH_3-\overset{O}{\overset{\|}{C}}-H \ \longrightarrow CH_3-\overset{O}{\overset{\|}{C}}-O-OH \ + \ CH_3-\overset{O}{\overset{\|}{C}}\cdot$

then (2), (3), (2), (3), etc.

Some of the peracetic acid decomposes with the help of the catalyst and the catalyst is regenerated by this process.

(4) $\quad CH_3-\overset{O}{\overset{\|}{C}}-O-OH \ + \ Mn\cdot^{+2} \ \longrightarrow CH_3-\overset{O}{\overset{\|}{C}}-O\cdot \ + \ Mn^{+3} \ + \ OH^-$

(5) $\quad CH_3-\overset{O}{\overset{\|}{C}}-O\cdot \ + \ CH_3-\overset{O}{\overset{\|}{C}}-H \ \longrightarrow \ CH_3-\overset{O}{\overset{\|}{C}}-OH \ + \ CH_3-\overset{O}{\overset{\|}{C}}\cdot$

then (2), (3), (2), (3), etc.

Most of the peracetic acid decomposes via a cyclic reaction with acetaldehyde to form two moles of acetic acid.

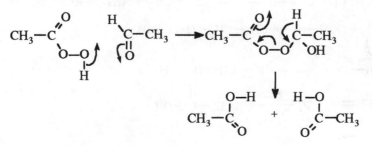

The second manufacturing method for acetic acid utilizes butane from the C_4 petroleum stream rather than ethylene. It is a very complex oxidation with avariety of products formed, but conditions can be controlled to allow a large percentage of acetic acid to be formed. Cobalt (best), manganese, or chromium acetates are catalysts with temperatures of 50-250°C and a pressure of 800 psi.

$$C_4H_{10} + O_2 \xrightarrow{Co(OAc)_2} 76\% \ CH_3-\overset{\overset{O}{\|}}{C}-OH$$

(95% *n*-butane)

$$6\% \ \ H-\overset{\overset{O}{\|}}{C}-OH$$

$$6\% \ \ CH_3-CH_2-OH$$

$$4\% \ \ CH_3-OH$$

$$8\% \ \ \text{Other}$$

The mechanism of this reaction involves first a free radical oxidation of butane to butane hydroperoxide catalyzed by metal ions. This is probably similar to the one on page 171 for acetaldehyde to peracetic acid. The butane hydroperoxide then decomposes to two moles of acetaldehyde via ß-scissions of a C—C bond followed by a C—H bond. The acetaldehyde is then further oxidized to acetic acid as discussed.

The third and now preferred (since 1970) method of acetic acid manufacture is the carbonylation of methanol (Monsanto process), involving reaction of methanol and carbon monoxide (both derived from methane) with rhodium and iodine as catalysts at 175°C and 1 atm. The yield of acetic acid is 99% based on methanol and 90% based on carbon monoxide.

$$CH_3-OH \ + \ CO \ \xrightarrow[I_2]{Rh} \ CH_3-\overset{\overset{\displaystyle O}{\|}}{C}-OH$$

The mechanism is well understood, involving complexation of the rhodium with iodine and carbon monoxide, reaction with methyl iodide (formed from the methanol by hydrogen iodide), insertion of CO in the rhodium—carbon bond, and hydrolysis to give product and regeneration of the complex and more hydrogen iodide.

$$CH_3OH \ + HI \ \longrightarrow CH_3I \ + \ H_2O$$

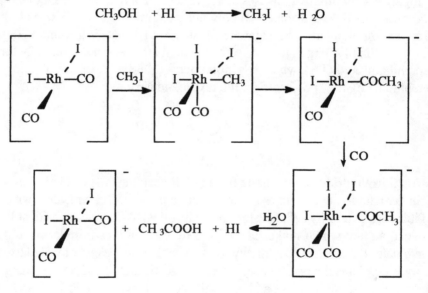

Although we have continued to treat acetic acid manufacture under ethylene derivatives, as you can see it is made from three of the seven basic organics: ethylene, C_4 hydrocarbons, and methane. Pure 100% acetic acid is sometimes called glacial acetic because when cold it will solidify into layered crystals similar in appearance to a glacier. It is a colorless liquid with a pungent, vinegar odor and sharp acid taste, bp 118°C, and mp 17°C.

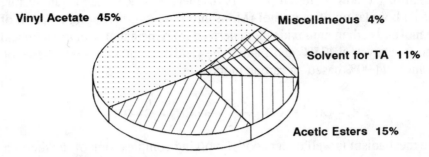

Figure 9.4 Uses of acetic acid. (*Source:* CP.)

Figure 9.4 summarizes the uses of acetic acid. Vinyl acetate is another top 50 chemical. Cellulose acetate is a polymer used mainly as a fiber in clothing. Another important use is in cigarette filters. Ethyl acetate is a common organic solvent. Finally, large amounts of acetic acid are used in the manufacture of terephthalic acid (TA), which is a monomer for the synthesis of poly (ethylene terephthalate), the "polyester" of the textile industry.

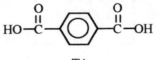

TA

Acetic anhydride can be made from either acetaldehyde oxidation or acetic acid dehydration and at times has appeared in the top 50 chemicals. Almost all of it is used to esterify cellulose for fibers. In 1989 BP Chemicals in Great Britain opened a plant which makes acetic acid by methanol carbonylation and also methylates the acetic acid to methyl acetate, which can be carbonylated to acetic anhydride. Thus the ratio of acetic acid and acetic anhydride can be changed as the market demands. If we add up all the acetaldehyde, acetic acid, and acetic anhydride manufactured the total amounts to about 5 billion lb, a considerable market.

$$\overset{O}{\overset{\|}{}}$$

VINYL ACETATE, CH_3-C-O-CH=CH$_2$ or AcO-CH=CH$_2$

Reference

CP, 4-17-89

Vinyl acetate is one of many compounds where classical organic chemistry has been replaced by a catalytic process. It is also an example of older acetylene chemistry becoming outdated by newer processes involving other basic organic building blocks. Up until recently the preferred manufacture of this important monomer was based on the addition of acetic acid to the triple bond of acetylene using zinc amalgam as the catalyst, a universal reaction of alkynes.

In 1969, 90% of vinyl acetate was manufactured by this process. By 1975 only 10% was made from acetylene, and in 1980 it was obsolete. Instead, a newer method based on ethylene replaced this old acetylene chemistry. A Wacker catalyst is used in this process similar to that for acetic acid. Since the acetic acid can also be made from ethylene, the basic raw material is solely ethylene, in recent years very economically advantageous as compared to acetylene chemistry. An older liquid-phase process has been replaced by a vapor-phase reaction run at 70-140 psi and 175-200°C. Catalysts may be (1) C—PdCl$_2$—CuCl$_2$, (2) PdCl$_2$—Al$_2$O$_3$, or (3) Pd—C, KOAc. The product is distilled; water, acetadehyde, and some polymer are separated. The acetaldehyde can be recycled to acetic acid. The pure colorless liquid is collected at 72°C. It is a lachrymator (eye irritant). The yield is 95%. The mechanism of this reaction is the same used as the Wacker process for ethylene to acetic acid, except that acetic acid attacks rather than water.

$$CH\equiv CH \ + \ HOAc \ \xrightarrow[\text{Ag}]{\text{Zn}} \ CH_2{=}CH{-}OAc$$

$$CH_2{=}CH_2 \ + \ CH_3{-}\overset{\overset{\displaystyle O}{\|}}{C}{-}OH \ + \ {}^1\!/_2O_2 \ \xrightarrow[\text{PdCl}_2]{\text{CuCl}_2} CH_3{-}\overset{\overset{\displaystyle O}{\|}}{C}{-}O{-}CH{=}CH_2 \ + \ H_2O$$

Fig. 9.5 gives the uses of vinyl acetate. Poly (vinyl acetate) is used primarily in adhesives, coatings, and paints. The shift to water-based coatings has certainly helped vinyl acetate production. Copolymers of poly (vinyl acetate) with poly (vinyl chloride) are used in flooring, phonograph records, and PVC pipe. Poly (vinyl alcohol) is used in textile sizing, adhesives, emulsifiers, and paper coatings. Poly (vinyl butyral) is the plastic inner liner of most safety glass.

Vinyl acetate is a good example of an ethylene chemical with a high percentage of exports. The United States now has a cost advantage in ethylene production and many ethylene derivatives have high export percentages.

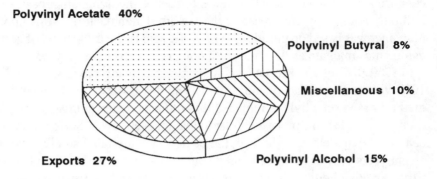

Polyvinyl Acetate 40%

Polyvinyl Butyral 8%

Miscellaneous 10%

Exports 27%

Polyvinyl Alcohol 15%

Figure 9.5 Uses of vinyl acetate. (*Source:* CP.)

ETHYLBENZENE

Reference

CP, 8-21-89

Despite the use of new catalysts for manufacturing some industrial organic chemicals, many well-known classical reactions still abound. The Friedel-Crafts alkylation is one of the first reactions studied in electrophilic aromatic substitution. It is used on a large scale for making ethylbenzene.

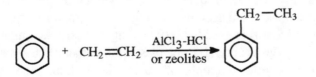

Note that ethylbenzene is a derivative of two basic organic chemicals, ethylene and benzene. A vapor-phase method with boron trifluoride, phosphoric acid, or alumina-silica as catalysts has given away to a liquid-phase reaction with aluminum chloride at 90°C and atmospheric pressure. A new Mobil-Badger zeolite catalyst at 420°C and 175-300 psi in the gas phase may be the method of choice for future plants to avoid corrosion problems. The mechanism of the reaction involves complexation of the ethylene with the Lewis acid catalyst, attack of the electrophilic carbon on the aromatic ring, loss of the proton to rearomatize, and desorption of the catalyst with subsequent protonation in the side chain.

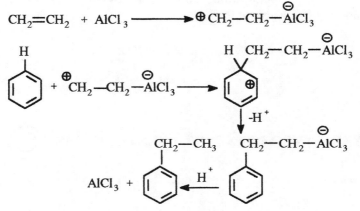

Excess benzene must be used. A common benzene: ethylene ratio is 1.0:0.6. This avoids the formation of di-and triethylbenzenes. The first ethyl group, being electron-donating inductively as compared to hydrogen, will activate the benzene ring toward electrophilic attack by stabilizing the intermediate

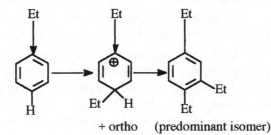

+ ortho   (predominant isomer)

carbocation. The benzene when in excess prevents this since it increases the probability of the attack on benzene rather than on ethylbenzene, but some

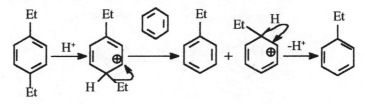

polyethylbenzenes are formed, and these can be separated in the distillation process and burned for fuel. Alternatively, disubstituted isomers can be transalkylated with benzene to give two moles of monosubstituted product.

The benzene is recycled. Ethylbenzene is a colorless liquid, bp 136°C. Despite the elaborate separations required, including washing with caustic and water and three distillation columns, the overall yield of ethylbenzene is economically feasible at 98%.

<u>Uses of Ethylbenzene</u>
Styrene	99%
Solvent	1

STYRENE; VINYLBENZENE, PHENYLETHENE

Reference

CP, 8-14-89

Nearly all, about 83%, of the styrene produced in the United States is made from ethylbenzene by dehydrogenation. This is a high-temperature reaction (630°C) with various metal oxides as catalysts, including zinc, chromium, iron, or magnesium oxides coated on activated carbon, alumina, or bauxite. Iron oxide on potassium carbonate is also used. Most dehydrogenations do not occur readily even at high temperatures. The driving force for this reaction is the extension in conjugation that results, since the double bond on the side chain is in conjugation with the ring. Conditions must be controlled to avoid polymerization of the styrene. Sulfur is added to prevent polymerization. The crude product has only 37% styrene but contains 61% ethylbenzene. A costly vacuum distillation through a 70-plate column at 90°C and 35 torr is needed to separate the two. The ethylbenzene is recycled. Usually a styrene plant is combined with an ethylbenzene plant when designed. The yield is 90%.

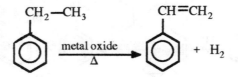

Because styrene readily polymerizes it is immediately treated with an antioxidant such as *p-t-* butylcatechol at l0 ppm.

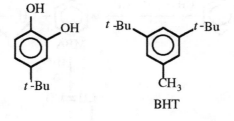

BHT

Many phenols, especially "hindered phenols" such as butylated hydroxy toluene (BHT), are good antioxidants. They act as radical scavengers by readily reacting with stray radicals to give very stable radicals via resonance. The alkyl radicals then cannot initiate the polymerization of substances such as styrene.

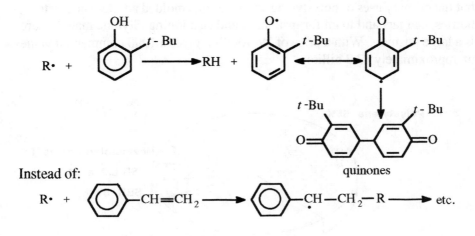

quinones

Instead of:

A new alternate method for the manufacture of styrene, called the oxirane process, now accounting for 17% of production, uses ethylbenzene. It is oxidized to the hydroperoxide and reacts with propylene to give phenylmethylcarbinol (or methyl benzyl alcohol, MBA) and propylene oxide, the latter being a top 50 chemical itself. The alcohol is then dehydrated at relatively low temperatures (180-400°C) using an acidic silica gel or titanium dioxide catalyst, a much cleaner and less energy-dependent reaction than the dehydrogenation. Other olefins besides propylene could be used in the epoxidation reaction, but it is chosen because of the high demand for the epoxide. Some acteophenone is separated and hydrogenated back to MBA.

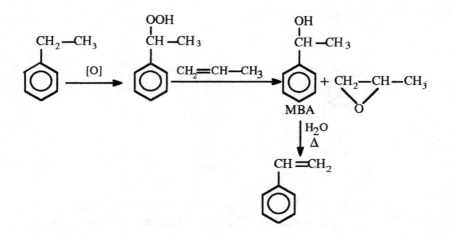

Fig. 9.6 shows the uses of styrene. These are dominated by polymer chemistry and involve polystyrene and its copolymers. We will study these in detail later, but the primary uses of polystyrene are in various molded articles such as toys, bottles, and jars and foam for insulation and cushioning. Styrene manufacture is a large business. With a price of 39-40¢/lb it gives a 1990 commercial value of approximately $3.2 billion.

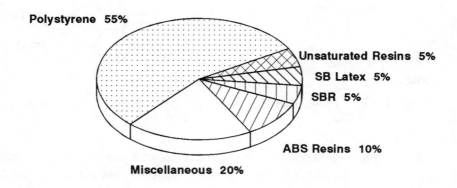

Figure 9.6 Uses of styrene. (*Source:* CP.)

ETHYLENE OXIDE $\overset{\displaystyle O}{\overset{\displaystyle /\backslash}{CH_2-CH_2}}$

Reference

CP, 5-12-90

Another example of a famous organic chemical reaction being replaced by a catalytic process is furnished by the manufacture of ethylene oxide. For many years it was made by chlorohydrin formation followed by dehydrochlorination to the epoxide. Although the chlorohydrin route is still used to convert propylene to propylene oxide, a more efficient air epoxidation of ethylene is used and the chlorohydrin process for ethylene oxide manufacture has not been used since 1972.

$$\text{old:}\quad CH_2{=}CH_2 \;+\; Cl_2 \;+\; H_2O \xrightarrow{10\text{-}50\,^{\circ}C} \underset{OH}{\overset{Cl}{CH_2{-}CH_2}} \;+\; HCl \xrightarrow[\text{or Ca(OH)}_2]{NaOH}$$

$$\underset{O}{CH_2{-}CH_2} \;+\; NaCl \,(\text{or } CaCl_2) \;+\; H_2O$$

$$\text{new: } 2CH_2{=}CH_2 \;+\; O_2 \xrightarrow{Ag} 2\,CH_2{-}CH_2\,({+}CO_2 \;+\; H_2O)$$

The higher yields (85%) in the chlorohydrin method are not enough to outweigh the waste of chlorine inherent in the process. Although the yields in the direct oxidation method (75%) are lower, the cheap oxidant atmospheric oxygen is hard to beat for economy. Some overoxidation to carbon dioxide and water occurs. Good temperature control at 270-290°C and pressures of 120-300 psi with a 1 sec contact time on the catalyst are necessary. Tubular reactors containing several thousand tubes of 20-50 mm diameter are used. Even though metallic silver is placed in the reactor, the actual catalyst is silver oxide under the conditions of the reaction. Ethylene oxide is a gas at room temperature with a bp of 14°C.

Fig. 9.7 lists the uses of ethylene oxide. Ethylene glycol is eventually used in two primary types of end products: polyesters and antifreeze. About half the

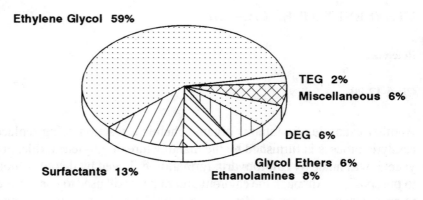

Figure 9.7 Uses of ethylene oxide. (*Source*: CP.)

ethylene glycol is used for each end product. Poly (ethylene terephthalate) is the leading synthetic fiber and has other important applications in plastic film and bottles. Ethylene glycol is a common antifreezing agent especially in automobiles.

As a hospital sterilant for plastic materials, ethylene oxide was ideal since radiation or steam cannot be used. It is in this application that recent evidence of high miscarriage rates (3x normal) of women on hospital sterilizing staffs has caused a lowering of the TLV to 1 ppm. The chemical is most often used in closed systems but in this application incidental exposures are said to go as high as 250 ppm. As of 1984, ethylene oxide is on the "suspect carcinogen list" and has not been used for this purpose. Ethylene oxide is important in the manufacture of many nonionic detergents to be discussed in a later chapter. It is a feedstock for synthesizing glycol ethers (solvents for paints, brake fluids) and ethanolamines (surfactants and gas scrubbing of refineries to remove acids). The manufacturing chemistry of these two materials is given on page 183.

Diethylene and triethylene glycol (DEG and TEG) are produced as byproducts of ethylene glycol (see page 183). DEG and TEG are used in polyurethane and unsaturated polyester resins and in the drying of natural gas. DEG is also used in antifreeze and in the synthesis of morpholine, a solvent, corrosion inhibitor, antioxidant, and pharmaceutical intermediate.

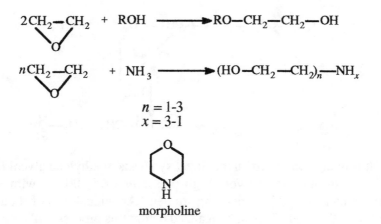

$$2\,CH_2{-}CH_2 \;+\; ROH \longrightarrow RO{-}CH_2{-}CH_2{-}OH$$

$$n\,CH_2{-}CH_2 \;+\; NH_3 \longrightarrow (HO{-}CH_2{-}CH_2)_n{-}NH_x$$

$$n = 1\text{-}3$$
$$x = 3\text{-}1$$

morpholine

ETHYLENE GLYCOL, HO-CH₂-CH₂-OH; ETHAN-1, 2-DIOL

Reference

CP, 3-16-87

The primary manufacturing method of making ethylene glycol is from acid or thermal-catalyzed hydration and ring opening of the oxide. Nearly all the glycol is made by this process. Either a 0.5-1.0% H_2SO_4 catalyst is used at 50-70°C for 30 min or, in the absence of the acid, a temperature of 195°C and 185 psi for 1 hr will form the diol. A 90% yield is realized when the ethylene oxide:water molar ratio is 1:5-8. The advantage of the acid-catalyzed reaction is no high pressure; the thermal reaction however needs no corrosion resistance and no acid separation step.

$$CH_2{-}CH_2 \;+\; H_2O \xrightarrow{\ H^+\text{ or }\Delta\ } HO{-}CH_2{-}CH_2{-}OH$$

+ some
HO—CH₂—CH₂—O—CH₂—CH₂—OH
diethylene glycol

+ some
HO—CH₂—CH₂—O—CH₂—CH₂—O—CH₂—CH₂—OH
triethylene glycol

The ethylene glycol, bp 198°C, is readily vacuum distilled and separated from the diethylene glycol, bp 246°C, and triethylene glycol, bp 288°C. The mechanism of the reaction follows the general scheme for acid-catalyzed ring openings of epoxides.

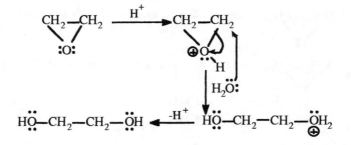

Research is being conducted on the direct synthesis of ethylene glycol from synthesis gas. In one process very high pressures of 5,000 psi with very expensive catalysts Rh_x (CO) are being studied. An annual loss of rhodium catalyst of only 0.000001% must be realized before this process will compete economically. There are at least five other alternate syntheses of ethylene glycol which bypass toxic ethylene oxide that are being researched.

Fig. 9.8 shows the use profile for ethylene glycol. In 1950 only 48% of antifreeze was ethylene glycol, whereas in 1962 it accounted for 95% of all antifreeze. Its ideal properties of low melting point, high boiling point, and unlimited water solubility make it a good material for this application. The 1991 price for ethylene glycol of 30¢/lb gives a commercial value of $1.5 billion.

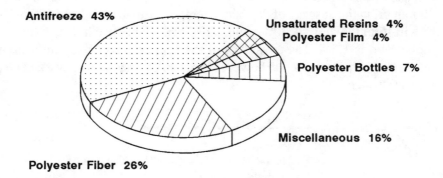

Figure 9.8 Uses of ethylene glycol. (*Source:* CP.)

ETHANOL, CH3-CH2-OH

Reference

CP, 3-25-91

Ethanol is also named ethyl alcohol, industrial alcohol, grain alcohol, and alcohol.

An ethylene derivative that sometimes makes the top 50, ethanol is an excellent example of the tremendous development that synthetic organic chemicals have had in the last few decades. In 1935 over 90% of all alcohol was made nonsynthetically via the fermentation of sugars. Waste syrup (molasses) after crystallization of sugar from sugar cane processing containing mostly sucrose can be treated with enzymes from yeast to give ethyl alcohol and carbon dioxide. The enzymes require 28-72 hr at 20-38°C and give a 90% yield. This process is not used to any great extent for industrial alcohol today.

$$C_{12}H_{22}O_{11} + H_2O \xrightarrow{\text{invertase}} 2C_6H_{12}O_6$$
sucrose

$$C_6H_{12}O_6 \xrightarrow{\text{zymase}} 2CH_3CH_2OH + 2CO_2$$
glucose and fructose

The next method of manufacture, the esterification/hydrolysis of ethylene, was the method of choice in the 1950s and 1960s. An important side product in this electrophilic addition to ethylene was diethyl ether.

$$CH_2{=}CH_2 + H_2SO_4 \longrightarrow CH_3{-}CH_2{-}O{-}\overset{\overset{\displaystyle O}{\|}}{\underset{\underset{\displaystyle O}{\|}}{S}}{-}OH \xrightarrow{H_2O}$$

$$CH_3{-}CH_2{-}OH + H_2SO_4$$

Now we use direct hydration of ethylene with a catalytic amount of phosphoric acid. In the early 1980s over 90% of ethanol was synthetic and was manufactured by direct hydration. Temperatures average 300-400°C with 1000 psi. Only 4% of the ethylene is converted to alcohol per pass, but this cyclic process eventually gives a net yield of 97%. Less diethyl ether byproduct is formed and no pollution problems of sulfuric acid are encountered. The mechanism involves electrophillic attack by the catalyst on the double bond, reaction of the carbocation with water, and deprotonation to give the product.

Reaction: $CH_2{=}CH_2 \;+\; H_2O \xrightarrow{\;H_3PO_4\;} CH_3CH_2OH$

Mechanism: $CH_2{=}CH_2 \xrightarrow{\;\overset{+}{H}\;} CH_3{-}\overset{\oplus}{CH_2} \xrightarrow{\;H_2\ddot{O}:\;} CH_3{-}CH_2{-}\overset{\oplus}{\underset{\cdot\cdot}{O}H_2} \xrightarrow{\;{-}\overset{+}{H}\;}$

$$CH_3{-}CH_2{-}\ddot{\underset{\cdot\cdot}{O}}H$$

The fermentation process is making a comeback for use in gasohol, a new automobile fuel which is a simple mixture of 90% gasoline and 10% alcohol claimed to increase mileage. The federal government is giving a special subsidy to farmers who produce alcohol by corn fermentation for this use. Corn fermentation was the source of 66% of all ethanol made in 1987 and 82% in 1988. Current technology yields about 2.5 gal of ethanol per bushel of corn. Brazil has a new process for making ethanol from sugarcane and it is now the world's largest producer of ethanol. It is exporting this technology to other developing, petroleum-poor countries.

Most alcohol is sold as 95% ethanol-5% water since it forms an azeotrope at that temperature. To obtain absolute alcohol a third component such as benzene must be added during the distillation. This tertiary azeotrope carries over the water and leaves the pure alcohol behind. Common industrial alcohol is "denatured." Additives are purposely included to make it nondrinkable and therefore not subject to the high taxes of the alcoholic beverage industry.

The use and production values for ethanol do not include that amount produced for most alcoholic beverages, which is a separate industry itself and will not be discussed here. However, one term that you should understand is "U.S. proof". Originally this term was applied to a test of strong alcoholic beverages done by pouring the sample onto gunpowder and lighting it. Ignition of the gunpowder was considered proof that the beverage was not diluted. A value of 100 proof now refers to 50% alcohol by volume (42.5% by weight). Thus 190 proof is 95% alcohol by volume (92.4% by weight).

Industrial uses of ethanol are shared by synthetic and fermentation alcohol in a 7:3 ratio. These uses include solvents, 59% (especially for toiletries and cosmetics, coatings and inks, and detergents and household cleaners), and chemical intermediates, 41% (especially ethyl acrylate, vinegar, ethylamines, and glycol ethers). Most corn fermentation alcohol is used in fuel, 86%; industrial solvents and chemicals, 7%; and beverages, 7%.

10

Chemicals from Propylene and Butylene

References

L & M, selected pages
W & R I, pp. 67-92
Kent, pp. 942-959
Wiseman, pp. 65-88
Szmant, pp. 265-378

INTRODUCTION

As we learned in Chapter 8, the official production of propylene is usually about half that of ethylene, only because nearly half the propylene is used by petroleum refineries internally to alkylate gasolines. This captive use is not reported. Of the propylene used for chemical manufacture, nearly one fourth is polymerized to polypropylene, to be discussed in a later chapter. Of the remaining amount of propylene, 6 chemicals from the top 50 are manufactured. These are listed in Table 10.1. Their industrial manufacturing methods are summarized in Fig. 10.1. Note that three of these chemicals, cumene, phenol, and acetone, are also derived from a second basic organic chemical, benzene.

TABLE 10.1 Propylene Derivatives in the Top 50

1990 Rank	Chemical
32	Cumene
35	Phenol
36	Propylene oxide
38	Acrylonitrile
43	Acetone
48	Isopropyl alcohol

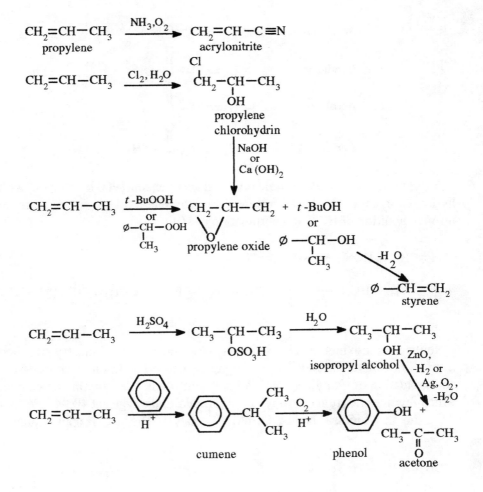

Figure 10.1 Synthesis of propylene derivatives.

ACRYLONITRILE, CH_2=CH—C≡N; 2-PROPENONITRILE

Reference

CP, 3-6-89

Acrylonitrile and other three-carbon analogs containing a double bond have a common name derived from the word *acrid*, meaning strong and disagreeable, in regard to the odor of most of these chemicals. Compounds in this family are given here with their common names.

Acrylonitrile	CH_2=CH—C≡N
Acrylic acid	CH_2=CH—$\overset{\overset{\displaystyle O}{\|\|}}{C}$—OH
Acrolein	CH_2=CH—$\overset{\overset{\displaystyle O}{\|\|}}{C}$—H
Acrylamide	CH_2=CH—$\overset{\overset{\displaystyle O}{\|\|}}{C}$—$NH_2$

Acrylonitrile was made completely from acetylene in 1960 by reaction with hydrogen cyanide. This was followed by ethylene oxide as a raw material through addition of HCN and elimination of H_2O.

$$CH≡CH + HCN \longrightarrow CH_2=CH_2—CN$$

$$CH_2—CH_2 + HCN \longrightarrow HO—CH_2—CH_2—CN \xrightarrow{\Delta} CH_2=CH—CN + H_2O$$

Neither of these methods is used today. Around 1970 the industry switched from C_2 raw materials and classical organic chemical addition reactions to the ammoxidation of propylene. Now all acrylonitrile is made by this procedure, which involves reaction of propylene, ammonia, and oxygen at 400-450°C and 0.5-2 atm in a fluidized bed $Bi_2O_3 \cdot nMoO_3$ catalyst. The yield is approximately 70%.

$$2CH_2=CH—CH_3 + 2NH_3 + 3O_2 \longrightarrow 2CH_2=CH—C≡N + 6H_2O$$

The mechanism is undoubtedly a free radical reaction that occurs very easily at the allyl site in propylene, forming the resonance stabilized allyl radical.

$$CH_2=CH-CH_2\cdot \longleftrightarrow \cdot CH_2-CH=CH_2$$

Byproducts of this reaction are acetonitrile, $CH_3-C≡N$, and hydrogen cyanide. This is now a major source of these two materials. Interestingly, the C_2 byproduct acetonitrile has a bp of 81.6°C, whereas acrylonitrile with three carbons has a lower bp of 77.3°C, quite an unusual reversal of this physical property's dependence on molecular weight. The TWA of acrylonitrile is 2 ppm and it is a "suspect carcinogen."

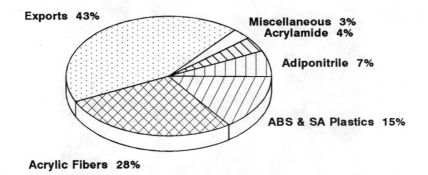

Figure 10.2 Uses of acrylonitrile. (*Source*: CP.)

Figure 10.2 outlines the use of acrylonitrile. The most important uses of acrylonitrile are in the polymerization to polyacrylonitrile. This substance and its copolymers make good synthetic fibers for the textile industry. Acrylic is the fourth largest produced synthetic fiber behind polyester, nylon, and polyolefin. It is known primarily for its warmth, similar to the natural and very expensive fiber, wool. Approximately 15% of the acrylonitrile is made into plastics, including the copolymer of styrene-acrylonitrile (SA) and the terpolymer of acrylonitrile, butadiene, and styrene (ABS). Fig. 10.3 shows a reactor used for the ammoxidation of propylene to acrylonitrile.

Adiponitrile is made by two different methods. One method is by the electrohydrodimerization of acrylonitrile. It is converted into hexamethyl-enediamine (HMDA) which is used to make nylon. The other adiponitrile synthesis is C_4 chemistry which will be discussed later.

Figure 10.3 Reactor for the ammoxidation of propylene to make acrylonitrile. (Courtesy of Du Pont, Beaumont, TX.)

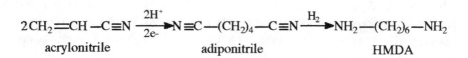

$$2\,CH_2\!=\!CH-C\!\equiv\!N \xrightarrow[\text{2e-}]{\text{2H}^+} N\!\equiv\!C-(CH_2)_4-C\!\equiv\!N \xrightarrow{H_2} NH_2-(CH_2)_6-NH_2$$

acrylonitrile   adiponitrile   HMDA

About one third of all adiponitrile is made from acrylonitrile. In the electrodimerization of acrylonitrile a two phase system is used containing a phase transfer catalyst tetrabutylammonium tosylate [(n-Bu)$_4$N$^+$OTs$^-$]. The head-to-head dimerization may be visualized to occur in the following manner:

$$2CH{=}CH{-}C{\equiv}N{:} \xrightarrow{2e^-} 2 \cdot CH_2{-}CH{=}C{=}\ddot{N}{:}^{\ominus} \longrightarrow$$

$$^{\ominus}{:}\ddot{N}{=}C{=}CH{-}CH_2{-}CH_2{-}CH{=}C{=}\ddot{N}{:}^{\ominus} \xrightarrow{2H^+}$$

$$H\ddot{N}{=}C{=}CH{-}CH_2{-}CH_2{-}CH{=}C{=}\ddot{N}H \longrightarrow$$

$$:N{\equiv}C{-}CH_2{-}CH_2{-}CH_2{-}CH_2{-}C{\equiv}N{:}$$

The byproduct of acrylonitrile manufacture, HCN, has its primary use in the manufacture of methyl methacrylate by reaction with acetone. This is covered in the section on acetone.

PROPYLENE OXIDE, CH_3-CH-CH$_2$; 1, 2-EPOXYPROPANE

Reference

CP, 2-2-87

There are two important methods for the manufacture of propylene oxide, each accounting for one half the total amount produced. The older method involves chlorohydrin formation from the reaction of propylene with chlorine water. Before 1969 this was the exclusive method. Unlike the analogous procedure for making ethylene oxide from ethylene which now is obsolete, this method for propylene oxide is still economically competitive. Many old ethylene oxide plants have been converted to propylene oxide synthesis.

$$CH_2{=}CH{-}CH_3 + Cl_2 + H_2O \longrightarrow \overset{Cl}{\underset{OH}{CH_2{-}CH{-}CH_3}} + HCl \xrightarrow[\text{or Ca(OH)}_2]{\text{NaOH}}$$

$$\underset{O}{CH_2{-}CH{-}CH_3} + NaCl \text{ (or CaCl}_2) + H_2O$$

The mechanism in the first step involves an attack of the electrophilic chlorine on the double bond of propylene to form a chloronium ion, which is attacked by a hydroxide ion to complete the first reaction. The dilute cholorohydrin solution is mixed with a 10% slurry of lime to form the oxide, which is purified by distillation, bp 34°C. The yield is 90%.

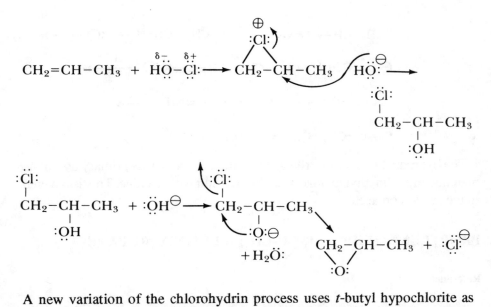

A new variation of the chlorohydrin process uses *t*-butyl hypochlorite as chlorinating agent. The waste brine solution can be converted back to chlorine and caustic by a special electrolytic cell to avoid the waste of chlorine.

The second manufacturing method for propylene oxide is via peroxidation of propylene, called the Halcon process after the company that invented it. Oxygen is first used to oxidize isobutane to *t*-butyl hydroperoxide(BHP) over a molybdenum naphthenate catalyst at 90°C and 450 psi. This oxidation occurs at the preferred tertiary carbon because a tertiary alkyl radical intermediate can be formed easily.

$$
4CH_3{-}\underset{\underset{\displaystyle CH_3}{|}}{CH}{-}CH_3 \; + \; 3O_2 \longrightarrow 2CH_3{-}\underset{\underset{\displaystyle O{-}OH}{|}}{\overset{\overset{\displaystyle CH_3}{|}}{C}}{-}CH_3 \; + \; 2CH_3{-}\underset{\underset{\displaystyle OH}{|}}{\overset{\overset{\displaystyle CH_3}{|}}{C}}{-}CH_3
$$

Mechanism: BHP

(1) $(CH_3)_3\,CH\;+X\cdot \longrightarrow (CH_3)_3\,C\cdot\;+\;HX$

(2) $(CH_3)_3\cdot\;+\;O_2 \longrightarrow (CH_3)_3\,COO\cdot$

(3) $(CH_3)_3\,COO\cdot\;+\;(CH_3)_3CH \longrightarrow (CH_3)_3COOH\;+\;(CH_3)_3C\cdot$

then (2), (3), (2), (3), etc.

The BHP is then used to oxidize propylene to the oxide. This reaction is ionic and its mechanism follows. The yield of propylene oxide from propylene is 90%.

Reaction:

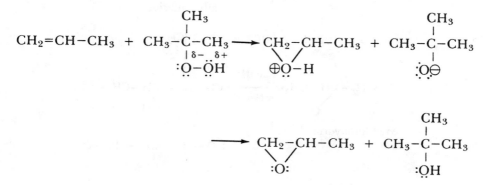

Mechanism:

The *t*-butyl alcohol can be used to increase the octane of unleaded gasoline or it can be made into methyl *t*-butyl ether (MTBE) for the same application. The alcohol can also be dehydrated to isobutylene, which in turn is used in alkylation to give highly branched dimers for addition to straight-run gasoline.

Since approximately 2.2 lb. of *t*-butyl alcohol would be produced per 1 lb of propylene oxide, an alternative reactant in this method is ethylbenzene hydroperoxide. This eventually forms phenylmethylcarbinol along with the propylene oxide. The alcohol is dehydrated to styrene. This chemistry was covered in the previous chapter as one of the syntheses of styrene. Thus the side product can be varied depending on the demand for substances such as *t*-butyl alcohol or styrene. Research is being done on a direct oxidation of propylene with oxygen, analogous to that used in the manufacture of ethylene oxide from ethylene and oxygen (see Chapter 9). But the proper catalyst and conditions have not yet been found. The methyl group is very sensitive to oxidation conditions.

As an aside to the manufacture of propylene oxide via the chlorohydrin process let us mention use of this type of chemistry to make epichlorohydrin. Although not in the top 50, it is an important monomer for making epoxy adhesives as well as glycerin (HO—CH_2—$CHOH$—CH_2—OH). Propylene is first chlorinated free radically at the allyl position at 500°C to give allyl chloride, which undergoes chlorohydrin chemistry as discussed previously to give epichlorohydrin. The student should review the mechanism of allyl-free radical substitution from a basic organic chemistry course and also work out the mechanism for this example of a chlorohydrin reaction.

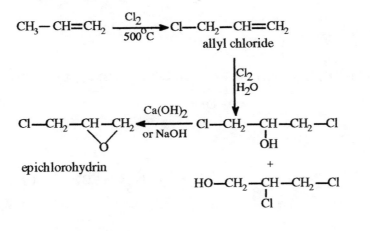

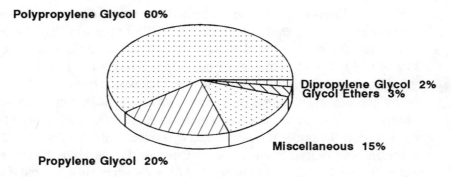

Polypropylene Glycol 60%

Dipropylene Glycol 2%
Glycol Ethers 3%

Miscellaneous 15%

Propylene Glycol 20%

Figure 10.4 Uses of propylene oxide. (*Source*: CP.)

Figure 10.4 summarizes the uses of propylene oxide. Propylene glycol is made by hydrolysis of propylene oxide. The student should develop the mechanism for this reaction which is similar to the ethylene oxide to ethylene glycol conversion (see Chapter 9). Propylene glycol is a monomer in the manufacture of unsaturated polyester resins, which are used for boat and automobile bodies, bowling balls, and playground equipment.

But an even larger use of the oxide is its polymerization to poly (propylene glycol), which is actually a polyether, although it has hydroxy end groups. These hydroxy groups are reacted with an isocyanate such as toluene diisocyanate (TDI) to form the urethane linkages in the high molecular weight polyurethanes, useful especially as foams for automobile seats, furniture, bedding, and carpets. Poly (propylene glycol) is used to make both flexible and rigid polyurethane in a 75:25 market ratio.

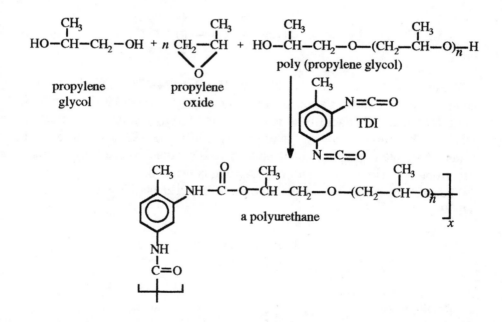

ISOPROPYL ALCOHOL, CH_3-CH-CH_3
$\qquad\qquad\qquad\qquad\qquad$ **OH**

Reference

CP, 9-3-90

Isopropyl alcohol (IPA) is also named 2-propanol, isopropanol, and rubbing alcohol.

One of the longest continuous running processes for an organic chemical used since World War I is the esterification/hydrolysis of propylene to isopropyl alcohol. Unlike ethanol, for which the esterification/hydrolysis has been replaced by direct hydration, the direct process for isopropyl alcohol has not become popular in this country, although it is used in Germany and the United Kingdom. It does not work well for crude propylene. In the esterification process only the propylene reacts and conditions can be maintained so that ethylene is inert. (Why is ethylene less reactive?)

$$CH_3{-}CH{=}CH_2 \ + \ H_2SO_4 \longrightarrow CH_3{-}\underset{\underset{OSO_3H}{|}}{CH}{-}CH_3$$

$$\Big\downarrow H_2O$$

$$H_2SO_4 \ + \ CH_3{-}\underset{\underset{OH}{|}}{CH}{-}CH_3$$

The esterification step occurs with 85% H_2SO_4 at 24-27°C. Dilution to 20% concentration is done in a separate tank. The isopropyl alcohol is distilled from the dilute acid. This acid is concentrated and returned to the esterification reactor. The isopropyl alcohol is originally distilled as a 91% azeotrope with water. Absolute isopropyl alcohol, bp 82.5°C, is obtained by distilling a tertiary azeotrope with isopropyl ether. A 95% yield is realized.

The mechanism for this process follows. *n*-Propyl alcohol is not formed here. Why?

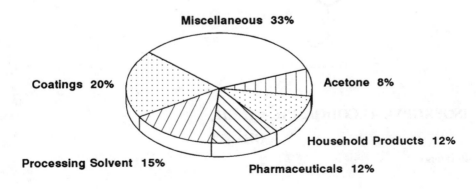

Figure 10.5 Uses of isopropyl alcohol. (*Source*: CP.)

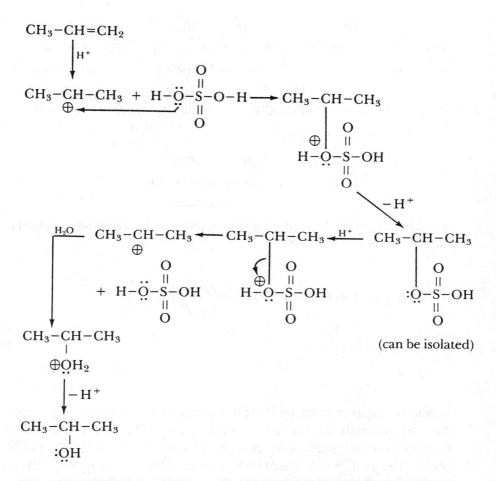

Fig. 10.5 lists the uses of isopropyl alcohol. Only 7% of all acetone is made from isopropyl alcohol. Only one such plant is left in the United States. The rest is made along with phenol from cumene as we will see subsequently. But since cumene is derived from propylene (and benzene) all the acetone produced commercially is propylene-based. When the phenol market is down then this alternate acetone process increases in percentage. Some of the chemicals derived from isopropyl alcohol going into the end uses of Fig. 10.5 are isopropyl ether, a good industrial extraction solvent; isopropyl acetate, a solvent for cellulose derivatives; isopropyl myristate, an emollient, lubricant, and blending agent in cosmetics, inks, and plasticizers; *t*-butylperoxy isopropyl carbonate, a poly-

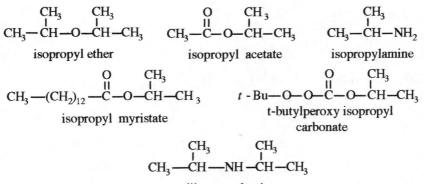

isopropyl ether   isopropyl acetate   isopropylamine

isopropyl myristate   t-butylperoxy isopropyl carbonate

diisopropylamine

merization catalyst and curing agent; and isopropylamine and diisopropylamine, convenient low-boiling bases.

CUMENE; ISOPROPYLBENZENE

Reference

CP, 7-23-90

Cumene is an important intermediate in the manufacture of phenol and acetone. The feed materials are benzene and propylene. This is a Friedel-Crafts alkylation reaction catalyzed by solid phosphoric acid at 175-225°C and 400-600 psi. The yield is 97% based on benzene and 90% on propylene. Excess benzene stops the reaction at the monoalkylated stage and prevents the polymer-

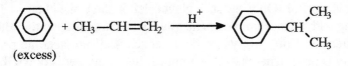

(excess)

ization of propylene. Interestingly, if benzene is left out similar conditions are used to manufacture the trimer and tetramer of propylene. The cumene is separated by distillation, bp 153°C.

    The mechanism of the reaction involves electrophilic attack of the catalyst on the double bond of propylene to form the more stable secondary cation, which reacts with the π cloud of benzene to give a delocalized ion. Deprotonation rearomatizes the ring.

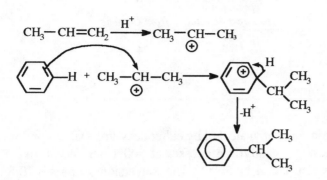

Almost all of the cumene is used to make phenol and acetone. Only 1% is used to make α-methylstyrene by dehydrogenation. This material is used in small amounts during the polymerization of styrene to vary the properties of the resulting copolymer.

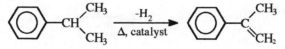

ACETONE, CH_3–$\overset{\displaystyle O}{\underset{\displaystyle \|}{C}}$– CH_3; 2-PROPANONE

Reference

CP, 7-3-90

Presently there are two processes that make acetone in large quantities. The feedstock for these is either isopropyl alcohol or cumene. In the last few years there has been a steady trend away from isopropyl alcohol and toward cumene, but isopropyl alcohol should continue as a precursor since manufacture of acetone from only cumene would require a balancing of the market with the coproduct phenol from this process.

This is not always easy to do, so an alternate acetone source is required. In fact, isopropyl alcohol may become attractive again since cumene can be used to increase octane ratings in unleaded gasoline, and phenol, as a plywood adhesive, is slumping with the housing industry. The percentage distribution of the two methods is given on page 202.

Year	Percent from Isopropyl Alcohol	Percent from Cumene
1959	80	20
1975	46	54
1980	30	70
1985	15	85
1990	10	90

Acetone is made from isopropyl alcohol by either dehydrogenation (preferred) or air oxidation. These are catalytic processes at 500°C and 40-50 psi. The acetone is purified by distillation, bp 56°C. The conversion per pass is 70-85% and the yield is over 90%.

dehydrogenation
$$CH_3 - \overset{OH}{\underset{|}{CH}} - CH_3 \xrightarrow[\text{or ZnO}]{Cu, -Zn} CH_3 - \overset{O}{\overset{\|}{C}} - CH_3 + H_2$$

air oxidation
$$2CH_3 - \overset{OH}{\underset{|}{CH}} - CH_3 + O_2 \xrightarrow[\text{or Cu}]{Ag} 2CH_3 - \overset{O}{\overset{\|}{C}} - CH_3 + 2H_2O$$

The formation of phenol and acetone from cumene hydroperoxide involves a fascinating rearrangement of cumene hydroperoxide where a phenyl group migrates from carbon to an electron-deficient oxygen atom. This was discovered by German chemists Hock and Lang in 1944 and commercialized in 1953 in the U.S. and U.K. The hydroperoxide is made by reaction of cumene with oxygen at 110-115°C until 20-25% of the hydroperoxide is formed. Concentration of the hydroperoxide to 80% is followed by the acid-catalyzed rearrangement at 70-80°C. The overall yield is 90-92%.

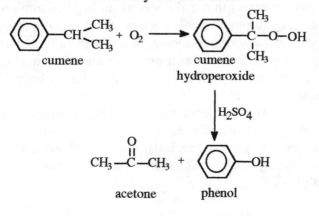

cumene

cumene hydroperoxide

acetone   phenol

Side products are acetophenone, 2-phenylpropan-2-ol, and α-methylstyrene. Acetone is distilled first at bp 56°C.

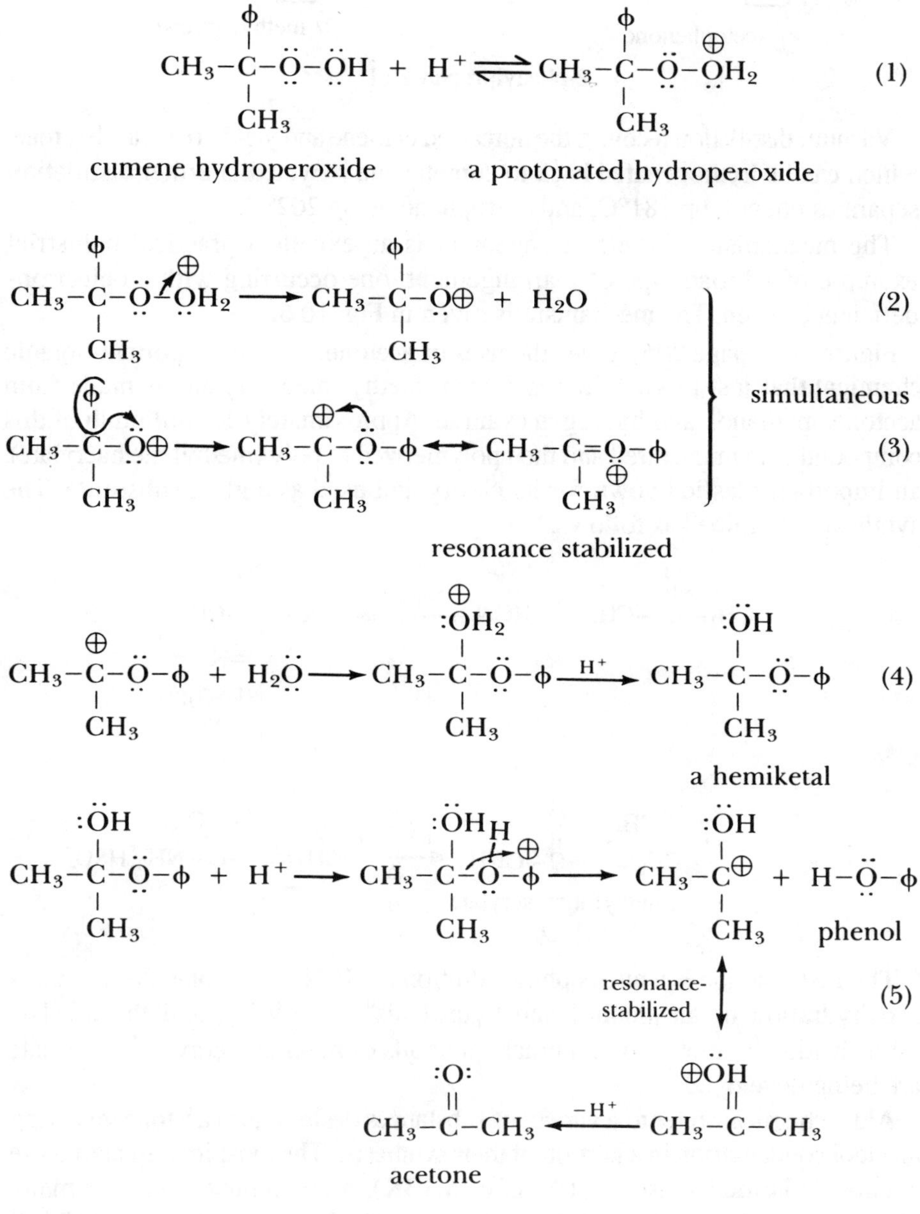

$$CH_3-\underset{\underset{CH_3}{|}}{\overset{\overset{\phi}{|}}{C}}-\ddot{\underset{..}{O}}-\ddot{\underset{..}{O}}H \;+\; H^+ \; \rightleftharpoons \; CH_3-\underset{\underset{CH_3}{|}}{\overset{\overset{\phi}{|}}{C}}-\ddot{\underset{..}{O}}-\overset{\oplus}{O}H_2 \qquad (1)$$

cumene hydroperoxide   a protonated hydroperoxide

$$CH_3-\underset{\underset{CH_3}{|}}{\overset{\overset{\phi}{|}}{C}}-\ddot{\underset{..}{O}}\overset{\oplus}{-}OH_2 \;\longrightarrow\; CH_3-\underset{\underset{CH_3}{|}}{\overset{\overset{\phi}{|}}{C}}-\ddot{O}\oplus \;+\; H_2O \qquad (2)$$

$$CH_3-\underset{\underset{CH_3}{|}}{\overset{\overset{\phi}{|}}{C}}-\ddot{O}\oplus \;\longrightarrow\; CH_3-\underset{\underset{CH_3}{|}}{\overset{\overset{\oplus}{}}{C}}-\ddot{\underset{..}{O}}-\phi \;\longleftrightarrow\; CH_3-\underset{\underset{CH_3}{|}}{C}=\overset{\oplus}{O}-\phi \qquad (3)$$

$$\left.\begin{array}{c} \text{simultaneous} \end{array}\right.$$

resonance stabilized

$$CH_3-\underset{\underset{CH_3}{|}}{\overset{\overset{\oplus}{}}{C}}-\ddot{\underset{..}{O}}-\phi \;+\; H_2\ddot{O} \;\longrightarrow\; CH_3-\underset{\underset{CH_3}{|}}{\overset{\overset{\overset{\oplus}{O}H_2}{|}}{C}}-\ddot{\underset{..}{O}}-\phi \;\xrightarrow{H^+}\; CH_3-\underset{\underset{CH_3}{|}}{\overset{\overset{\ddot{O}H}{|}}{C}}-\ddot{\underset{..}{O}}-\phi \qquad (4)$$

a hemiketal

$$CH_3-\underset{\underset{CH_3}{|}}{\overset{\overset{\ddot{O}H}{|}}{C}}-\ddot{\underset{..}{O}}-\phi \;+\; H^+ \;\longrightarrow\; CH_3-\underset{\underset{CH_3}{|}}{\overset{\overset{\ddot{O}H}{|}}{C}}-\overset{\overset{H}{\oplus}}{O}-\phi \;\longrightarrow\; CH_3-\underset{\underset{CH_3}{|}}{\overset{\overset{\ddot{O}H}{|}}{C}}\oplus \;+\; H-\ddot{\underset{..}{O}}-\phi \qquad (5)$$

phenol

resonance-stabilized

$$CH_3-\underset{}{\overset{\overset{:O:}{||}}{C}}-CH_3 \;\xleftarrow{-H^+}\; CH_3-\overset{\overset{\oplus\ddot{O}H}{||}}{C}-CH_3$$

acetone

Figure 10.6 Mechanism of the cumene hydroperoxide rearrangement.

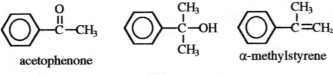

acetophenone

2-phenylpropan-2-ol

α-methylstyrene

Vacuum distillation recovers the unreacted cumene and yields α-methylstyrene, which can be hydrogenated back to cumene and recycled. Further distillation separates phenol, bp 181°C, and acetophenone, bp 202°C.

The mechanism of the rearrangement is an excellent practical industrial example of a broad type of rearrangement, one occurring with an electron-deficient oxygen. The mechanism is given in Fig. 10.6.

Figure 10.7, page 206, gives the uses of acetone. A very important organic chemical that just missed the top 50 list, methyl methacrylate, is made from acetone, methanol, and hydrogen cyanide. Approximately 1.2 billion lb of this compound is manufactured and then polymerized to poly (methyl methacrylate), an important plastic known for its clarity and used as a glass substitute. The synthesis is outlined as follows.

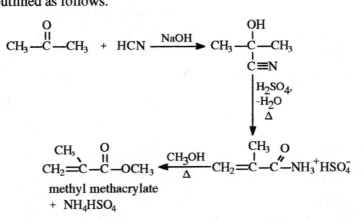

methyl methacrylate
+ NH₄HSO₄

The first reaction is a nucleophilic addition of HCN to a ketone, the second is a dehydration of an alcohol and hydrolysis of a nitrile, and the third is esterification by methanol. Alternate methods of making methyl methacrylate are being developed.

Aldol chemicals refer to a variety of substances desired from acetone involving an aldol condensation in a portion of their synthesis. The most important of these chemicals is methyl isobutyl ketone (MIBK), a common solvent for many plastics, pesticides, adhesives, and pharmaceuticals. Approximately 0.17 billion lb of MIBK were made in 1989. The synthesis is outlined here.

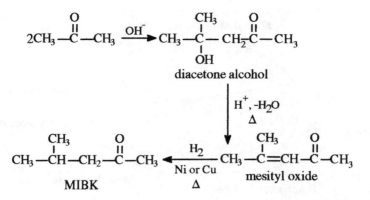

diacetone alcohol

$H^+, -H_2O$
Δ

$\begin{array}{c} CH_3 \\ | \\ CH_3-CH-CH_2-\overset{O}{\overset{||}{C}}-CH_3 \end{array}$ $\xleftarrow[\text{Ni or Cu}\;\Delta]{H_2}$ $\begin{array}{c} CH_3 \\ | \\ CH_3-C=CH-\overset{O}{\overset{||}{C}}-CH_3 \end{array}$

MIBK   mesityl oxide

Diacetone alcohol is a solvent used in hydraulic fluids and printing inks. Recall that the aldol condensation is an example of a variety of carbanion reactions used to make large molecules from smaller ones. An aldehyde or a ketone with at least one hydrogen on the carbon next to the carbonyl will react to give the aldol condensation. The mechanism is given as follows.

$$CH_3-\overset{:O:}{\overset{||}{C}}-CH_3 \; + :\!\overset{..}{\underset{..}{O}}H \rightleftharpoons \overset{\ominus}{C}H_2-\overset{:O:}{\overset{||}{C}}-CH_3 \; + \; H_2\overset{..}{\underset{..}{O}}:$$

$$CH_3-\overset{:O:}{\overset{||}{C}}-CH_3 \; + \overset{\ominus}{C}H_2-\overset{:O:}{\overset{||}{C}}-CH_3 \rightleftharpoons CH_3-\overset{:\overset{..}{O}:\ominus}{\underset{CH_3}{\overset{|}{C}}}-CH_2-\overset{:O:}{\overset{||}{C}}-CH_3$$

$$CH_3-\overset{:\overset{..}{O}:\,\ominus}{\underset{CH_3}{\overset{|}{C}}}-CH_2-\overset{:O:}{\overset{||}{C}}-CH_3 \; + \; H_2\overset{..}{\underset{..}{O}}: \rightleftharpoons CH_3-\overset{:\overset{..}{O}H}{\underset{CH_3}{\overset{|}{C}}}-CH_2-\overset{:O:}{\overset{||}{C}}-CH_3 \; + :\!\overset{..}{\underset{..}{O}}H$$

Finally, bisphenol A is manufactured by a reaction between phenol and acetone, the two products from the cumene hydroperioxide rearrrangement. Bisphenol A is an important diol monomer used in the synthesis of polycarbonates and epoxy resins as we will see later. Over a billion pounds per year of this chemical is made.

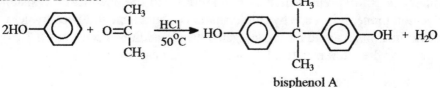

bisphenol A

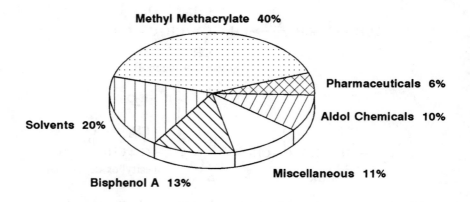

Figure 10.7 Uses of acetone. (*Source*: CP.)

The student should develop the mechanism of this reaction using the following stepwise information: (1) protonation of the carbonyl; (2) electrophilic attack on the aromatic ring; (3) rearomatization by proton loss; (4) another protonation, but then loss of a water molecule; and (5) electrophilic attack and rearomatization. Small amounts of the o, o' and p, p' isomers are also found.

CHEMICALS FROM THE C_4 FRACTION

Chemicals obtained from petroleum having four carbons are manufactured at a considerably lower scale than ethylene or propylene derivatives Only three C_4 compounds, butadiene, acetic acid, and methyl t- butyl ether (MTBE), appear in the top 50. Butadiene's manufacture, as well as the separation of other C_4 compounds from petroleum, is described in Chapter 8. Acetic acid was discussed primarily as a derivative of ethylene in Chapter 9. A few important derivatives of butadiene, isobutylene, and n-butene will be briefly mentioned here as well as MTBE.

Butadiene Derivatives

Besides butadiene, another important monomer for the synthetic elastomer industry is chloroprene which is polymerized to the chemically resistant polychloroprene. It is made by chlorolination of butadiene followed by dehydrochlorination. As with most conjugated dienes, addition occurs either 1,2 or 1,4 because the intermediate allyl carbocation is delocalized.

Reaction:

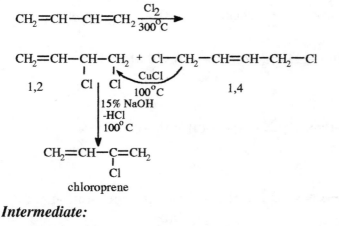

CH₂=CH—C=CH₂
          |
          Cl
chloroprene

Intermediate:

$$CH_2=CH-CH-CH_2-Cl \longleftrightarrow \oplus CH_2-CH=CH-CH_2-Cl$$
$$\qquad\qquad\quad \oplus$$

The 1,4-isomer can be isomerized to the 1,2-isomer by heating with cuprous chloride.

Another derivative of butadiene, hexamethylenediamine (HMDA), is used in the synthesis of nylon. We have already met this compund earlier in this chapter since it is made from acrylonitrile through adiponitrile.Approximately two thirds of all adiponitrile is made from 1,3-butadiene and 2 moles of hydrogen cyanide. This is an involved process chemically and it is summarized in Fig. 10.8. Butadiene first adds one mole of HCN at 60°C with a nickel catalyst via both 1,2- and 1,4-addition to give respectively 2-methyl-3-butenonitrile (2M3BN) and 3-pentenonitrile (3PN) in a 1:2 ratio. The 1,2-addition is the usual Markovnikov addition with a secondary carbocation intermediate being preferred. Fig. 10.9 shows an ADN reactor. Next isomerization of the 2M3BN to 3PN takes place at 150°C. Then more HCN, more catalyst, and a triphenylboron promotor react with 3PN to form 5% methylglutaronitrile (MGN) and mostly adiponitrile (ADN). The ADN is formed from 3PN probably through isomerization of 3PN to 4-pentenonitrile and then anti-Markovnikov addition of HCN to it. The nickel catalyst must play a role in this last unusual mode of addition, and a steric effect may also be operating to make CN⁻ attack at the primary carbon rather than a cationically preferred secondary carbon. A complicated set of extractions and distillations is necessary to obtain pure ADN. Even then the hexamethylenediamine (HMDA) made by hydrogenation of ADN must also be distilled through seven columns to purify it before polymerization to nylon. Fig. 10.10 pictures some HMDA distillation units.

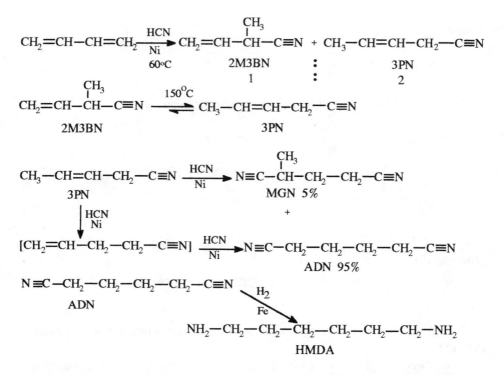

Figure 10.8 Manufacture of adiponitrile and hexamethylenediamine from 1,3-butadiene.

Figure 10.9 Reactors used in the conversion of 1,3-butadiene and HCN to adiponitrile. (Courtesy of Du Pont.)

Isobutylene Derivatives

Isobutylene can be obtained from the C_4 fraction of petroleum. Large amounts can be made from isobutane by dehydrogenation. It can also be made by dehydration of *t*-butyl alcohol. Two important derivatives of isobutylene deserve mention. *t*-Butyl alcohol can be made by hydration of isobutylene. The favored tertiary carbocation intermediate limits the possible alcohols produced to only this one.

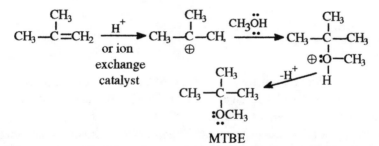

isobutylene   *t*-butyl alcohol

In 1984 methyl *t*-butyl ether (MTBE) broke into the top 50 for the first time with a meteoric rise in production from 0.8 billion lb in 1983 to 1.47 billion lb in 1984 to be ranked 47th. In 1990 it was 24th with production over 6 billion lb. The demand for MTBE in the U.S. could rise to 23 billion lb/yr. Some discussion of MTBE was given in Chapter 7, under gasoline additives. It is an important tetraethyl lead substitute. MTBE is manufactured by the acid catalyzed electrophilic addition of methanol to isobutylene.

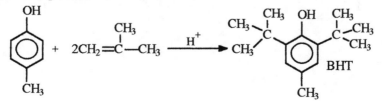

MTBE

An important antioxidant for many products is butylated hydroxytoluene (BHT), more properly named 4-methyl-2,6-di-*t*-butylphenol. Acid-catalyzed electrophilic aromatic substitution of a *t*-butyl cation at the activated positions ortho to the hydroxy group of *p*-cresol yields this product. *p*-Cresol is obtained from coal tar or petroleum.

Figure 10.10 Distillation columns associated with the purification of hexamethylenediamine. (Courtesy of Du Pont.)

n-Butene Derivatives

For many years maleic anhydride (MA) was made from benzene by oxidation and loss of two moles of CO_2. Even as late as 1978 83% of maleic anhydride was made from benzene. However, the new OSHA standards for benzene plants required modifications in this process, and butane is also cheaper than benzene. As a result since 1989 all maleic anhydride is now made from butane. This has been a very rapid and complete switch in manufacturing method.

$$CH_3-CH_2-CH_2-CH_3 \ + \ O_2 \ \xrightarrow[\Delta]{\text{Zn, P, V salts}} \ \underset{\text{MA}}{\text{⬡}} \ + \ H_2O$$

The uses of maleic anhydride are summarized in Fig. 10.11. Unsaturated polyester resins are its prime use area. Food acidulants include fumaric and malic acids. Malic acid competes with citric acid as an acidulant for soft drinks, and it is added to products that contain aspartame, the artificial sweetener, because it makes aspartame taste more like sugar. Agricultural chemicals made from MA include daminozide (Alar®), a growth regulator for apples which in 1989 was found to be carcinogenic because of a breakdown product, unsymmetric dimethylhydrazine (UDMH). Alar® is needed to keep the apple on the tree, to make a more perfectly shaped, redder, firmer apple, and to maintain firmness in stored apples by reducing ethylene production.

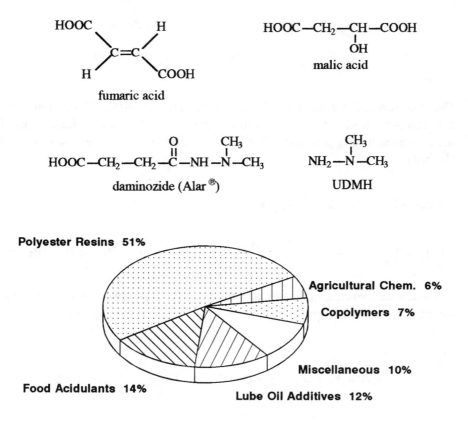

Figure 10.11 Uses of maleic anhydride. (*Source*: CP.)

Methyl ethyl ketone (MEK) is an important coating solvent for many polymers and is made at the 0.5 billion lb/yr level. Either 1- or 2-butene can be sulfated and hydrated to *sec*-butyl alcohol, which is then dehydrogenated to the ketone. This is the major route (87%) to MEK, though 13% is made by butane oxidation.

$$CH_2{=}CH{-}CH_2{-}CH_3$$
$$\text{or} \qquad \xrightarrow[\text{2) } H_2O,\ 100^\circ C]{\text{1) } H_2SO_4,\ 50^\circ C} CH_3{-}\underset{\underset{OH}{|}}{CH}{-}CH_2{-}CH_3$$
$$CH_3{-}CH{=}CH{-}CH_3$$

sec-butyl alcohol

$$\Big\downarrow \ \begin{array}{l} ZnO \\ \text{or} \\ Zn{-}Cu \end{array}$$

$$H_2\ +\ CH_3{-}\underset{\underset{O}{\|}}{C}\ {-}CH_2{-}CH_3$$

MEK

The "oxo process" ("hydroformylation") is used to convert the C_4 fraction to C_5 derivatives (as well as C_3 to C_4). Synthesis gas (CO + H_2) is catalytically reacted with 1-butene to give pentanal which can be hydrogenated to 1-pentanol (*n*-amyl alcohol).

$$CH_3{-}CH_2{-}CH{=}CH_2 \xrightarrow[\text{catalyst}]{CO,\ H_2}$$

$$CH_3{-}CH_2{-}CH_2{-}CH_2{-}\underset{\underset{O}{\|}}{C}{-}H \xrightarrow{H_2} CH_3{-}CH_2{-}CH_2{-}CH_2{-}CH_2{-}OH$$

The classic oxo catalyst is octacarbonyldicobalt. This reacts with hydrogen to give hydridotetracarbonyl cobalt, the active catalyst in the oxo process.

$$2Co + 8CO \longrightarrow Co_2(CO)_8 \xrightarrow{H_2} 2\ \underset{CO}{\overset{CO}{\underset{|}{\overset{|}{CO{\diagdown}\underset{\diagup}{Co}{-}H}}}}$$

The mechanism of carbonylation/hydrogenation involves addition of the alkene to form a pi complex, followed by alternating additions of H, CO, and H.

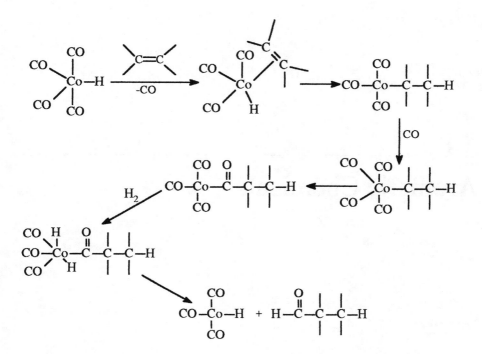

The oxo process is also important in making *n*-butyraldehyde and isobutyraldehyde from propylene in yields of 67% and 15% respectively. Newer rhodium catalysts can be used to allow the reactions to be run at lower temperatures and pressures, but no catalyst can be lost here because of its expense. The C_4 aldehydes can be hydrogentaed to the alcohols. Thus some C_4 compounds are derivatives of propylene by the oxo process.

$$CH_3-CH=CH_2 \xrightarrow{CO, H_2} CH_3-CH_2-CH_2-\overset{\overset{O}{\|}}{C}-H + CH_3-\overset{CH_3}{\underset{|}{C}H}-\overset{\overset{O}{\|}}{C}-H$$

n-butyraldehyde : isobutyraldehyde
67 :15

Many other processes utilizing the C_4 fraction of petroleum are possible. The C_4 supplies may increase as it becomes more popular to crack heavier fractions of petroleum, yielding larger amounts of C_4 material. The chemical industry may see an increase in the versatile C_4 chemistry if this becomes a reality.

11

Aromatic Chemicals

References

L & M, selected pages
W & RI, pp. 92-108
Kent, pp. 960-971
Wiseman, pp. 101-140
Szmant, pp. 407-574

BENZENE DERIVATIVES

There are eight chemicals in the top 50 that are manufactured from benzene. These are listed in Table 11.1 according to rank. Two of these, ethylbenzene and styrene, have already been discussed in Chapter 9 since they are also derivatives of ethylene. Two others, cumene and acetone, were covered in Chapter 10 when propylene derivatives were studied. Although the three carbons of acetone do not formally come from benzene, its primary manufacturing method is from cumene, which is made by reaction of benzene and propylene. These compounds need not be discussed further at this point. That leaves phenol, cyclohexane, adipic acid, and caprolactam. We will also briefly mention a few other important industrial chemicals of high production volumes that are made from benzene but do not quite make the top 50 list. Figure 11.1 summarizes the synthesis of important chemicals made from benzene.

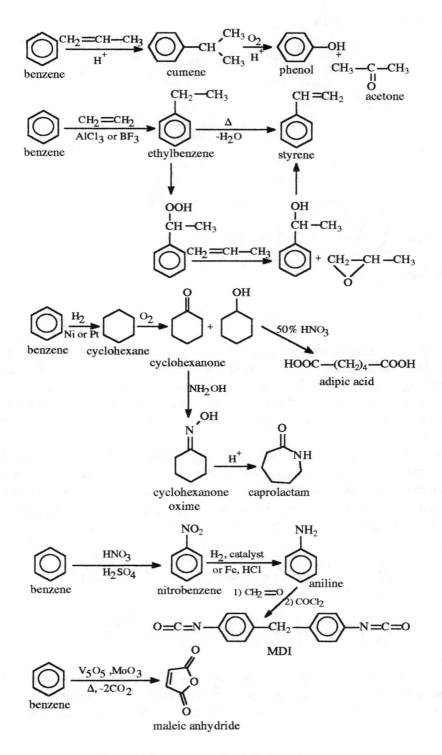

Figure 11.2 Synthesis of benzene derivatives.

TABLE 11.1 Benzene Derivatives in the Top 50

1990 Rank	Chemical
19	Ethylbenzene
20	Styrene
32	Cumene
35	Phenol
41	Cyclohexane
43	Acetone
46	Adipic acid
48	Caprolactam

Phenol 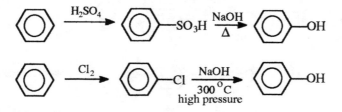 —OH   **Carbolic Acid**

Reference

CP, 9-14-87

The major manufacturing process for making phenol was discussed in the previous chapter, since it is the coproduct with acetone from the acid-catalyzed rearrangement of cumene hydroperoxide. The student should review this process. It accounts for 97% of the total phenol production and has dominated phenol chemistry since the early 1950s. But a few other syntheses deserve some mention.

A historically important method, first used about 1900, is sulfonation of benzene followed by desulfonation with caustic. This is classic aromatic chemistry. In 1924 a chlorination route was discovered. Both the sulfonation and chlorination reactions are good examples of electrophilic aromatic substitution on an aromatic ring. These routes are no longer used.

The remaining route, which now accounts for 2% of phenol, takes advantage of the usual surplus of toluene from petroleum refining. Oxidation with a number of reagents gives benzoic acid. Further oxidation to *p*-hydroxybenzoic acid and

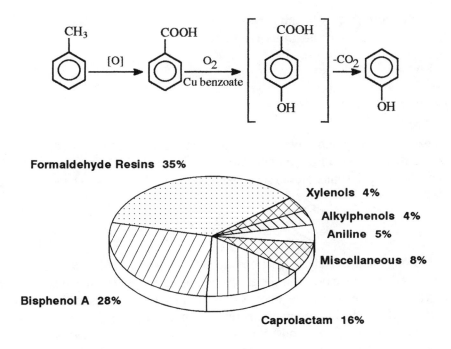

Figure 11.2 Uses of phenol. (*Source:* CP.)

decarboxylation yields phenol. Here phenol competes with benzene manufacture, also made from toluene when the surplus is large as seen in Chapter 8. The last 1% of phenol comes from distillation of petroleum.

Figure 11.2 outlines the uses of phenol. We will consider the details of phenol uses in later chapters. Phenol-formaldehyde polymers (phenolics) have a primary use as the adhesive in plywood formulations. We studied the synthesis of bisphenol A from phenol and acetone in Chapter 10. Phenol's use in detergent synthesis to make alkylphenols will be discussed later. Caprolactam and aniline are mentioned in the following sections

Although phenol ranked only thirty-fifth in 1990, it is still the highest ranked derivative of benzene other than those using ethylene or propylene along with benzene. Its 1990 price was 33 ¢/lb. That gives a total commercial value of $1.2 billion for the 3.51 billion lb produced.

Cyclohexane 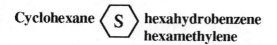 hexahydrobenzene
hexamethylene

Reference

CP, 10-23-89

Benzene can be quantitatively transformed into cyclohexane by hydrogenation over either a nickel or platinum catalyst. This reaction is carried out at 210°C and 350-500 psi, sometimes in several reactors placed in series. The yield is over 99%.

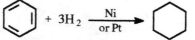

Although many catalytic reactions are not well understood, a large amount of work has been done on hydrogenations of double bonds. The metal surface acts as a source of electrons. The π bonds as well as hydrogen atoms are bound to this surface. Then the hydrogen atoms react with the complexed carbons one at a time to form new C—H bonds. No reaction occurs without the metal surface. The metal in effect avoids what would otherwise have to be a free radical mechanism that would require considerably more energy. The mechanism is outlined as follows.

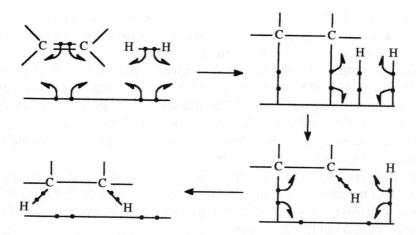

Fig. 11.3 shows the main uses of cyclohexane. Adipic acid is used to manufacture nylon 6,6, the major nylon used currently in the U.S. Caprolactam is the monomer for nylon 6, for which there is a growing market.

$$\text{O} \qquad\qquad \text{O}$$
$$\text{Adipic Acid HO-C-CH}_2\text{-CH}_2\text{-CH}_2\text{-CH}_2\text{-C-OH}$$
1,6-Hexandioic Acid

Reference

CP, 10-9-89

Nearly all the adipic acid manufactured, 98%, is made from cyclohexane by oxidation. Air oxidation of cyclohexane with a cobalt or manganese(II) naphthenate or acetate catalyst at 125-160°C and 50-250 psi pressures gives a mixture of cyclohexanone and cyclohexanol. Benzoyl peroxide is another possible catalyst. The yield is 75-80% because of some ring opening and other further oxidation which takes place. The cyclohexanone/cyclohexanol mixture (sometimes referred to as ketone-alcohol, KA mixture, or "mixed oil" is further oxidized with 50% nitric acid with ammonium vanadate and copper present as catalysts at 50-90°C and 15-60 psi for 10-30 min.

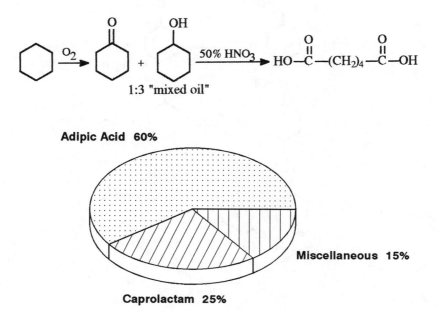

Figure 11.3 Uses of cyclohexane. (*Source:* CP.)

The mechanism of cyclohexane oxidation involves cyclohexane hydroperoxide as a key intermediate. Its formation is similar to the oxidation of isobutane in Chapter 10.

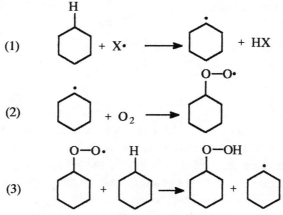

then (2), (3), (2), (3), etc.

The cyclohexane hydroperoxide then undergoes a one-electron transfer with cobalt or manganese (II).

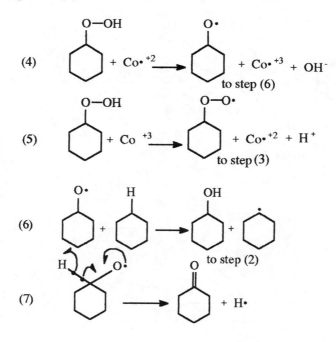

Chain transfer of the cyclohexyloxyl radical gives cyclohexanol or β-scisson gives cyclohexanone.

Fig. 11.4 shows a cyclohexane oxidation reactor. The further oxidation of the ketone and alcohol to adipic acid is very complex but occurs in good yield, 94%, despite some succinic and glutaric acid byproducts being formed because the adipic acid can be preferentially crystallized and centrifuged.

Figure 11.4 The large tower on the right is the cyclohexane oxidation chamber and purification unit to convert cyclohexane to the hydroperoxide and then to cyclohexanone/cyclohexanol. An elevator leads to the top platform of this narrow tower, where an impressive view of this and other surrounding plants can be obtained. (Courtesy of Du Pont.)

A small amount of adipic acid, 2%, is made by hydrogenation of phenol with a palladium or nickel catalyst (150°C, 50 psi) to the mixed oil, then nitric acid oxidation to adipic acid. If palladium is used, more cyclohexanone is formed. Although the phenol route for making adipic acid is not economically advantageous because phenol is more expensive than benzene, the phenol conversion to greater cyclohexanone percentages can be used successfully for caprolactam manufacture, where cyclohexanone is necessary.

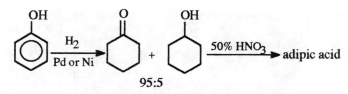

95:5

Fig. 11.5 gives the uses of adipic acid. As will be seen later, the nylon 6,6 has large markets in textiles, carpets, and tire cords. It is made by reaction of HMDA and adipic acid.

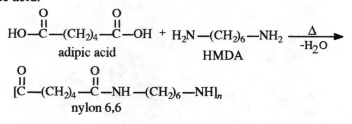

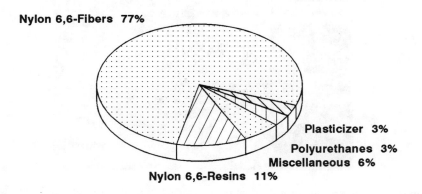

Figure 11.5 Uses of adipic acid. (*Source:* CP.)

Two routes to adipic acid from C_4 chemistry are being studied. Recently BASF in Germany has developed an alternate synthesis of adipic acid but it is not yet commercial.

$$CH_2{=}CH{-}CH{=}CH_2 \ + \ 2CO \ + \ 2CH_3OH \ \longrightarrow \ CH_3O{-}\overset{O}{\overset{\|}{C}}{-}(CH_2)_4{-}\overset{O}{\overset{\|}{C}}{-}OCH_3$$

$$2H_2O \Big\downarrow$$

$$2CH_3OH \ + \ HO{-}\overset{O}{\overset{\|}{C}}{-}(CH_2)_4{-}\overset{O}{\overset{\|}{C}}{-}OH$$

Monsanto is developing a new route to adipic acid. It involves dicarbonylation of 1,4-disubstituted 2-butenes with palladium chloride.

$$X{-}CH_2{-}CH{=}CH{-}CH_2{-}X \ + \ CO$$

$$\searrow PdCl_2$$

$$X{-}\overset{O}{\overset{\|}{C}}{-}CH_2{-}CH{=}CH{-}CH_2{-}\overset{O}{\overset{\|}{C}}{-}X$$

where $X = CH_3COO$, CH_3O, OH, or Cl

Polar, aprotic, and nonbasic solvents are preferred to avoid unwanted side products from hydrogenolysis or isomerization.

Caprolactam

Reference

CP, 10-16-89

Caprolactam is the most recent edition to the top 50. It replaced sodium tripolyphosphate in 1987, although for many years it has been made at the billion lb/yr level. The unusual name for this chemical comes from the original name for the C_6 carboxylic acid, caproic acid. Caprolactam is the cyclic amide (lactam) of 6-aminocaproic acid Its manufacture is from cyclonexanone, made usually from cyclohexane (58%), but also available from phenol (42%). Some of the cyclohexanol in cyclohexanone/cyclohexanol mixtures can be converted to cyclohexanone by a ZnO catalyst at 400°C. Then the cyclohexanone is converted into the oxime with hydroxylamine. The oxime undergoes a very famous

acid-catalyzed reaction called the Beckmann rearrangement to give caprolactam. Sulfuric acid at 100-120°C is common but phosphoric acid is also used,
since after treatment with ammonia the byproduct becomes ammonium phosphate which can be sold as a fertilizer. The caprolactam can be extracted and
vacuum distilled, bp 139° at 12 mm. The overall yield is 90%.

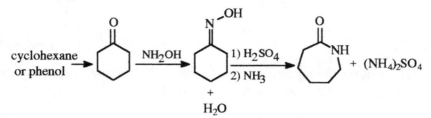

The first reaction, formation of the oxime, is a good example of a nucleophilic
addition to a ketone followed by subsequent dehydration. Oximes are common
derivatives of aldehydes and ketones because they are solids that are easily
purified.

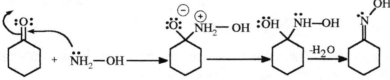

In the rearrangement of cumene hydroperoxide we saw an industrial example of
a rearrangement to electron-deficient oxygen. The Beckmann rearrangement of
caprolactam is a successful large-scale example of a rearrangement to electron-
deficient nitrogen. Protonation of the hydroxyl followed by loss of a water

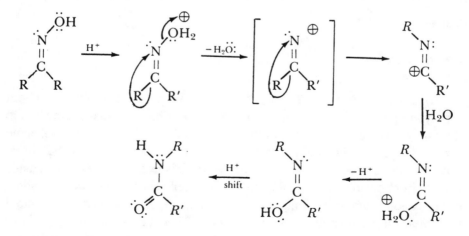

molecule forms the positive nitrogen, but the R group can migrate while the water leaves, so the nitrenium ion may not be a discreet intermediate. Attack of water on the rearranged ion and a proton shift to form the amide completes the process.

The student should adapt this general mechanism and work through the specific cyclic example of cyclohexanone oxime to caprolactam. Note that the result of the shift is an expansion of the ring size in the final amide product with the incorporation of the nitrogen atom as part of the ring.

All of the caprolactam goes into nylon 6 manufacture, especially fibers (83%) and plastic resin and film (7%). Although nylon 6,6 is still the more important nylon in this country (about 2:1) and in the U.K., nylon 6 is growing rapidly, especially in certain markets such as nylon carpets. In other countries, for example, Japan, nylon 6 is more predominant. Nylon 6 is made directly from caprolactam by heating with a catalytic amount of water.

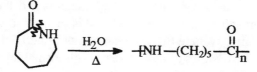

In 1990 1.38 billion lb of caprolactam were made at 89¢/lb for a commercial value of $1.2 billion.

Aniline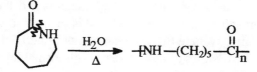

Reference

CP, 9-17-90

Aniline is an important derivative of benzene that can be made in two steps by nitration to nitrobenzene and either catalytic hydrogenation or acidic metal reduction to aniline. Both steps occur in excellent yield. Almost all nitrobenzene manufactured (97%) is directly converted into aniline. The nitration of benzene with mixed acids is an example of an electrophilic aromatic substitution

Reaction:

Mechanism:

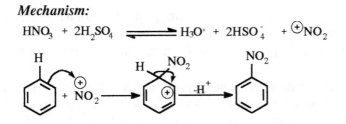

$$HNO_3 + 2H_2SO_4 \rightleftharpoons H_3O^+ + 2HSO_4^- + {}^{\oplus}NO_2$$

involving the nitronium ion as the attacking species. The hydrogenation of nitrobenzene has replaced the iron-acid reduction process for many years. At one time the special crystalline structure of the Fe_3O_4 formed as a byproduct in the latter process made it unique for use in pigments. But the demand for this pigment was not great enough to justify continued use of this older method of manufacturing aniline.

The uses of aniline are given in Fig. 11.6. Aniline's use in the rubber industry is in the manufacture of various vulcanization accelerators and age resistors. By far the most important and growing use for aniline is in the manufacture of *p,p'*-methylene diphenyl diisocyanate (MDI), which is polymerized with a diol to give a polyurethane. Two moles of aniline react with formaldeyde to give *p,p'*-methylenedianiline (MDA), which reacts with phosgene to give MDI.

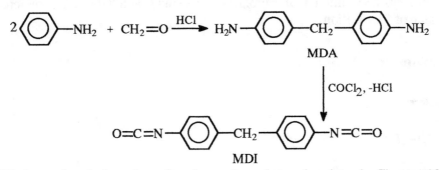

We have already been introduced to polyurethane chemistry in Chapter 10, where we used toluene diisocyanate (TDI) reacting with a diol to give a polyurethane. Polyurethanes derived from MDI are more rigid than those from TDI. New applications for these rigid foams are in home insulation and exterior autobody parts. The MDI market could grow 8 to 15% a year if the construction industry is healthy. Nearly 1 billion lb/yr of MDI is now being produced. However, the intermediate MDA was added to the suspect carcinogen list in 1984 and the effect of this action on the market for MDI remains to be seen. The TLV-TWA values for MDA and MDI are some of the lowest of the chemicals we have discussed, being 0.1 and 0.005 ppm respectively.

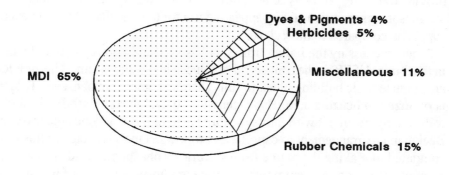

Figure 11.6 Uses of aniline. (*Source:* CP.)

TOLUENE DERIVATIVES

Other than benzene, 30% of which is made from toluene by the hydrodealkylation processes described in Chapter 8, there are no other top 50 chemicals derived from toluene in large amounts. However, we wish to mention a few important processes and chemicals made from toluene. As we learned earlier in this chapter, a very small amount of phenol is made from toluene. Toluene also reactions

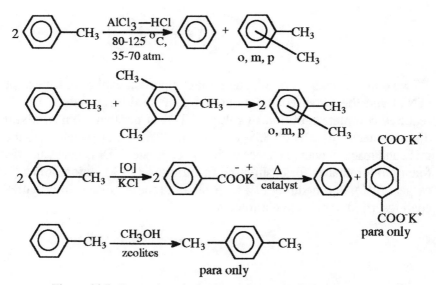

Figure 11.7 Conversion of toluene to other aromatic compounds.

provides an alternate source that is becoming more popular for the xylenes, especially *p*-xylene. These routes are indicated in Fig. 11.7. The last two provide routes respectively to terephthalic acid and *p*-xylene without the need for an isomer separation, a very appealing use for toluene that is often in excess supply as compared to the xylenes.

A new process by the Italian firm Snia Viscosa, and also reported to be used in the former USSR, is the conversion of toluene into caprolactam. This provides an alternate basic building block for this chemical other than benzene. Toluene is oxidized to benzoic acid. Hydrogenation to cyclohexanecarboxylic acid is followed by treatment with nitrosylsulfuric acid to form cyclohexanone oxime. Beckmann rearrangement to caprolactam follows. One advantage of this route compared to making the oxime from sulfuric or phosphoric acids is that large amounts of a byproduct ammonium sulfate or phosphate are not formed. This process is not yet used in the U.S.

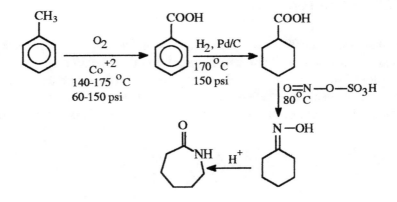

Two other derivatives of toluene are the important explosive trinitrotoluene (TNT) and the polyurethane monomer toluene diisocyanate (TDI). TNT requires complete nitration of toluene. TDI is derived from a mixture of dinitrotoluenes (usually 65-85% *o,p* and 35-20% *o,o*) by reduction to the diamine and reaction with phosgene to the diisocyanate. TDI is made into flexible foam polyurethanes for cushioning in furniture (28%), automobiles (16%), carpets (8%), and bedding (7%). A small amount is used in polyurethane coatings, rigid foams and elastomers.

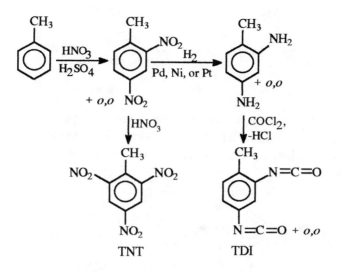

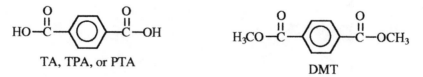

XYLENE DERIVATIVES

There are only two top 50 chemicals, terephthalic acid and dimethyl terephthalate, derived from *p*-xylene and none from *o*- or *m*-xylene. But we also wish to discuss phthalic anhydride, which is made in large amounts from *o*-xylene.

Terephthalic Acid and Dimethyl Terephthalate

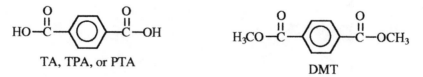

TA, TPA, or PTA

DMT

Reference

CP, 7-7-89

Terephthalic acid is commoly abbreviated TA or TPA. The abbreviation PTA (P = pure) is reserved for the product of 99% purity for polyester manufacture. For many years polyesters had to be made from dimethyl terephthalate (DMT) because the acid could not be made pure enough economically. Now either can be used. TA is made by air oxidation of *p*-xylene in acetic acid as a solvent in the presence of cobalt and manganese salts of heavy metal bromides as catalysts at 200°C and 400 psi. TA of 99.6% purity is formed in 90% yield. This is called the Amoco process.

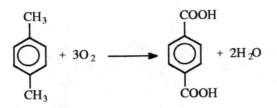

Catalysts for the Amoco process include cobalt, manganese, and bromide ions. A partial mechanism with some intermediates is given below. Details are similar to other oxidation processes we have discussed.

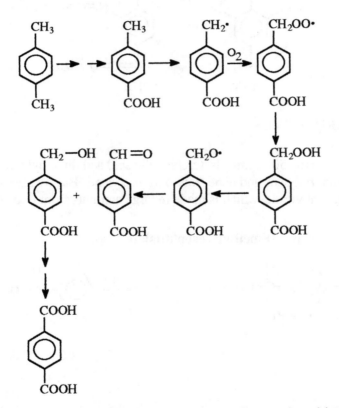

The crude TA is cooled and crystallized. The acetic acid and xylene are evaporated and the TA is washed with hot water to remove traces of the catalyst and acetic acid. Some *p*-formylbenzoic acid is present as an impurity from incomplete oxidation. This is most easily removed by hydrogenation to *p*-methylbenzoic acid and recrystallization of the TA to give 99.9% PTA, which is a polyester-grade product, mp > 300°C.

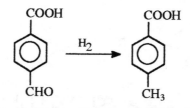

p-formylbenzoic acid   p-methylbenzoic acid

DMT can be made from crude TA or from *p*-xylene directly. Esterification of TA with methanol occurs under sulfuric acid catalysis. Direct oxidation of *p*-xylene with methanol present utilizes copper and manganese salt catalysis.

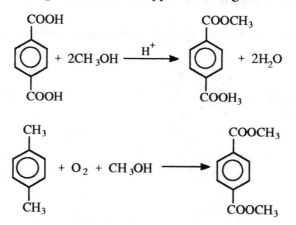

The DMT must be carefully purified via a five-column distillation system, bp 288°C, mp 141°C. The present distribution of the TA/DMT market in the U.S. is 44:56.

Fig. 11.8 shows the uses of TA/DMT. TA or DMT is usually reacted with ethylene glycol to give poly (ethylene terephthalate) (90%) but sometimes it is combined with 1,4-butanediol to yield poly (butylene terephthalate). Polyester fibers are used in the textile industry. Films find applications as magnetic tapes, electrical insulation, photographic film, and packaging. Polyester bottles, especially in the soft drink market, are growing rapidly in demand.

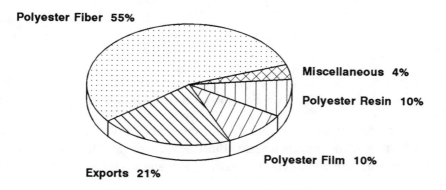

Figure 11.8 Uses of terephthalic acid and dimethyl terephthalate. (*Source:* CP.)

Phthalic Anhydride

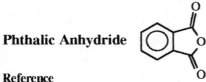

Reference

CP, 7-24-89

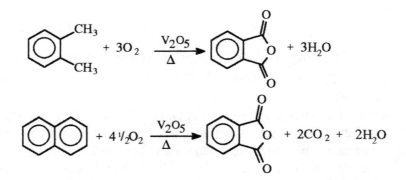

The manufacturing method of making phthalic anhydride has been changing rapidly similar to the switchover in making maleic anhydride. In 1983 28% of phthalic anhydride came from naphthalene, 72% from *o*-xylene. No naphthalene-based plants were open in 1989. Despite the better yield in the naphthalene process, energetic factors make this less favorable economically compared to the *o*-xylene route.

The uses of phthalic anhydride include plasticizers (48%), unsaturated polyester resins (23%), and alkyd resins (19%).

Phthalic anhydride reacts with alcohols such as 2-ethylhexanol to form liquids that impart great flexibility when added to many plastics without hurting their strength. Most of these plasticizers, about 80%, are for poly(vinyl chloride) flexibility. Dioctyl phthalate is a common plasticizer. The structure of polyesters and alkyds incorporating phthalic anhydride will be discussed in a later chapter. In 1985 high doses of DEHP were found to cause liver cancer in rats and mice. Certain plasticizer applications, such as those in infants' pacifiers and squeeze toys, may be affected in the years ahead.

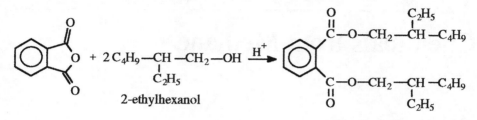

di-(2-ethylhexyl) phthalate, DEHP
(dioctyl phthalate, DOP)

12

Chemicals from Methane

References

L & M, selected pages
W & R I, pp. 108-121
Kent, pp. 917-923
Wiseman, pp. 148-155
Szmant, pp. 61-187

In previous discussions we studied a large percentage of important chemicals derived from methane. Those in the top 50 are listed in Table 12.1 and their syntheses are summarized in Fig. 12.1.

As we learned in Chapters 3 and 4, many inorganic compounds, not just ammonia, are derived from synthesis gas, made from methane by steam-reforming. This includes carbon dioxide, ammonia, nitric acid, ammonium nitrate, and urea in the top 50. No further mention need be made of these important processes. One of the manufacturing methods for acetic acid that is growing in importance is the carbonylation of methanol. This was covered in Chapter 9 since acetic acid is still made from ethylene (and butane). We discussed MTBE in Chapters 7 and 10 since it is also a C_4 derivative. In Chapter 11 we discussed dimethyl terephthalate. Review these pertinent sections. That leaves only two chemicals, methanol and formaldehyde, as derivatives of methane that have not been discussed in detail.

TABLE 12.1 Derivatives of Methane in the Top 50	
1990 Rank	*Chemical*
6	Ammonia (and other inorganics)
21	Methanol
22	Dimethyl terephthalate
23	Formaldehyde
24	Methyl *t*-butyl ether
33	Acetic acid

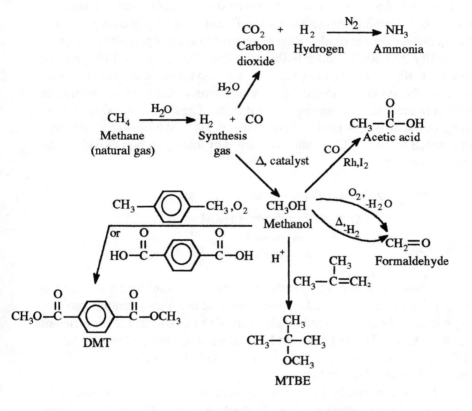

Figure 12.1 Synthesis of methane derivatives.

METHANOL, CH₃OH; Wood Alcohol, Methyl Alcohol

Reference

CP, 9-25-89

Before 1926 all methanol was made by distillation of wood. Now it is all synthetic. Methanol is obtained from synthesis gas under appropriate conditions. This includes zinc, chromium, manganese, or aluminum oxides as catalysts, 300°C, 250-300 atm (3000-5000 psi), and most importantly a 1:2 ratio of $CO:H_2$. Newer copper oxide catalysts require lower temperatures and pressures, usually 200-300°C and 50-100 atm (750-1500 psi). A 60% yield of methanol is realized. As seen in Chapter 3, many synthesis gas systems are set to maximize the amount of hydrogen in the mixture so that more ammonia can be made from the hydrogen reacting with nitrogen. The shift conversion reaction aids the attainment of this goal. When synthesis gas is to be used for methanol manufacture, a 1:2 ratio $CO:H_2$ ratio is obtained by adding carbon dioxide to the methane and water.

$$3CH_4 + 2H_2O + CO_2 \longrightarrow 4CO + 8H_2$$

$$CO + 2H_2 \xrightarrow{\Delta} CH_3OH$$

$$\text{pressure}$$
$$\text{metal oxides}$$

Thus methanol and ammonia plants are sometimes combined since carbon dioxide, which must be removed from hydrogen to use it for ammonia production, can in turn be used as feed to adjust the $CO:H_2$ ratio to 1:2 for efficient methanol synthesis. The methanol can be condensed and purified by distillation, bp 65°C. Unreacted synthesis gas is recycled. Other products include higher boiling alcohols and dimethyl ether.

Fig. 12.2 gives the uses for methanol. The percentage of methanol used in the manufacture of formaldehyde has been fluctuating. It was 42% in 1981. It has decreased in part because of recent toxicity scares of formaldehyde. The percentage of methanol used in acetic acid manufacture is up from 7% in 1981 because the carbonylation of methanol has become the preferred acetic acid manufacturing method. MTBE is the octane enhancer and is synthesized directly from isobutylene and methanol. It is the fastest growing use for methanol. Many other important chemicals are made from methanol, although they have not quite made the top 50 list. Some of these can be found in the next

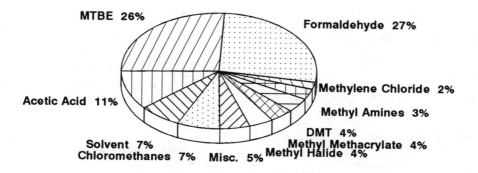

Figure 12.2 Uses of methanol. (*Source:* CP.)

chapter where the second 50 chemicals are summarized. With a U.S. production of 7.99 billion lb and a price of 8.7¢/lb, the commercial value of methanol is $0.7 billion.

Not mentioned in the table is the direct use of methanol as fuel for automobiles. It is added in small amounts to gasoline, sometimes as a blend with other alcohols such as *t*-butyl alcohol, to increase octane ratings and lower the price of the gasoline. Experimentation is even being done on vehicles that burn pure methanol. This fuel use is usually captive but a good estimate is that it may account for almost 10% of the methanol produced.

$$\underset{\text{FORMALDEHYDE, H-C-H; METHANAL}}{\overset{\displaystyle \overset{\text{O}}{\|}}{}}$$

FORMALDEHYDE, H-C-H; METHANAL

Reference

CP, 9-18-89

Formaldehyde is produced solely from methanol. The process can be air oxidation or simple dehydrogenation. Since the oxidation is exothermic and the dehydrogenation is endothermic, usually a combination is employed where the heat of reaction of oxidation is used for the dehydrogenation. Various metal oxides or silver metal are used as catalysts. Temperatures range from 450-900°C and there is a short contact time of 0.01 sec. Formaldehyde is stable only

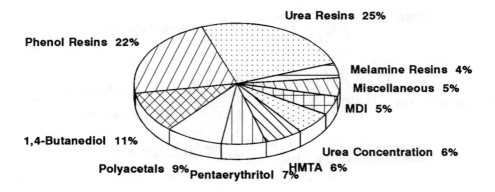

Figure 12.3 Uses of formaldehyde. (*Source:* CP.)

oxidation $2CH_3OH + O_2 \xrightarrow{\Delta} CH_2{=}O + 2H_2O$

dehydrogenation $CH_3OH \xrightarrow{\Delta} CH_2{=}O + H_2$

in water solution, commonly 37-56% formaldehyde by weight. Methanol (3-15%) may he present as a stabilizer. Formaldehyde in the pure form is a gas with a bp of - 21°C but is unstable and readily trimerizes to trioxane or polynerizes to paraformaldehyde.

trioxane

$HO{-}(CH_2{-}O)_n{-}H$

paraformaldehyde
$n = 8{-}50$

Fig. 12.3 summarizes the uses of formaldehyde. Two important thermosetting plastics, urea- and phenol-copolymers, take nearly one half the formaldehyde manufactured. 1,4-Butanediol is made for some polyesters and is an example of acetylene chemistry that has not yet been replaced. Tetrahydrofuran (THF) is

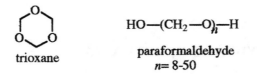

$HC{\equiv}CH + 2CH_2O \longrightarrow HO{-}CH_2{-}C{\equiv}C{-}CH_2{-}OH \xrightarrow{2H_2}$

$HO{-}CH_2{-}CH_2{-}CH_2{-}CH_2{-}OH \xrightarrow{-H_2O}$

THF

a common solvent that is made by dehdration of 1,4-butanediol. Pentaerythritol finds end-uses in polyesters and explosives (pentaerythritol tetranitrate). To appreciate this synthesis, the student should review two condensation reactions, the crossed aldol and the crossed Canizzarro. Acetaldehyde reacts with 3 mol of formaldehyde in three successive aldol condensations. This product then undergoes a Canizzarro reaction with formaldehyde.

$$CH_3-\overset{\overset{\displaystyle O}{||}}{C}-H \;+\; 3CH_2O \;\xrightarrow{OH^-}\; HO-CH_2-\underset{\underset{\displaystyle OH}{\overset{\displaystyle |}{CH_2}}}{\overset{\overset{\displaystyle OH}{\overset{\displaystyle |}{CH_2}}}{C}}-C\overset{\nearrow O}{\underset{\searrow H}{}} \;\xrightarrow[OH^-]{CH_2O}$$

$C(CH_2OH)_4 + \; HCOOH$
pentaerythritol

Hexamethylenetetramine (HMTA) has important uses in modifying phenolic resin manufacture and is an intermediate in explosive manufacture. Although it is a complex three-dimensional structure, it is easily made by the condensation of formaldehyde and ammonia.

$$6CH_2O \;+\; 4NH_3 \;\longrightarrow\; \text{[HMTA structure]} \;+\; 6H_2O$$

Debate is continuing on the safety and toxicity of formaldehyde and its products, especially urea-formaldehyde foam used as insulation in construction and phenol-formaldehyde as a plywood adhesive. Presently the TLV-TWA of formaldehyde is 1 and the TLV-STEL is 2. Conflicting studies of formaldehyde as an irritant and possible carcinogen abound in the literature. Although the Consumer Product Safety Commission banned urea-formaldehyde foam insulation, in 1983 a U.S. Circuit Court of Appeals overturned this decision. The final story is still to be written. Formaldehyde is now a "suspect carcinogen."

CHLOROFLUOROCARBONS (CFCs)

References

CP, 3-13-89
C & E News, selected articles, 1982-1990

Because of the growing importance of CFCs in environmental chemistry, a basic understanding of the chemistry and uses of this diverse chemical family is necessary. Together they represent a production of over 1 billion lb/yr which, at $1.00/lb, is a large commercial value.

This industry segment uses common abbreviations and a numbering system for CFCs and related compounds. The original nomenclature developed in the 1930s at Du Pont is still employed and uses three digits. When the first digit is 0, it is dropped. The first digit is the number of carbons minus 1, the second digit is the number of hydrogens plus 1, and the third digit is the number of fluorines. All other atoms filling the four valences of each carbon are chlorines. Important nonhydrogen-containing CFCs are given below. Originally these were called Freons®.

CCl_2F_2	CCl_3F	CCl_2FCClF_2
CFC-12	CFC-11	CFC-113

When some of the chlorines are replaced by hydrogens, CFCs become HCFCs, the now more common nomenclature for those chlorofluorocarbons containing hydrogen. The numbering is the same. When more than one isomer is possible, the most symmetrically substituted compound has only a number; letters a and b are added to designate less symmetrical isomers.

$CHClF_2$	CF_3CHCl_2	CF_3CHClF
HCFC-22	HCFC-123	HCFC-124
	CCl_2FCH_3	$CClF_2CH_3$
	HCFC-141b	HCFC-142b

When there is no chlorine and the chemical contains only hydrogen, fluorine, and carbon, they are called HFCs.

CH_2FCF_3	CH_3CHF_2
HFC-134a	HFC-152a

Halons, a closely related type of chemical which also contains bromine, are used as fire retardants. Numbering here is more straightforward: first digit, no. of carbons; second digit, no. of fluorines; third digit, no. of chlorines; and fourth digit, no. of bromines. Common Halons are the following: Halon 1211, CF_2BrCl; Halon 1301, CF_3Br; and Halon 2402, C_2F_4Br.

Most CFCs are manufactured by combining hydrogen fluoride and either carbon tetrachloride or chloroform. The hydrogen fluoride comes from fluorspar, CaF_2, reacting with sulfuric acid. The chlorinated methanes are manufactured from methane. Important reactions in the manufacture of CFC-11 and -12 and HCFC-22 are given in Fig. 12.4.

$$CCl_4: \quad CH_4 + 4S \longrightarrow CS_2 + 2H_2S$$
$$CS_2 + 3Cl_2 \longrightarrow CCl_4 + S_2Cl_2$$
$$\text{or } CH_4 + 4Cl_2 \longrightarrow CCl_4 + 4HCl$$
$$HF: \quad CaF_2 + H_2SO_4 \longrightarrow 2HF + CaSO_4$$
$$CFCs: 3HF + 2CCl_4 \xrightarrow{SbCl_5} CCl_2F_2 + CCl_3F + 3HCl$$
$$2HF + CHCl_3 \longrightarrow CHClF_2 + 2HCl$$

Figure 12.4

The current use pattern of CFCs is shown in Fig. 12.5. The classic CFCs that have been used for refrigeration and air conditioning are mostly CFC-11 and -12, with some -114 and -115. A large portion of this 37% is now for automobile air conditioning. Foam blowing agents use mainly CFC-11 and -12. Solvent use, especially for cleaning of electronic circuit boards, employs CFC-113. A large previous use of CFCs was in aerosols and propellants. This has now been outlawed. An estimated 3 billion aerosol cans/yr used CFCs in the early 1970s.

What are the properties of CFCs that make them unique for certain applications? Propellants for aerosols need high volatility and low boiling points. Interestingly, compared to the same size hydrocarbons, fluorocarbons have

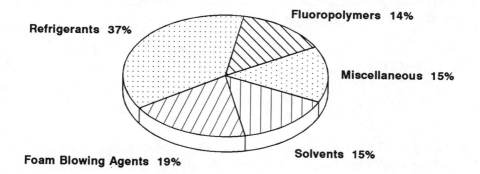

Figure 12.5 Uses of chlorofluorocarbons. (*Source:* CP.)

higher volatility and lower boiling points, unusual for halides. They are less reactive, more compressible, and more thermally stable than hydrocarbons. They also have low flammability, toxicity, and odor. They are used in air conditioners and refrigerators because they have high specific heats, high thermal conductivities, and low viscosities. Their nonflammability and low toxicity are also attractive in these applications.

What's the problem with CFCs? In the mid-1970s CFCs were determined to photodissociate in the stratosphere to form chlorine atoms. These chlorine atoms then react with ozone to deplete this protective layer in our atmosphere. The mechanism is a typical free radical chain process. Initiation in step (1) involves breaking a carbon-chlorine bond, weaker than a carbon-fluorine bond. Two propagation steps then can rapidly deplete ozone by reaction with the chlorine atoms.

$$(1) \qquad CCl_2F_2 \xrightarrow{\ h\nu\ } Cl\cdot + \cdot CClF_2$$

$$\text{or} \quad CCl_3F \xrightarrow{\ h\nu\ } Cl\cdot + \cdot CCl_2F$$

$$(2) \qquad Cl\cdot + O_3 \longrightarrow ClO\cdot + O_2$$

$$(3) \qquad ClO\cdot + O \longrightarrow Cl\cdot + O_2$$

then (2), (3), (2), (3), etc.

Net reaction, (2) + (3): $O_3 + O \longrightarrow 2O_2$

Long-range effects of having less ozone in the stratosphere are difficult to determine, but involve greater ultraviolet sunlight transmission, alteration of weather, and an increased risk of skin cancer. The ozone depletion potential for CFCs and other fluorocarbons have been measured and are given below relative to CFC-11 and -12. Notice that the HCFCs with lower chlorine content have lower depletion potentials than the CFCs, and the one HFC studied shows no depletion potential because it contains no chlorine.

CFC-11	1.0	HCFC-123	0.016
CFC-12	1.0	HCFC-141b	0.081
CFC-113	0.8	HCFC-22	0.053
CFC-114	1.0	HFC-134a	0
CFC-115	0.6		

HCFCs and HFCs, because of the hydrogen in the molecule, react with hydroxyl groups in the lower atmosphere. The HCFCs are being pushed as possible temporary replacements in some applications of CFCs, though HCFCs will be phased out early in the 21st century.

The manufacturing picture and the recent past and future of CFCs are rapidly changing. In 1988 annual CFC consumption was 2.5 billion lb. In the U.S. about 5,000 businesses at 375,000 locations produced goods and services valued in excess of $28 billion. More than 700,000 jobs were supported by these businesses. Obviously the CFC phaseout must be done properly to minimize the effects on these businesses and individuals. The following brief chronology will give the student an idea of the situation as of this writing.

1978 -The EPA outlawed CFC-11 and -12 in aerosol and propellant applications because of fear of ozone depletion.

1984 -An ozone hole over Antarctica was discovered with especially low concentrations of ozone above that continent in their spring (Northern Hemisphere's fall). This may be linked to CFCs.

1988 -Du Pont, the largest producer of CFCs, called for a total CFC production phaseout.

-A possible arctic ozone hole was studied.

-Du Pont suggested HFC-134a as a substitute for CFC-12 in refrigerators and auto air conditioners. However, the cost is 3-5 times higher for production, and the product is less energy efficient and requires expensive redesign of equipment.

-Pennwalt Corporation marketed a blend of HCFC-22 and HCFC-142b for refrigeration and air conditioning.

-Du Pont used HCFC-22 to replace CFC-12 as a blowing agent for plastic foams. It also tested HCFC-113 and HCFC-141b for this application.

-The EPA called for a total ban of CFCs.

1989 -The Montreal Protocol was completed. This asked for a worldwide production freeze at the 1986 levels, a 20% cut by 1993, and another 30% lowering of production by 1998 for CFC-11, -12, -113, -114, and -115.

1990 -The EPA proposed a nationwide recycling program for CFCs. It is estimated that two thirds of all CFCs in the U.S. can be recycled, especially for air conditioning, refrigeration, and solvent cleaning.

-Du Pont finished building its HFC-134a plant in Corpus Christi, TX.

-Du Pont suggested a blend of HCFC-22, HCFC-124, and HFC-152a to replace CFC-12 in refrigeration and air conditioning, with better energy efficiency than HFC-134a.

13

The Second Fifty Industrial Chemicals

References

Chenier, P. J.; Artibee, D. S. "The Second 50 Industrial Chemicals, Part 1." *J. Chem. Educ.* **1988**, *65,* 244-50.
Chenier, P. J.; Artibee, D. S. "The Second 50 Industrial Chemicals, Part 2." *J. Chem. Educ.* **1988**, *65,* 433-36.

INTRODUCTION

The basis of most of our study of industrial chemicals thus far has been the top 50 list published each year by *Chemical and Engineering News.* These chemicals are produced utilizing some fascinating processes on a large scale, and an understanding and appreciation of these chemicals, their manufacture and uses, provides the student with a solid background in industrial chemicals. We might ask ourselves what comes next? What are some other important large-scale chemicals that, though not made quite at the volume of the top 50, nevertheless in their own right make important contributions to the chemical industry?

To our knowledge, no recent attempt previous to our list first published in 1988 has identified these second 50 chemicals arranged in order by annual U.S. production. This is a valuable reference to those interested in studied commodity chemicals in greater detail than what the top 50 list offers. Indeed in previous chapters we have included some of these chemicals in our discussions, especially those which are monomers for important polymers to be mentioned later. Nearly half of the second 50 have been noted already in passing. This chapter will present these compounds as a unit, discuss the second 50 list in general terms, and summarize briefly each of the 50 chemicals in regard to manufacturing methods and use.

Production figures are sometimes very difficult to obtain for chemicals. A number of sources have been examined and at least one figure for each chemical has been found for production amounts from the last three years. We make no claim that the order is exact; nor do we suggest that no chemicals have been overlooked. Specialized trade literature has not been examined. Keeping this in mind, we present to you these 50 chemicals in Table 13.1, page 248.

Generally, the guidelines for selection are similar to those for the top 50 list. A single chemical, or a commercially useful, closely related family of chemicals often used as a single entity, is included. One might argue against the families being listed, as is done for linear alpha-olefins, fluorocarbons, *n*-paraffins, linear alkylbenzenes, and ethanolamines. However, these families are important in the industry and it is instructive to include them, even though they are not

Figure 13.1 The picture shows a portion of a horizontal cylindrical rotating reactor approximately 150 ft long in which hydrofluoric acid is made from fluorspar and sulfuric acid. HF comes out the cold end, $CaSO_4$ out the hot end. (Courtesy of Du Pont, La Porte, TX.)

chemically homogeneous and may not have made the list if their components were considered separately. No polymers are included. Sometimes chemicals do not appear on the list or are lower than what they might be due to a large captive use and unofficially reported production, e.g., adiponitrile is not listed even though it is the precursor of hexamethylenediamine. Production figures are not readily available for these chemicals and we hesitated to estimate their amount. Nevertheless, we feel the list as developed is quite functional.

CHARACTERISTICS OF THE SECOND 50 CHEMICALS

The most immediate characteristic of the second 50 list that is very striking is the dominance of organic chemicals, a total of 42, leaving only 8 inorganics, though some might claim phosgene and hydrogen cyanide as being borderline inorganics. In contrast, the top 50 have 29 organics and 21 inorganics, a more balanced distribution, and eight of the top 10 are inorganic. Organics in the second 50 account for 40 of a total of 45 billion lb of chemical production for all 50; in the top 50 the inorganics win in total production at 409 billion vs. organics at 217 billion lb, due mainly to those big eight inorganics. In grand total production, the second 50 at 45 billion lb is only a fraction of the top 50's 626 billion lb, barely 7% of the higher list. While the top 50 varies in production from sulfuric acid at 88.56 billion lb to caprolactam at 1.38 billion lb, the second 50 extends from phosgene at 2.28 billion lb to hydrogen peroxide and methylene chloride at the 0.48 billion lb level. As is obvious from this mention, we believe five chemicals might be candidates for the top 50 list, since they exceed 1.38 billion lb: phosgene, butyraldehyde, acetic anhydride, linear alpha-olefins and fluorocarbons. However, *C & E News* probably weeds out captive production and may have better production figures than those more readily available.

Also included in Table 13.1 are long-term growth patterns for each of the second 50 chemicals. The percent average annual growth is given, in most cases for approximately a ten-year duration, the exact years depending on the available source used. Positive growth is evident for 41 chemicals with only nine showing a net annual decrease in production over this period. Double digit positive percent annual growth rates are recorded for two chemicals, butraldehyde (52) and linear alpha-olefins (54). The largest decreases in production are exhibited by perchloroethylene (97) and sodium tripolyphosphate (65). The average annual growth rate for all the second 50 chemicals is +3.4%. The average prices of commercial quantities in ¢/lb are given in Table 13.1. Three chemicals are over $1.00/lb, HMDA (59), MDI (69), and TDI (75). The cheapest of the second 50 chemicals is isobutane (68), with hexanes (80) and lignosulfonic acid salts (76) close to it. The average price for all the second 50 chemicals is 46¢/lb.

Of interest are some other chemicals that were investigated in this study whose recent production figures did not qualify them for the second 50. However, due to their close proximity future figures may show that they have indeed passed up others. Those closest, with total production in billion lb, are maleic anhydride, 0.45; 2-butoxyethanol, 0.41; 1,4-butanediol, 0.40; tallow acids, sodium salt, 0.40; and ethoxylated nonylphenol, 0.39.

DERIVATIVES OF THE SEVEN BASIC ORGANIC CHEMICALS

The preponderance of organic chemicals in the list has prompted us to separate them according to the seven organic chemicals (ethylene, propylene, the C_4 stream, benzene, toluene, xylene, and methane) on which they are based. Table 7.1, page 114, lists the recent breakdown of the 29 organics in the 50 as being eight from ethylene, six from propylene, four from the C_4 stream, eight from benzene, one from toluene, three from xylene, and seven from methane. Double counting is, of course, necessary since some derivatives are made from more than one basic organic. Table 13.2, page 249, shows the organic chemicals in the second 50 separated by basic source. The list includes nine from ethylene, eleven from propylene, five from the C_4 stream, six from benzene, three from toluene, two from xylene, and sixteen from methane. Larger numbers of chemicals in the second 50 are derived from propylene and methane than in the top 50. Examining this list more closely, we see that derivatives are not always obvious from a simple count of carbons. The manufacturing method must be considered. For instance, butyraldehyde is not made from the C_4 stream, but by the oxo process with propylene and synthesis gas (carbon monoxide/hydrogen). Thus, it is derived from propylene and methane.

Unlike the organics in the top 50, the second 50 organics have some alternative sources other than the seven basic organics. Hexanes and *n*-paraffins are from various fractions of petroleum; nonene is made primarily by dehydrogenation of *n*-paraffins; linear alkylbenzenes, linear alpha olefins, and linear alcohols, ethoxylated are made from *n*-paraffins in part; and lignosulfonic acid salts come to us from the pulp and paper industry.

TABLE 13.1 The Second 50 Industrial Chemicals

Rank	Chemical	Production Billion lb	Average % Annual Growth	Average Price ¢/lb
51	Phosgene	2.28	4.2	61
52	Butyraldehyde	1.78	12.8	20
53	Acetic anhydride	1.75	1.1	46
54	Linear alpha olefins (LAO)	1.54	11.0	54
55	Chlorofluorocarbons (CFC)	1.44	2.9	94
56	Acetone cyanohydrin	1.31	4.2	na
57	n-Butyl alcohol	1.28	5.1	40
58	Methyl methacrylate (MMA)	1.24	5.0	71
59	Hexamethylenediamine (HMDA)	1.19	3.8	111
60	Hydrogen cyanide	1.18	4.8	60
61	Isobutylene	1.16	1.1	24
62	Bisphenol A	1.13	8.0	93
63	Cyclohexanone	1.06	8.3	77
64	Aniline	1.03	5.0	61
65	Sodium tripolyphosphate (STPP)	1.03	-3.5	36
66	Acrylic acid	1.01	5.5	67
67	2,4- and 2,6-Dinitrotoluene	0.98	4.8	37
68	Isobutane	0.97	-1.9	8
69	Methylene diphenyl diisocyanate (MDI)	0.96	7.2	120
70	Nitrobenzene	0.96	1.3	34
71	Phthalic anhydride	0.95	0.1	42
72	Propylene glycol	0.86	6.2	58
73	o-Xylene	0.86	-1.5	17
74	n-Paraffins	0.82	-0.4	28
75	Toluene diisocyanate (TDI)	0.78	2.6	129
76	Lignosulfonic acid, salts	0.74	1.0	10
77	Acetaldehyde	0.73	4.0	46
78	1,1,1-Trichloroethane	0.71	-0.3	41
79	Phosphorus	0.70	-2.1	91
80	Hexanes	0.68	5.5	10
81	Linear alkylbenzenes (LAB)	0.68	0.9	49
82	Sodium bicarbonate	0.66	3.0	19
83	Carbon tetrachloride	0.65	-0.7	24
84	Ethanolamines	0.65	4.6	63
85	2-Ethylhexanol	0.65	5.5	48
86	Linear alcohols, ethoxylated	0.64	3.3	49
87	Hydrofluoric acid	0.63	2.0	69
88	Potassium hydroxide	0.59	1.9	14
89	Nonene	0.58	3.5	22
90	1-Butene	0.57	16.0	30
91	Ethanol (synthetic)	0.56	1.2	31
92	Sodium chlorate	0.56	6.0	22
93	Butyl acrylate	0.55	9.6	53
94	Methyl chloride	0.55	1.7	26
95	Chloroform	0.53	3.8	35
96	Diethylene glycol	0.53	1.4	30
97	Perchloroethylene	0.50	-4.9	31
98	Sulfur dioxide	0.50	1.0	12
99	Hydrogen peroxide	0.48	7.2	48
100	Methylene chloride	0.48	-2.7	28

na=not available

Table 13.2 The Second 50 Chemicals as Derivatives of the Seven Basic Organics

Ethylene
    Acetaldehyde
    Acetic anhydride
    Diethylene glycol
    Ethanol
    Ethanolamines
    Linear alcohols, ethoxylated
    Linear alpha olefins
    Perchloroethylene
    1,1,1-Trichloroethane

Propylene
    Acetone cyanohydrin
    Acrylic acid
    Bisphenol A
    Butyl acrylate
    n-Butyl alcohol
    Butyraldehyde
    2-Ethylhexanol
    Hexamethylenediamine
    Hydrogen cyanide
    Methyl methacrylate
    Propylene glycol

C4 Stream
    Acetic anhydride
    1-Butene
    Hexamethylenediamine
    Isobutane
    Isobutylene

Other Sources
    Hexanes
    Lignosulfonic acid, salts
    Linear alcohols, ethoxylated
    Linear alkylbenzenes
    Linear alpha olefins
    Nonene
    n-Paraffins

Benzene
    Aniline
    Bisphenol A
    Cyclohexanol
    Linear alkylbenzenes
    Methylene diphenyl diisocyanate
    Nitrobenzene

Toluene
    Dinitrotoluene
    Toluene diisocyanate
    o-Xylene

Xylene
    Phthalic anhydride
    o-Xylene

Methane
    Acetic anhydride
    Alkylamines
    Butyl acrylate
    n-Butyl alcohol
    Butyraldehyde
    Carbon tetrachloride
    Chlorofluorocarbons
    Chloroform
    2-Ethylhexanol
    Hydrogen cyanide
    Methyl chloride
    Methylene chloride
    Methylene diphenyl diisocyanate
    Methyl methacrylate
    Phosgene
    Toluene diisocyanate

TABLE 13.3 Second 50 Chemical Pairs

Precursor	Derivative
Acetaldehyde	Acetic anhydride
Acetone cyanohydrin	Methyl methacrylate
Acrylic acid	Butyl acrylate
Aniline	Methylene diphenyl diisocyanate
n-Butyl alcohol	Butyl acrylate
Butyraldehyde	2-Ethylhexanol
Carbon tetrachloride	Chlorofluorocarbons
Dinitrotoluene	Toluene diisocyanate
2-Ethylhexanol	Dioctyl phthalate
Hydrogen cyanide	Acetone cyanohydrin
Isobutane	Isobutylene
Linear alpha olefins	Linear alcohols, ethoxylated
Methyl chloride	Methylene chloride
Methylene chloride	Chloroform
Nitrobenzene	Aniline
Nitrobenzene	Methylene diphenyl diisocynate
Nonene	Linear alkylbenzenes
n-Paraffins	Linear alkylbenzenes
n-Paraffins	Linear alpha olefins
n-Paraffins	Nonene
Phosgene	Methylene diphenyl diisocyanate
Phosgene	Toluene diisocyanate
o-Xylene	Phthalic anhydride

SECOND 50 CHEMICAL PAIRS

The second 50 chemicals commonly are made with another representative of this same list as one of its precursors, while only originally being derived from a more basic chemical. Table 13.3 gives 23 such paired chemicals. These second 50 pairs are one reason why an exact list of ranking may not be possible, since so many production amounts are very close.

MANUFACTURE AND USES

The following sections briefly mention the one or two processes used to make the chemical on a large scale. Then the chemical's uses are given with approximate percentages where found. A close study of this chemistry uncovers many interesting relationships among all 100 top chemicals in the United States.

51. PHOSGENE

$$CO + Cl_2 \longrightarrow Cl-\overset{\overset{\textstyle O}{\|}}{C}-Cl$$

Phosgene is manufactured by reacting chlorine gas and carbon monoxide in the presence of activated carbon. Much of the market is captive. The merchant market is small.

Uses of phosgene include the manufacture of toluene diisocyanate (45%), methylene diisocyanate (35%), and polycarbonate resins (10%).

52. BUTYRALDEHYDE

$$CH_3-CH=CH_2 + CO + H_2 \longrightarrow CH_3-CH_2-CH_2-\overset{\overset{\textstyle O}{\|}}{C}-H$$

$$+ \quad CH_3-\overset{\overset{\textstyle CH_3}{|}}{CH}-\overset{\overset{\textstyle O}{\|}}{C}-H$$

Butyraldehyde is made by the reaction of propylene, carbon monoxide, and hydrogen at 130-175°C and 250 atm over a cobalt carbonyl catalyst. The reaction is referred to as the oxo process, and a second product of the reaction is isobutyraldehyde. The ratio of *n*/iso is 4:1. A new rhodium catalyst can be used at lower temperatures and pressures and gives a ratio of 16:1.

Butyraldehyde is used most in the production of *n*-butyl alcohol.

53. ACETIC ANHYDRIDE

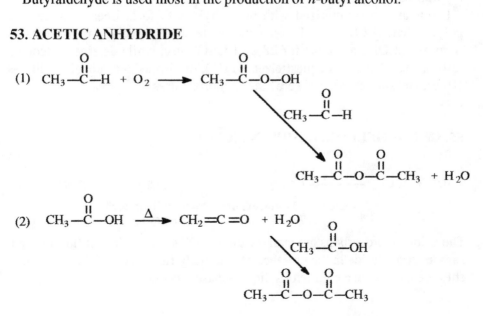

$$(3) \quad CH_3\overset{\overset{\displaystyle O}{\|}}{C}-O-CH_3 + CO \longrightarrow CH_3\overset{\overset{\displaystyle O}{\|}}{C}-O-\overset{\overset{\displaystyle O}{\|}}{C}-CH_3$$

Acetic anhydride may be produced by three different methods. The first procedure involves the *in situ* production from acetaldehyde of peracetic acid, which in turn reacts with more acetaldehyde to yield the anhydride. In the preferred process, acetic acid (or acetone) is pyrolyzed to ketene, which reacts with acetic acid to form acetic anhydride. A new process to make acetic anhydride involves CO insertion into methyl acetate. This may be the process of the future.

Approximately 80% of acetic anhydride is used as a raw material in the manufacture of cellulose acetate.

54. LINEAR ALPHA OLEFINS (LAO)

$$CH_2{=}CH_2 \longrightarrow CH_3-(CH_2)_n-CH{=}CH_2$$

$$n = 3 \text{ to } 15$$

Linear hydrocarbons with a double bond at the end of the chain are made by oligomerization of ethylene. Compounds with 6-18 carbons are the most popular. Ziegler catalysts are used in this process. Note that certain olefins such as nonene and dodecene can also be made by cracking and dehydrogenation of *n*-paraffins.

LAOs are copolymerized with polyethylene to form linear low density polyethylene (LLDPE). 1-Hexene and 1-octene are especially useful for this purpose. LLDPE accounts for 28% of LAO's use, while detergent alcohols (20%), oxo alcohols for plasticizers (11%), lubricants and lubeoil additives (10%), and surfactants (6%) are other important uses.

55. CHLOROFLUOROCARBONS (CFC)

$$3HF + 2CCl_4 \xrightarrow{SbCl_5} CCl_2F_2 \quad + \quad CCl_3F \quad + 3HCl$$

$$\text{dichlorodifluoromethane} \quad \text{trichlorofluoromethane}$$
$$\text{CFC-12} \quad \text{CFC-11}$$

The chlorofluorocarbons are manufactured by reacting hydrogen fluoride and carbon tetrachloride in the presence of a partially fluorinated antimony pentachloride catalyst in a continuous, liquid-phase process.

Common uses for the fluorocarbons are as refrigerants (37%), foam blowing agents (19%), solvents (15%), and fluoropolymers (14%).

56. ACETONE CYANOHYDRIN

$$CH_3-\overset{\overset{\displaystyle O}{\|}}{C}-CH_3 \ + \ HCN \ \longrightarrow \ CH_3-\overset{\overset{\displaystyle OH}{|}}{\underset{\underset{\displaystyle C\equiv N}{|}}{C}}-CH_3$$

Acetone cyanohydrin is manufactured by the direct reaction of hydrogen cyanide with acetone catalyzed by base, generally in a continuous process.

Acetone cyanohydrin is an intermediate in the manufacture of methyl methacrylate.

57. *n*-BUTYL ALCOHOL

(1) $C_6H_{12}O_6 \longrightarrow CH_3-CH_2-CH_2-CH_2-OH \ + \ CH_3-\overset{\overset{\displaystyle O}{\|}}{C}-CH_3 \ +$

$$CH_3-CH_2-OH \ + \ CO_2 \ + \ H_2$$

(2) $CH_3-CH=CH_2 \ + \ CO_2 \ + \ H_2 \ \longrightarrow \ CH_3-CH_2-CH_2-\overset{\overset{\displaystyle O}{\|}}{C}-H \overset{H_2}{\longrightarrow}$

$$CH_3-CH_2-CH_2-CH_2-OH$$

Butanol can be obtained from carbohydrates (such as molasses and grain) by fermentation. Acetone and ethanol are also produced. Synthetic processes account for the majority of current-day production. Propylene and synthesis gas give *n*-butyl alcohol. Isobutyl alcohol is a byproduct.

n-Butyl alcohol is used for butyl acrylate and methacrylate (30%), glycol ethers (25%), solvent (11%), butyl acetate (11%), and plasticizers (4%).

58. METHYL METHACRYLATE (MMA)

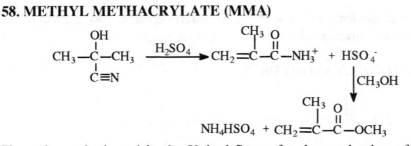

The only method used in the United States for the production of methyl methacrylate is the acetone cyanohydrin process. Acetone cyanohydrin (from the reaction of acetone with hydrogen cyanide) is reacted with sulfuric acid to yield methacrylamide sulfate, which is further hydrolyzed and esterified in a continuous process. Other processes using different raw materials have been tried in the United States and abroad, but the acetone cyanohydrin process has prevailed over the years. It is being made in Japan by oxidation of isobutene or *t*-butyl alcohol.

Methyl methacrylate is polymerized to poly(methyl methacrylate), which is used in cast and extruded sheet (24%), molding powder and resins (21%), surface coatings (18%), impact modifiers (10%), and emulsion polymers (8%).

59. HEXAMETHYLENEDIAMINE (HMDA)

$$2CH_2{=}CH{-}C{\equiv}N$$

$$CH_2{=}CH{-}CH{=}CH_2 \xrightarrow[\substack{2H^+ \\ 2e^- \\ 2HCN \\ Ni^{+2}}]{} N{\equiv}C{-}(CH_2)_4{-}C{\equiv}N \xrightarrow{H_2} NH_2{-}(CH_2)_6{-}NH_2$$

Hexamethylenediamine is produced from adiponitrile by hydrogenation. Adiponitrile comes from electrodimerization of acrylonitrile (32%) or from anti-Markovnikov addition of 2 moles of hydrogen cyanide to butadiene (68%).

HMDA is used exclusively in the production of nylon 6,6.

60. HYDROGEN CYANIDE

(1) $2CH_4 + 2NH_3 + 3O_2 \longrightarrow 2HCN + 6H_2O$

(2) $CH_2{=}CH{-}CH_3 + 2NH_3 + 3O_2 \longrightarrow 2CH_2{=}CH{-}C{\equiv}N + 6H_2O(+HCN)$

Approximately 80% of all hydrogen cyanide is manufactured by the reaction of air, ammonia, and natural gas over a platinum or platinum-rhodium catalyst at elevated temperature. The reaction is referred to as the Andrussow process. Hydrogen cyanide is also available as a by-product from acrylonitrile manufacture by ammoxidation (20%).

Methyl methacrylate production accounts for 33% of hydrogen cyanide use, adiponitrile for 43%. Other uses include cyanuric chloride (6%) and chelating agents (5%).

61. ISOBUTYLENE

$$C_nH_{2n+2} \xrightarrow{\Delta} CH_2{=}\underset{\underset{CH_3}{|}}{C}{-}CH_3 + H_2$$

Isobutylene is made largely by the catalytic and thermal cracking of hydrocarbons. Other C_4 products formed in the reaction include other butylenes, butane, isobutane, and traces of butadiene. The most widely used process is the fluid catalytic cracking of gas oil; other methods are delayed coking and flexicoking.

Isobutylene is used as a raw material in the production of methyl *t*-butyl ether (MTBE, gasoline additive) and butylated hydroxytoluene (an important antioxidant). It is also used in the alkylation of gasoline.

62. BISPHENOL A (BPA)

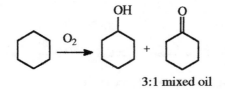

Bisphenol A is made by reacting phenol with acetone in the presence of an acid catalyst. The temperature of the reaction is maintained at 50°C for about 8-12 hours. A slurry of BPA is formed, which is neutralized and distilled to remove excess phenol.

The only major uses of BPA are in the production of epoxy (25%) and polycarbonate resins (55%).

63. CYCLOHEXANOL/CYCLOHEXANONE

3:1 mixed oil

Cyclohexanol and cyclohexanone are made by the air oxidation of cyclohexane with a cobalt(II) naphthenate or acetate or benzoyl peroxide catalyst at 125-160°C and 50-250 psi. Also used in the manufacture of cyclohexanol is the hydrogenation of phenol at elevated temperatures and pressures, in either the liquid or vapor phase and with a nickel catalyst. The former production method accounts for 80% of cyclohexanol made in the United States.

Cyclohexanol is used primarily in the production of adipic acid (90%), which is further used as a raw material in nylon 6,6 production, and in methylcyclohexanol production (7%).

64. ANILINE

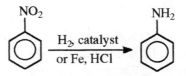

Aniline is made by the reduction of nitrobenzene (83%) by either catalytic hydrogenation or acidic metal reduction. The reaction of ammonia and phenol is a newer process that shows promise and is being used (17%).

Major uses of aniline include *p-p'*-methylene diphenyl diisocyanate (MDI) (65%) and rubber chemicals (15%) production. It is also used to a smaller extent in herbicides (5%), dyes and pigments (4%), and specialty fibers (2%).

65. SODIUM TRIPOLYPHOSPHATE (STPP)

$$2H_3PO_4 + Na_2CO_3 \longrightarrow 2NaH_2PO_4 + H_2O + CO_2$$

$$4H_3PO_4 + 4Na_2CO_3 \longrightarrow 4Na_2HPO_4 + 4H_2O + 4CO_2$$

$$NaH_2PO_4 + 2Na_2HPO_4 \longrightarrow 2Na_5P_3O_{10} + 2H_2O$$

STPP is used mainly as a builder for detergents: home laundry detergents, 44%; industrial and institutional detergents, 19%; dishwashing detergents, 18%; and food uses, 4%.

66. ACRYLIC ACID

(1) $CH_2{=}CH{-}CH_3 \xrightarrow{O_2} CH_2{=}CH{-}CHO \xrightarrow{O_2} CH_2{=}CH{-}COOH$

(2) $CH_2{=}CH{-}CN \xrightarrow[H^+]{H_2O} CH_2{=}CH{-}COOH$

Acrylic acid is made by the oxidation of propylene to acrolein and further oxidation to acrylic acid. Another common method of production is acrylonitrile hydrolysis.

Acrylic acid and its salts are raw materials for an important range of esters, including methyl, ethyl, butyl, and 2-ethylhexyl acrylates. The acid and its esters are used in polyacrylic acid and salts (23%, including superabsorbent polymers, detergents, water treatment chemicals, and dispersants), surface coatings (15%), adhesives and sealants (11%), textiles and non-wovens (10%), and plastic modifiers (5%).

67. 2,4- AND 2,6-DINITROTOLUENE

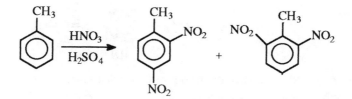

The dinitrotoluenes are manufactured by the classical nitration of toluene. This gives a mixture of 65-80% 2,4-dinitro derivative and 35-20% 2,6-dinitro compound. If the pure 2,4-compound is required, mononitration followed by separation of pure *p*-nitrotoluene from the ortho isomer, and then further nitration of *p*-nitrotoluene gives the pure 2,4-dinitro isomer.

Nearly all the dinitrotoluenes are hydrogenated to diamines, which are converted into diisocyanates to give toluene diisocyanate, a monomer for polyurethanes. A small amount of the dinitrotoluenes are further nitrated to 2,4,6-trinitrotoluene (TNT), the famous explosive.

68. ISOBUTANE

$$C_4 \xrightarrow[\text{distillation}]{\text{extraction}} CH_3-\overset{\overset{\displaystyle CH_3}{|}}{CH}-CH_3$$

Isobutane can be isolated from the petroleum C_4 fraction or from natural gas by extraction and distillation (see Chapter 8, p. 145).

There are two major uses of isobutane. Dehydrogenation to isobutylene is a large use. The isobutylene is then converted into the gasoline additive methyl *t*-butyl ether. Isobutane is also oxidized to the hydroperoxide and then reacted with propylene to give propylene oxide and *t*-butyl alcohol. The *t*-butyl alcohol can be used as a gasoline additive, or dehydrated to isobutylene.

69. METHYLENE DIPHENYL DIISOCYANATE (MDI)

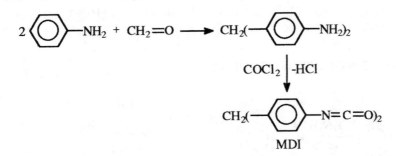

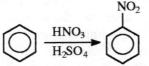

MDI

Aniline is condensed with formaldehyde; reaction with phosgene gives MDI. Rigid polyurethane foams account for 60% of MDI use, especially for construction, refrigeration and packaging.

70. NITROBENZENE

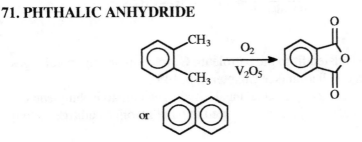

Nitrobenzene is made by the direct nitration of benzene using a nitric acid-sulfuric acid mixture. The nitrator of the reaction is a specially built cast-iron steel kettle.

About 98% of nitrobenzene is used in the production of aniline. The other 2% goes towards the production of acetaminophen.

71. PHTHALIC ANHYDRIDE

or

In 1983 about 72% of the phthalic anhydride made in the United States came from the reaction of *o*-xylene with air. The rest was made from naphthalene, which was isolated from coal tar and petroleum. In 1989 all plants used *o*-xylene.

Plasticizers, such as dioctyl phthalate (48%), unsaturated polyester resins (23%), and alkyd resins (19%), account for the majority of phthalic anhydride use.

72. PROPYLENE GLYCOL

$$CH_3-CH-CH_2 + H_2O \xrightarrow[\text{or } \Delta]{H^+} CH_3-CH-CH_2-OH$$

Propylene glycol is produced by hydration of propylene oxide in a process similar to that for the production of ethylene glycol by hydration of ethylene oxide.

Unsaturated polyester resins account for the majority of the commercial use of propylene glycol (41%). Other uses include food, cosmetics, and pharmaceuticals (11%), pet food (7%), and tobacco humectants (4%).

73. *o*-XYLENE

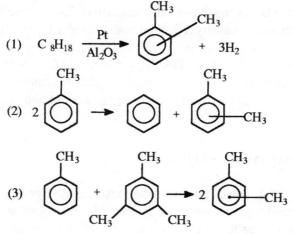

(1)　C_8H_{18} $\xrightarrow[\text{Al}_2\text{O}_3]{\text{Pt}}$ ... + $3H_2$

There are two methods of manufacture of the xylenes. The major one is from petroleum by catalytic reforming with a platinum-alumina catalyst. The second method (which has been developed recently) is by processes involving the disproportionation of toluene or the transalkylation of toluene with trimethylbenzenes. The ortho isomer is separated from the meta and para isomer by fractional distillation.

o-Xylene is used almost exclusively as feedstock for phthalic anhydride manufacture.

74. *n*-PARAFFINS

$$C_nH_{2n+2}$$

The production of the *n*-paraffins, especially C_{10}-C_{14}, involves the use of zeolites to separate straight chain compounds from the kerosene fraction of petroleum.

The main use of *n*-paraffins is in the production of linear alkylbenzenes (68%) for the detergent industry. The other uses are for linear alcohols (9%), solvents (4%), and chlorinated paraffins (2%).

75. TOLUENE DIISOCYANATE (TDI)

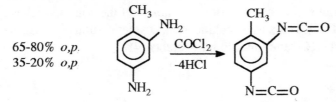

65-80% *o,p.*
35-20% *o,p*

Toluene diisocyanate (TDI) is made from the reaction of 2,4-toluenediamine and phosgene. The diamine is made by reduction of dinitrotoluene, which in turn is manufactured by nitration of toluene.

Polyurethanes account for the use of TDI. Approximately 65% of this goes toward flexible polyurethane foams (28% furniture, 16% transportation, 8% carpet underlay, 7% bedding), 4% toward coatings, 2% in rigid foams, and 2% in elastomers.

76. LIGNOSULFONIC ACID, SALTS

Lignosulfonates are metal salts of the sulfonated products of lignin and are a byproduct of the pulp and paper industry (see Chapter 22). Pulping of wood with a sulfite solution dissolves the lignin portion, leaving behind the cellulose fibers which are processed into paper. The sulfite solution is concentrated and the resulting solids sold as lignosulfonates. Only a small amount, about 15%, of the spent sulfite liquor solids is used each year. The rest is burned to recover heat. A molecular weight of 250 for the lignosulfonate monomer is approximate, but the product may contain material with a molecular weight as high as 100,000. Liquor from a kraft alkaline sulfate pulping process can be concentrated to give a solid called alkali lignin, but little of this is used except as fuel.

The uses of lignosulfonates include binders and adhesives (35%), surfactants (32%), animal feed additives (17%), and manufacture of vanilla (16%).

77. ACETALDEHYDE

$$2\,CH_2{=}CH_2 \; + \; O_2 \; \xrightarrow[\text{PdCl}_2]{\text{CuCl}_2} \; 2CH_3{-}\overset{\displaystyle O}{\overset{\|}{C}}{-}H$$

Acetaldehyde may be made (1) from ethylene by direct oxidation with the Wacker-catalyst containing copper (II) and palladium (II) salts; (2) from ethanol by vapor-phase oxidation or dehydrogenation; or (3) from butane by vapor-phase oxidation. The direct oxidation of ethylene is the most commonly used process, accounting for 80% of acetaldehyde production.

The main use of acetaldehyde (70%) is in acetic acid and acetic anhydride production; other uses include pyridine bases (8%), pentaerythritol (7%), peracetic acid (6%), and 1,3-butylene glycol (2%).

78. 1,1,1-TRICHLOROETHANE

$$CH_2{=}CH{-}Cl \; + \; HCl \; \longrightarrow CH_3{-}CHCl_2$$

$$CH_3{-}CHCl_2 \; + \; Cl_2 \; \longrightarrow CH_3{-}CCl_3 \; + \; HCl$$

1,1,1-Trichloroethane is made primarily from vinyl chloride by the hydrochlorination-chlorination (60%) process shown above; however, it may also be made from vinylidene chloride by hydrochlorination (30%) or from ethane by chlorination (10%).

Uses of 1,1,1-trichloroethane are in vapor degreasing (34%), cold cleaning (12%), aerosols (10%), adhesives (8%), intermediates (7%), and coatings (5%).

79. PHOSPHORUS

$$2Ca_3(PO_4)_2 \; + \; 10C \; + \; 6SiO_2 \; \longrightarrow P_4 \; + \; 6CaSiO_3 \; + \; 10CO$$

Yellow phosphorus (known also as white phosphorus) is produced by reducing phosphate rock (calcium phosphate or calcium fluorophosphate) with carbon in the presence of silica as flux; heat of reaction is furnished by an electric-arc furnace.

Phosphorus is used for the manufacture of thermal phosphoric acid (83%) and other chemicals (12%), including phosphorus trichloride, pentasulfide, and pentoxide.

80. HEXANES

$$C_6H_{14}$$

Commercial hexanes are produced by two-tower distillation of straight-run gasolines that have been distilled from crude oil or natural gas liquids.

Motor oil is the major use of this chemical. It is also used as a solvent for oil-seed extraction and as a medium for various polymerization reactions.

81. LINEAR ALKYLBENZENES (LAB)

(1) $C_{12}H_{26} \xrightarrow{-H_2} C_{12}H_{24}$ + ⟨⟩ $\xrightarrow{HF}$ $C_{12}H_{25}$—⟨⟩

(2) $C_{12}H_{26} \xrightarrow[hv]{Cl_2} C_{12}H_{25}Cl$ + ⟨⟩ $\xrightarrow{AlCl_3}$ $C_{12}H_{25}$—⟨⟩

Linear alkylbenzenes are made from *n*-paraffins (C_{10}-C_{14}) by either partial dehydrogenation to olefins and addition to benzene with HF as catalyst (60%) or chlorination of the paraffins and Friedel-Crafts reaction with benzene and an aluminum chloride catalyst.

The major uses of linear alkylbenzenes are in the manufacture of linear alkyl sulfonates, LAS (89%), which are divided into household detergents (74%) and industrial cleaners (15%).

82. SODIUM BICARBONATE

$$Na_2CO_3 + CO_2 + H_2O \longrightarrow 2NaHCO_3$$

Sodium bicarbonate can be made by treating soda ash with carbon dioxide and water. Sodium bicarbonate is called bicarbonate of soda or baking soda. It is also mined from certain ores called nahcolite.

The uses of sodium bicarbonate include food preparation (34%), animal feeds (24%), industrial chemicals (13%), and soap and detergents (10%).

83. CARBON TETRACHLORIDE

$$CS_2 + 3Cl_2 \longrightarrow S_2Cl_2 + CCl_4$$
$$CS_2 + 2S_2Cl_2 \longrightarrow 6S + CCl_4$$
$$6S_4 + 3C \longrightarrow 3CS_2$$

Carbon tetrachloride may be made from the reaction of carbon disulfide and chlorine (accounting for 85% of carbon tetrachloride), with sulfur monochloride as an important intermediate. Elemental sulfur can be reconverted to carbon disulfide by reaction with coke. Chlorination of methane and higher aliphatic hydrocarbons accounts for 15% of the carbon tetrachloride produced.

$$CH_4 + 4Cl_2 \xrightarrow[400^\circ C]{250-} CCl_4 + 4HCl$$

The majority of carbon tetrachloride goes toward the making of fluorocarbons 11 and 12 (dichlorodifluoromethane and trichlorofluoromethane, respectively); approximately 91% of all carbon tetrachloride is used this way.

84. ETHANOLAMINES

$$\overset{O}{\overset{\displaystyle \triangle}{CH_2-CH_2}} + NH_3 \longrightarrow MEA + DEA + TEA$$

Monoethanolamine (MEA)	$HO-CH_2-CH_2-NH_2$
Diethanolamine (DEA)	$(HO-CH_2-CH_2)_2NH$
Triethanolamine (TEA)	$(HO-CH_2-CH_2)_3N$

Ethanolamines are made by reacting ethylene oxide and excess ammonia, followed by separation of unreacted ammonia and the three ethanolamines. The proportion of the three products depends on reaction conditions.

The breakdown of use of ethanolamines is surfactants (23%), natural gas conditioning and petroleum use (15%), and metal working (7%).

85. 2-ETHYLHEXANOL

$$2 CH_3—CH_2—CH_2—CH=O + H_2$$

$$\xrightarrow{Ni}$$

$$CH_3—CH_2—CH_2—CH_2—\overset{\overset{\displaystyle CH_2—CH_3}{|}}{CH}—CH_2—OH$$

2-Ethylhexanol is produced by aldol condensation of butylaldehyde followed by reduction.

Plasticizers account for 50% of the use of 2-ethylhexanol (especially dioctyl phthalate, 34%; dioctyl adipate, 5%; and trioctyl trimellitate, 4%). About 11% goes toward the making of 2-ethylhexyl acrylate for adhesives and coatings, and 5% for 2-ethylhexylnitrate.

86. LINEAR ALCOHOLS, ETHOXYLATED

$$C_{14}H_{29}OH + n CH_2—CH_2 \longrightarrow C_{14}H_{29}—O(—CH_2—CH_2—O)_n—H$$

Ethoxylated linear alcohols can be made by the reaction of straight-chain alcohols, usually C_{12}-C_{14}, with 3-7 moles of ethylene oxide. The resulting alcohols are one type of many alcohols used for detergents. The linear alcohols can come from *n*-paraffins via the alpha olefins or the chlorides. Or they can be made from alpha olefins formed from Ziegler oligomerization of ethylene.

By sulfonation and sodium salt formation these alcohols are converted into AES detergents for shampoos and dishwashing (Chapter 24).

87. HYDROFLUORIC ACID

$$CaF_2 + H_2SO_4 \xrightarrow{250\text{-}300\ ^O C} 2HF + CaSO_4$$

Fluorspar (CaF_2), 20% oleum, and sulfuric acid are heated in a horizontal rotating drum.

Uses are for fluorocarbons (67%), including fluoropolymers and CFCs; chemical intermediates (14%), including fluoroborates, surfactants, herbicides, and electronic chemicals; aqueous HF (8%), including stainless steel, glass, and metal processing; petroleum alkylation (5%); and uranium processing (4%).

88. POTASSIUM HYDROXIDE

$$2KCl + 2H_2O \xrightarrow{e^-} 2KOH + H_2 + Cl_2$$

Potassium hydroxide is produced by the electrolysis of potassium chloride solutions.

The breakdown of the use of potassium hydroxide is as follows: potassium carbonate, 23; tetrapotassium pyrophosphate and other potassium phosphates, 10%; other potassium compounds, 33%; liquid fertilizers, 10%; and soaps, 4%.

89. NONENE

$$C_9H_{20} \xrightarrow[\text{or catalyst}]{540\text{-}565\ ^{\circ}C} C_9H_{18} + H_2$$

Originally made by the trimerization of propylene to give a branched nonene, this product now has limited use for detergents because of nonbiodegradability. Cracking and dehydrogenation of *n*-paraffins is now the preferred method, giving very linear chains. With good linear wax, an olefin product containing as much as 90% linear alpha olefins can be prepared.

Nonene is used in the manufacture of nonylphenol (30%) and ethoxylated nonylphenol nonionic surfactants. It is also used in the oxo process to make isodecyl alcohol (34%) for esters as plasticizers.

90. 1-BUTENE

$$C_nH_{2n+2} \xrightarrow{\Delta} CH_3-CH_2-CH=CH_2 + H_2$$

The steam-cracking of naphtha and catalytic cracking in the refinery produce the C_4 stream, which includes butane, 1-butene (butylene), *cis-* and *trans*-2-butene, isobutylene, and butadiene. 1-Butene can be separated by extracting the isobutylene with sulfuric acid and distilling the 1-butene away from butane and butadiene. About 56% is made this way. It is also made by Ziegler ethylene oligomerization with other longer linear alpha olefins (26%). Shell uses a proprietary non-Ziegler oligomerization (18%).

1-Butene is used as a comonomer to make polyethylene. About 80% is used for both LLDPE and HDPE (93:7). It is also used to make valeraldehyde (pentanal) by the oxo process (8%), polybutene-1 (9%), and butylene oxide (1%).

91. ETHANOL (SYNTHETIC)

$$CH_2=CH_2 + H_2O \longrightarrow CH_3-CH_2-OH$$

Synthetic ethanol is made by the hydration of ethylene over a phosphoric acid-on-celite catalyst and accounts for 18% of all ethanol. The predominant method of ethanol manufacture, at one time, was by fermentation of sugars; this method went out of use in the 1930s. However, corn fermentation is now a source of 82% of all ethanol and is used for gasohol, a 10% alcohol:90% gasoline blend used for automobile fuel.

Industrial uses for ethanol are shared by synthetic and fermentation alcohol in a 7:3 ratio and include solvents (59%) and chemical intermediates (41%).

92. SODIUM CHLORATE

$$2NaCl + 2H_2O \xrightarrow{e^-} 2NaOH + H_2 + Cl_2$$

$$Cl_2 + 2NaOH \longrightarrow NaOCl + NaCl + H_2O$$

$$3NaOCl \longrightarrow NaClO_3 + 2NaCl$$

Sodium chlorate is readily produced in solution by the electrolysis of sodium chloride brine in a cell that is very similar to a diaphragm chlor-alkali cell, except that it has no diaphragm. By allowing the chlorine and sodium hydroxide to react to form hypochlorite and by providing an additional vessel where the cell liquor can be kept hot, the hypochlorite disproportionates to chloride and chlorate.

Almost all the sodium chlorate manufactured (88%) goes to the pulp and paper industry for bleaching. Other uses are in the manufacture of other chlorates, perchlorates, and chlorites (8%) and herbicides (2%).

93. BUTYL ACRYLATE

(1) $\quad CH\equiv CH + ROH + CO \xrightarrow[HCl]{Ni(CO)_4} CH_2=CH-COOR$

(2) $\quad CH_2=CH-CH_3 \xrightarrow{O_2} CH_2=CH-COOH \xrightarrow[H^+]{ROH} CH_2=CH-COOR$

Acrylates are still produced by a modified Reppe process that involves the reaction of actylene, the appropriate alcohol (in the case of butyl acrylate, butyl alcohol is used), and carbon monoxide in the presence of an acid. The process is continuous and a small amount of acrylates is made this way. The most economical method of acrylate production is that of the direct oxidation of propylene to acrylic acid, followed by esterification.

Acrylates find major use in coatings (45%), textiles (25%), and fibers, polishes, paper and leather (15% collectively).

94. METHYL CHLORIDE

$$(1) \quad CH_3OH + HCl \longrightarrow CH_3Cl + H_2O$$

$$(2) \quad CH_4 + Cl_2 \longrightarrow CH_3Cl + HCl$$

The major method (65%) for the production of methyl chloride is by the reaction of methanol and hydrogen chloride, with the aid of a catalyst and either in the vapor or liquid phases. Approximately 35% is made by the chlorination of methane.

The uses of methyl chloride are as follows: silicones, 74%; agricultural chemicals, 7%; methyl cellulose, 6%; and quaternary amines, 5%.

95. CHLOROFORM

$$CH_4 + Cl_2 \longrightarrow CH_3Cl + HCl$$

$$CH_3Cl + Cl_2 \longrightarrow CH_2Cl_2 + HCl$$

$$CH_2Cl_2 + Cl_2 \longrightarrow CHCl_3 + HCl$$

Chloroform is produced by the chlorination of methylene chloride, which in turn is made by the chlorination of methyl chloride and methane.

The main use of chloroform is in the manufacture of chlorofluorocarbon-22 (90%, 70:30 refrigerants: polymers).

96. DIETHYLENE GLYCOL

$$CH_2{-}CH_2 + H_2O \xrightarrow[or\ \Delta]{H^+} HO{-}CH_2{-}CH_2{-}OH$$

$$+$$

$$HO{-}CH_2{-}CH_2{-}O{-}CH_2{-}CH_2{-}OH$$

Diethylene glycol is produced as a byproduct in the manufacture of ethylene glycol from hydrolysis of ethylene oxide. It is separated from the ethylene glycol by vacuum distillation.

Breakdown of diethylene glycol use is as follows: polyurethane and unsaturated polyester resins, 40%; antifreeze blending, 15%; triethylene glycol, 12%; morpholine, 9%; and natural gas dehydration, 5%. Much of the market is captive. The merchant market is small.

97. PERCHLOROETHYLENE (perc)

$$2C_2H_4Cl_2 + 5Cl_2 \longrightarrow C_2H_2Cl_4 + C_2HCl_5 + 5HCl$$

$$C_2H_4Cl_2 + C_2HCl_5 \longrightarrow C_2HCl_5 + 2HCl + C_2Cl_4 \text{ (perc)}$$

$$7HCl + 1.75O_2 \longrightarrow 3.5H_2O + 3.5Cl_2$$

$$2C_2H_4Cl_2 + 1.5Cl_2 + 1.75O_2 \longrightarrow C_2HCl_3 + C_2Cl_4 + 3.5H_2O$$

There are three processes used in the making of perchloroethylene. Approximately 85% of the manufacture is from ethylene dichloride, as shown above.

Perchloroethylene and trichloroethylene are produced in a single-stage oxy chlorination process from ethylene dichloride and chlorine. The other 15% comes from either the chlorination of hydrocarbons (propane, for example) or from acetylene and chlorine via trichloroethylene.

The main use of perchloroethylene is in dry cleaning and textile processing (50%); other uses are as a chemical intermediate (28%) and in industrial metal cleaning (9%).

98. SULFUR DIOXIDE

$$S + O_2 \longrightarrow SO_2$$

Sulfur dioxide is made as part of the contact process for making sulfuric acid (Chapter 2, pp. 40-44). Sulfur and oxygen are burned at 1000°C. Sulfur dioxide can be made by oxidation of various metal sulfides or hydrogen sulfide, or it can be made from calcium sulfate or used sulfuric acid as well.

paper (20%); food and agriculture, especially corn processing (16%); water and waste treatment (10%); and metal and ore refining (6%).

99. HYDROGEN PEROXIDE

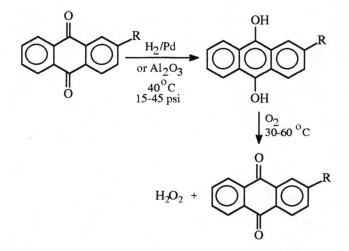

The most important method of making hydrogen peroxide is by reduction of anthraquinone to the hydroquinone, followed by reoxidation to anthraquinone by oxygen and formation of the peroxide. R is usually ethyl but *t*-butyl and *sec*-amyl have also been used.

Hydrogen peroxide is used in pulp and paper (38%); chemical synthesis (18%); environmental control, including municipal and industrial water treatment (17%); and textiles (11%).

100. METHYLENE CHLORIDE

$$CH_4 + Cl_2 \longrightarrow CH_3Cl + HCl$$
$$CH_3Cl + Cl_2 \longrightarrow CH_2Cl_2 + HCl$$

Methylene chloride is produced by the chlorination of methyl chloride, which in turn is made by the chlorination of methane.

The main uses of methylene chloride are in paint remover (28%), aerosols (18%), chemical processing (11%), urethane foam blowing agents (9%), metal degreasing (8%), and electronics (7%).

14

Basic Polymer Chemistry: Thermoplastics

Reference

W & R I, pp. 162-219

DEFINITIONS AND CLASSES

To begin our discussion of polymers we introduce some of the words used to describe different types of polymers. These terms will be used throughout our discussions of this subject, which will be quite detailed. The polymer industry stands out above all others as a consumer of heavy organic chemicals. The U.S. polymer industry produces over 60 billion lb of polymers as compared to the figure of 300 billion lb for the output of the U.S. organic chemical and polymer industry, which is about one fifth. However, the 300 billion lb figure includes values of all isolated chemicals, and of course some chemicals are used to make others and there is much double counting. Actually one estimate is that 60 billion lb of polymers consume about 130 billion lb of chemicals—really one half. It is also true that 50% of industrial chemists work with polymers. Thus we can see the importance of being acquainted with the polymers used in industry.

Polymers can be subdivided into a number of types. They may be specified as thermoplastic or thermoset, as linear or cross-linked depending on their structure. They may be step growth or chain growth, addition or condensation polymers depending on their mechanism of formation. They may be classed as block, graft, regular, random, and isotactic, syndiotactic, or atactic by their structures. Similarly, polymer processes may be free radical, cationic or anionic, metal complex or metal oxide catalyzed. The procedure or technique by which they are made may be bulk, solution, suspension, or emulsion polymerization. Finally, they may be classified by their end properties and uses as plastics, fibers, elastomers, coatings, or adhesives. In this and the next chapter we will try to clarify all these terms as we study polymers and give you numerous examples. Then we will be in the position to study their end-uses in detail by taking a separate look at plastics, fibers, elastomers, coatings, and adhesives.

To begin, *polymers* may be defined as substances that have repeating units and high molecular weight. *Polymerization* is the joining together of many small molecules to form very large ones with these repeating units. Perhaps the most important subdivision of polymerization is into chain growth or addition polymerization and step growth or condensation polymerization. The older designation of addition and condensation are not quite so accurate as chain and step growth.

Chain growth polymerization is characterized by the fact that the intermediates in the process—free radicals, ions, or metal complexes—are transient and cannot be isolated. Once a chain is initiated, monomer units add on to growing chains very quickly, and the molecular weight of that unit builds up in a fraction of a second. Consequently, the monomer concentration decreases steadily throughout the reaction. Prolonged reaction time has little effect on molecular weight but does provide higher yields. At any given time the reaction mixture contains unchanged reactant and "fully grown" polymer chains but a low concentration of growing chains. Chain polymerization often involves monomers containing a carbon-carbon double bond, although cyclic ethers such as ethylene and propylene oxides and aldehydes such as formaldehyde polymerize this way. There is no net loss of atoms in the polymer.

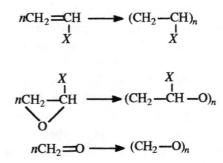

Step growth polymerization occurs because of reactions between molecules containing functional groups. This can be stopped at any time and low molecular weight products can be isolated (oligomers). The monomer does not decrease steadily in concentration; rather, it disappears early in the reaction because of the ready formation of oligomers. Long reaction times gradually build up the molecular weight. After the early stages of the reaction there is neither much reactant nor a great deal of "fully grown" polymer present. Instead, there is a wide distribution of slowly growing oligomers. Usually in step polymerization a small molecule such as water is lost as two monomers combine, but this is not always so. Common examples of step growth or condensation polymers are polyamides (nylons) and polyesters.

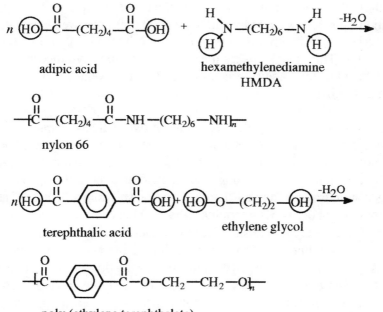

adipic acid

hexamethylenediamine
HMDA

nylon 66

terephthalic acid

ethylene glycol

poly (ethylene terephthalate)

The polymerization of caprolactam to nylon 6 is an example of a step polymerization that does not lose a molecule of water. Oligomers can be isolated at any time, which is clearly a step reaction. If we recall that it is actually the polymerization of 6-aminocaproic acid, then we can see that it is indeed a step polymerization.

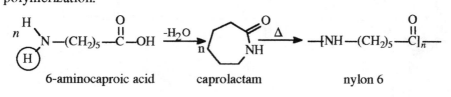

6-aminocaproic acid

caprolactam

nylon 6

First, let us treat in more detail the different types of chain or addition polymerizations and then later discuss as a unit the step or condensation polymerizations.

CHAIN GROWTH POLYMERIZATION

Free Radical Initiation. Many polymerizations are initiated by free radicals, especially alkoxy radicals formed by thermal decomposition of peroxides. A general mechanism for olefin free radical polymerization with initiation, propagation, and termination is given in Fig. 14.1.

After the initial reaction of a radical with the first monomer unit, a series of propagation steps follows, rapidly building up the molecular weight and degree of polymerization. The important part of this mechanism is therefore the (3), (3), etc. noted. This is what makes the polymer! With unsymmetrical monomers the "head-to-tail" addition is preferred because whatever it is in the R group that stabilized the radical once will do so each time a propagation step happens.

Chain termination can occur via coupling of two radicals. It may occur by disproportionation, that is, a hydrogen atom transfer from a carbon neighboring one radical site to another radical site, forming one saturated and one unsaturated end group. It may also be brought about by a chain transfer. This is simply a hydrogen atom transfer from an "internal" carbon site from a "finished" chain. If this happens, not only does it terminate the growing chain, but it also induces a branch in what was the "finished" chain. Reaction of this new radical will therefore occur nonlinearly. Branching can have a marked effect on polymer properties. It can also occur by hydrogen atom abstraction from a carbon atom in the same chain as the radical site, provided a stable, nonstrained six-membered ring transition state can be maintained. Low-density polyethylene is therefore characterized by C_4 branches. Finally, chain transfer is undesirable except when it is used intentionally to limit molecular weight by adding good chain transfer agents such as carbon tetrachloride. Here transfer of a chlorine

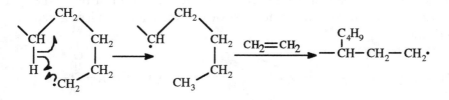

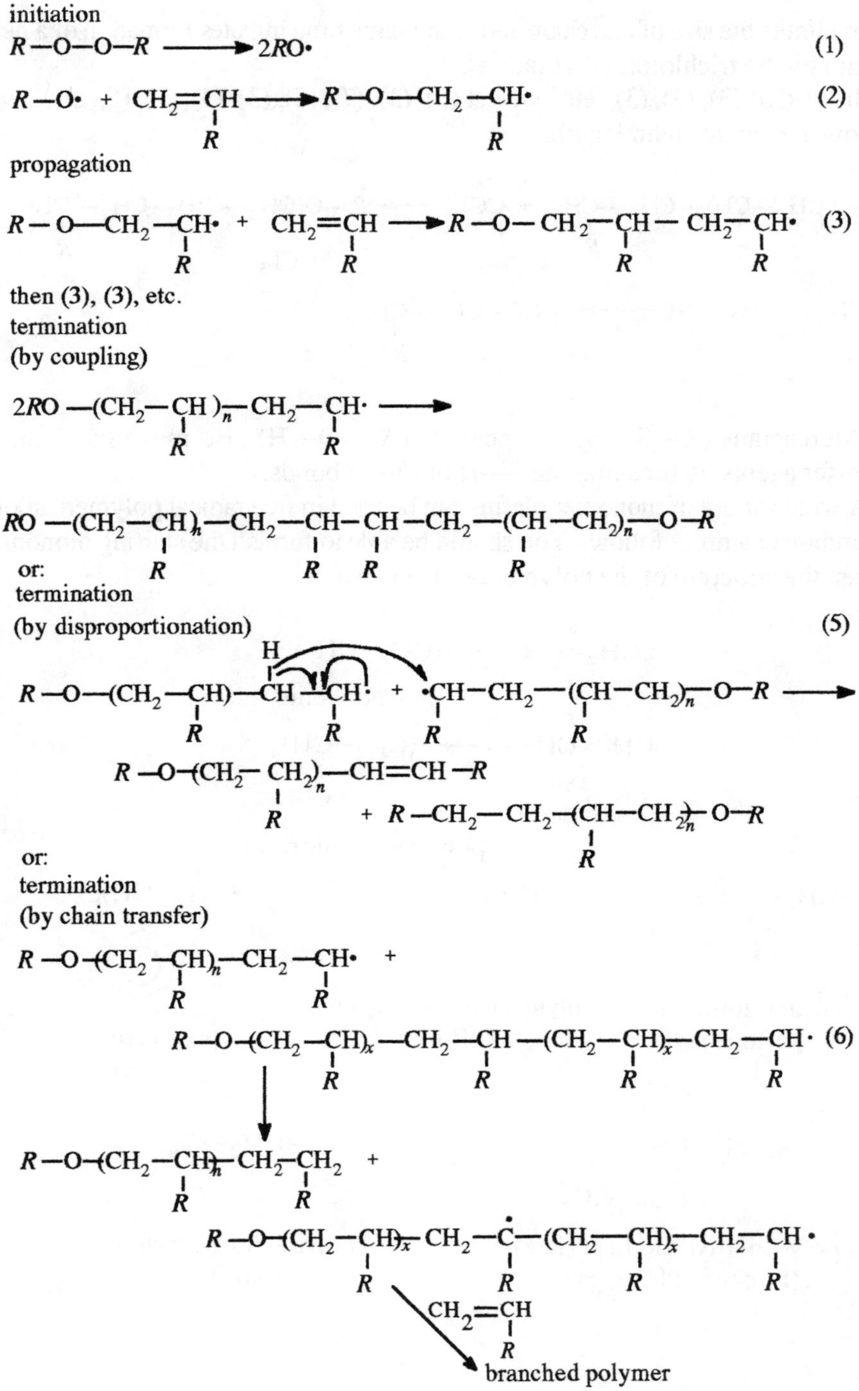

initiation

$$R-O-O-R \longrightarrow 2RO\cdot \qquad (1)$$

$$R-O\cdot \; + \; CH_2{=}\underset{R}{CH} \longrightarrow R-O-CH_2-\underset{R}{CH\cdot} \qquad (2)$$

propagation

$$R-O-CH_2-\underset{R}{CH\cdot} \; + \; CH_2{=}\underset{R}{CH} \longrightarrow R-O-CH_2-\underset{R}{CH}-CH_2-\underset{R}{CH\cdot} \qquad (3)$$

then (3), (3), etc.

termination
(by coupling)

$$2RO-(CH_2-\underset{R}{CH})_n-CH_2-\underset{R}{CH\cdot} \longrightarrow$$

$$RO-(CH_2-\underset{R}{CH})_n-CH_2-\underset{R}{CH}-\underset{R}{CH}-CH_2-(\underset{R}{CH}-CH_2)_n-O-R$$

or:

termination
(by disproportionation) $\qquad (5)$

$$R-O-(CH_2-\underset{R}{CH})-\underset{R}{CH}\overset{H}{-}\underset{R}{CH\cdot} \; + \; \underset{R}{\cdot CH}-CH_2-(\underset{R}{CH}-CH_2)_n-O-R \longrightarrow$$

$$R-O-(CH_2-CH_2)_n-\underset{R}{CH}-CH{=}CH-R$$

$$+ \; R-CH_2-CH_2-(\underset{R}{CH}-CH_2)_n-O-R$$

or:

termination
(by chain transfer)

$$R-O-(CH_2-\underset{R}{CH})_n-CH_2-\underset{R}{CH\cdot} \; +$$

$$R-O-(CH_2-\underset{R}{CH})_x-CH_2-\underset{R}{CH}-(CH_2-\underset{R}{CH})_x-CH_2-\underset{R}{CH\cdot} \qquad (6)$$

$$\downarrow$$

$$R-O-(CH_2-\underset{R}{CH})_n-CH_2-CH_2 \; +$$

$$R-O-(CH_2-\underset{R}{CH})_x-CH_2-\underset{R}{\overset{\cdot}{C}}-(CH_2-\underset{R}{CH})_x-CH_2-\underset{R}{CH\cdot}$$

$$\overset{CH_2{=}\underset{R}{CH}}{\searrow}$$

branched polymer

Figure 14.1 Mechanism of olefin free radical polymerization.

atom limits the size of one chain and at the same time initiates formation of a new chain by the trichloromethyl radical.

Instead of (3), (3), (3), etc., we get (3), (3), (7), (8), (3), (3), (7), (8), etc., with a lower average chain length.

$$R-O(CH_2-CH)_{\overline{n}}-CH_2-CH \cdot \ + \ CCl_4 \longrightarrow R-O(CH_2-CH)_{\overline{n}}-CH_2-CHCl \quad (7)$$

$$\underset{R}{\qquad} \underset{R}{\qquad} \qquad\qquad + \cdot CCl_3 \quad \underset{R}{\qquad} \underset{R}{\qquad}$$

$$\cdot CCl_3 \ + \ CH_2{=}CH \longrightarrow Cl_3C-CH_2-CH\cdot \qquad\qquad\qquad (8)$$

$$\underset{R}{\qquad} \qquad\qquad \underset{R}{\qquad}$$

Mercaptans (R—S—H) and phenols (Ar—O—H) also make good chain transfer agents by breaking the S—H or O—H bonds.

A wide variety of monomer olefins can be used in free radical polymerization. Common examples follow. You should be able to furnish the starting monomer given the structure of the polymer or vice versa.

$$nCH_2{=}CH_2 \longrightarrow (CH_2-CH_2)_{\overline{n}}$$

polyethylene

$$nCH_2{=}CH-Cl \longrightarrow (CH_2-CH)_{\overline{n}}$$
$$\qquad\qquad\qquad\qquad\qquad\quad | $$
$$\qquad\qquad\qquad\qquad\qquad Cl$$

poly (vinyl chloride)

$$(CH_2-CH)_{\overline{n}} \qquad\qquad (CF_2-CF_2)_{\overline{n}} \qquad\qquad (CH_2-CH)_{\overline{n}}$$
$$\quad | \qquad\qquad\qquad\qquad\qquad\qquad\qquad\qquad\qquad\qquad |$$
$$\quad C{\equiv}N$$

polyacrylonitrile   polytetrafluoroethylene   polystyrene
(Orlon®)   (Teflon®)

$$\qquad\qquad CH_3 \qquad\qquad\qquad\qquad\qquad\qquad\qquad\qquad Cl$$
$$\qquad\qquad | \qquad\qquad\qquad\qquad\qquad\qquad\qquad\qquad\quad |$$
$$(CH_2-C)_{\overline{n}} \qquad\qquad\qquad\qquad\qquad (CH_2-C)_{\overline{n}}$$
$$\qquad\qquad | \qquad\qquad\qquad\qquad\qquad\qquad\qquad\qquad\quad |$$
$$\qquad\quad CO_2-CH_3 \qquad\qquad\qquad\qquad\qquad\qquad Cl$$

poly (methyl methacrylate)   polydichloroethylene
(Lucite®, Plexiglas®)   (Saran®)

Free Radical Polymerization of Dienes

Conjugated dienes such as 1,3-butadiene very readily polymerize free radically. The important thing to remember here is that there are double bonds still present in the polymer. This is especially important in the case of elastomers (synthetic rubbers) because some cross-linking with disulfide bridges (vulcanization) can occur in the finished polymer at the allylic sites still present to provide elastic properties to the overall polymers. Vulcanization will be discussed in detail later. The mechanism shown here for simplicity demonstrates only the 1,4-addition of butadiene. 1,2-addition also occurs, and the double bonds may be *cis* or *trans* in their stereochemistry. Only with the metal complex catalysts will the stereochemistry be regular.

Reaction: $nCH_2=CH-CH=CH_2 \longrightarrow +CH_2-CH=CH-CH_2)_{\overline{n}}$

1,3-butadiene   polybutadiene

Mechanism:

initiation $ROOR \longrightarrow 2RO^{\cdot}$ (1)

$RO^{\cdot} + CH_2=CH-CH=CH_2 \longrightarrow RO-CH_2-CH=CH-CH_2\cdot$

                                          (2)

$RO-CH_2-CH=CH-CH_2\cdot + CH_2=CH-CH=CH_2 \longrightarrow$

$RO-CH_2-CH=CH-CH_2-CH_2-CH=CH-CH_2\cdot$ (3)

then (3), (3), (3), etc.

Vulcanization: polybutadiene $\xrightarrow[\text{heat}]{S}$ $+CH_2-CH=CH-CH)_{\overline{n}}$

                                                 S

                                               S

                      $+CH_2-CH=CH-CH)_{\overline{n}}$

Other examples:

        Cl CH_3

        | |

$+CH_2-C=CH-CH_2)_{\overline{n}}$ $+CH_2-C=CH-CH_2)_{\overline{n}}$

   polychloroprene   polyisoprene

  (Neoprene®, Duprene®)

Ionic Initiation

Although free radical initiation is by far the most conmon type of catalysis, accounting for about half all polymerizations, other types of initiation are commonly employed, since some monomers cannot be polymerized well free radically. For instance, propylene cannot be free radically polymerized to a high molecular weight because of its reactive allylic hydrogens which easily undergo chain transfer. As a general rule olefins containing an electron-withdrawing group can, in addition to free radical polymerization, use anionic initiation. Examples of anionic initiators commonly employed are *n*-butyllithium, sodium amide, and sodium or potassium metal in liquid ammonia or naphthalene. The mechanism for polymerization of acrylonitrile using *n*-butyllithium is given. We can see that the electron-witlhdrawing cyano group by its inductive effect is able to stabilize the intermediate negative charges on the neighboring carbon for each propagation step and aid the polymerization process. Although this mechanism is an oversimplification, it does give the basic idea. Chain termination is more complicated than in free radical polymerization. Coupling and disproportionation are not possible since two negative ions cannot easily come together. Termination may result from a proton transfer from a solvent or weak acid, such as water, sometimes present in just trace amounts. Actually it is well-known that ionic polymerization need not terminate.

$$Bu^-Li^+ \; + \; CH_2{=}CH{-}C{\equiv}N \longrightarrow Bu{-}CH_2{-}\overset{|}{\underset{C{\equiv}N}{\overline{C}H}}Li^+ \tag{1}$$

$$Bu{-}CH_2{-}\overset{|}{\underset{C{\equiv}N}{\overline{C}H}}Li^+ \; + \; CH_2{=}CH{-}C{\equiv}N \longrightarrow Bu{-}CH_2{-}\overset{|}{\underset{C{\equiv}N}{CH}}{-}CH_2{-}\overset{|}{\underset{C{\equiv}N}{\overline{C}H}}Li^+ \tag{2}$$

then (2), (2), etc.

$${-}(CH_2{-}\overset{|}{\underset{X}{CH}})_{\overline{n}}CH_2{-}\overset{|}{\underset{X}{\overline{C}H}}Li^+ \; + \; H_2O \longrightarrow$$

$${-}(CH_2{-}\overset{|}{\underset{X}{CH}})_{\overline{n}}CH_2{-}\overset{|}{\underset{X}{CH_2}} \quad + \; Li^+ \; + \; OH^-$$

They have been termed "living" polymers. If further monomer is added, weeks or months later there will be a further molecular weight increase as the polymer chains grow longer. As long as the counterion is present (lithium in the preceding case), the anionic end group is perfectly stable.

Ionic polymerization may also occur with cationic initiations such as protonic acids like HF and H_2SO_4 or Lewis acids like BF_3, $AlCl_3$ and $SnCl_4$. The polymerization of isobutylene is a common example. Note that the two inductively donating methyl groups stabilize the carbocation intermediate. Chain termination, if it does occur, usually proceeds by loss of a proton to form a terminal double bond. This regenerates the catalyst.

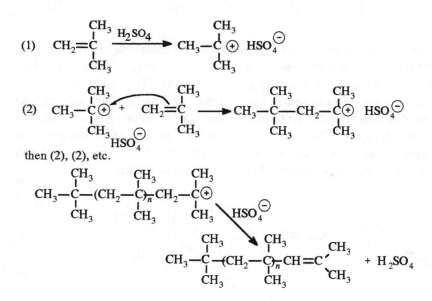

then (2), (2), etc.

Metal Complex Initiation (Ziegler-Natta Catalysis)

In the early 1950s Karl Ziegler in Germany and Giulio Natta in Italy found catalysts that polymerized olefins and dienes with stereoregularity and with mild polymerization conditions. For this revolutionary discovery they both won the Nobel prize. Let us take the example of propylene, which we have already said is not easily polymerized free radically. Not only was high molecular weight polypropylene obtained, but it was isotactic, with the methyls arranged stereoregularly. This is to be contrasted to atactic (random) or syndiotactic (alternating) structures.

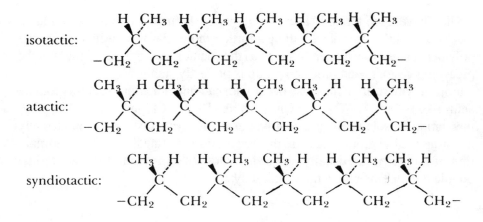

Ziegler-Natta catalysts are primarily complexes of a transition metal halide and an organometallic compound whose structure is not completely understood for all cases. Let us use as an example $TiCl_4$ and R_3Al. The mechanism of the polymerization catalysis is somewhat understood. The titanium salt and the organometallic compound react to give a penta-coordinated titanium complex with a sixth empty site of the octahedral configuration. The monomer alkene is then complexed with the titanium and finally inserts between the titanium and alkyl group, leaving a new empty site for repetition of the process.

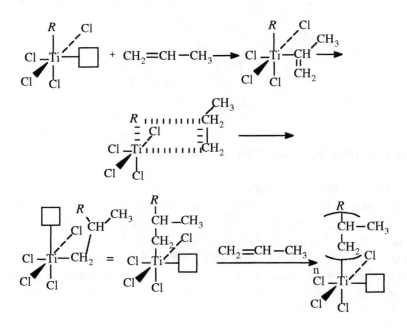

The versatility of Ziegler-Natta catalysis is shown in the polmerization of butadiene. Polybutadiene may have either a 1, 2 or 1,4-configuration. The 1,4 polymer has a double bond as part of the main chain and this can be atactic, isotactic, or syndiotactic. Thus five different polybutadienes can be made and all of them have been made with the aid of Ziegler-Natta catalysts.

Metal Oxide Initiation

Researchers for Standard Oil of Indiana have developed a molybdenum oxide catalyst and for Phillips Petroleum a chromic oxide catalyst for the polymerization of polyethylene with very few branches due to cyclic hydrogen atom transfer (see earlier discussion). This is a much stiffer polymer and has properties substantially different from polyethylene with branches. Completely linear polyethylene formed from this type of catalysis is called high-density polyethylene. More branched polyethylene has a much lower density because the chains cannot come as close together or be packed as tightly. An advantage over Ziegler-Natta is that this catalyst is not flammable. The main propagation step in this polymerization is a chromium-ethylene complex formation, followed by insertion of the two CH_2 units into the existing chromium-carbon bond. Chromium d orbitals are involved in the process. It is also known that chromium is attached to a silica surface through Cr-O-Si bonds.

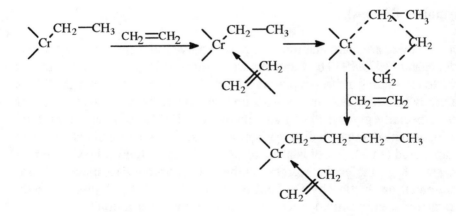

Chain termination and initiation with methyl formation can be pictured as shown below.

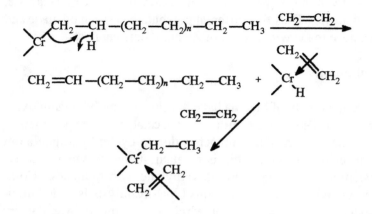

STEP GROWTH POLMERIZATION

At the beginning of this chapter we described step growth polymerization chiefly to contrast it with chain polymerization. We now consider this subject in more detail and discuss various types of step growth polymers.

Polyamides (Nylons)

Although there are many naturally occurring polyamides (proteins), synthetic work began in 1929-1930 by Carothers who worked at Du Pont. They were first interested in finding a cheap replacement for silk in women's stockings. Silk is a naturally occurring polyamide made up of a mixture of amino acid monomer units, especially glycine (44%) and alanine (40%). Wool is also a protein, keratin, which contains 18 different amino acids, the highest percentage being glutamic acid (14%). Wool is also cross-linked with sulfur bridges. The first successful high molecular weight synthetic polyamide was made in 1935. Commercial production by Du Pont began in 1940. This polymer was poly (hexamethyleneadipamide), now commonly referred to as nylon 66.

This is an example of a common way of making nylons: reaction of a dicarboxylic acid and a diamine. The first number of the nylon nomenclature refers to the number of carbon atoms in the amine, the second to the number of carbons in the acid. Nylon 66 was soon found to have higher strength than any

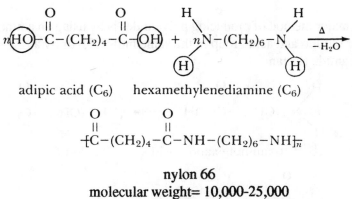

adipic acid (C_6)   hexamethylenediamine (C_6)

$$\begin{array}{ccc} & O & & O \\ & \parallel & & \parallel \\ \{C-(CH_2)_4-\overset{}{C}-NH-(CH_2)_6-NH\}_n \end{array}$$

nylon 66
molecular weight= 10,000-25,000
n= 40-110

natural fiber. It has good chemical stability and a high melting point (265°C) due to hydrogen bonding of the carbonyls of one chain with the N—H groups of another.

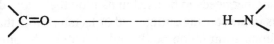

The tensile strength (how much pressure can be applied on the ends of a fiber before it breaks) of polymers is very dependent on the molecular weight and, although nylon 66 was made ten years earlier, the technical production problem of obtaining good molecular weight had to be overcome before it was used as a substitute for silk.

$$\begin{array}{ccc} O & & O \\ \parallel & & \parallel \\ nHO-C-(CH_2)_8-C-OH & + & nNH_2-(CH_2)_6-NH_2 \end{array} \xrightarrow[-H_2O]{\Delta}$$

sebacic acid

$$\begin{array}{ccc} O & & O \\ \parallel & & \parallel \\ \{C-(CH_2)_8-C-NH-(CH_2)_6-NH\}_n \end{array}$$

nylon 610
(used as bristles in brushes)

Another major method of producing polyamides is by using an amino acid as a monomer (amino and acid group in the same molecule) or by a ring opening of a cyclic amide (lactam).

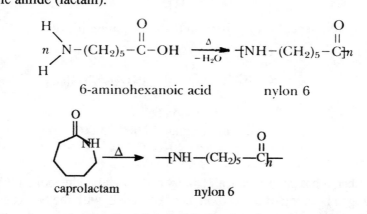

6-aminohexanoic acid   nylon 6

or

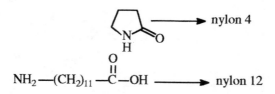

caprolactam   nylon 6

Here only one number needs to be used in naming the nylon, designating the number of carbons in the starting amino acid or lactam. Note that the structure of nylon 6 is different from nylon 66. In nylon 6 all of the amine groups are "facing the same way." However, both polymers have relatively similar physical properties. Nylon 6 is not quite so strong or as high melting (mp 215°C) as nylon 66 . But it is cheaper. It has found use in tire cords, carpet and brush fibers, and various molded articles.

Other examples:

$$NH_2—(CH_2)_{11}—\overset{\overset{\displaystyle O}{\|}}{C}—OH \longrightarrow \text{nylon 12}$$

It is interesting to note that the strength and melting point of polyamides is decreased as the number of carbons in the monomer is increased. This has to do with the number of amide linkages (and hydrogen bonds) per unit of weight in the polymer. The fewer the hydrogen bonds there are between chains, the freer the molecules are to move.

Polyesters

Polyesters are made in one of two ways: by either direct reaction of a diacid and a diol or ester interchange of a diester and a diol. By far the most commerically useful polyester is poly (ethylene terephthalate). Both methods are illustrated here.

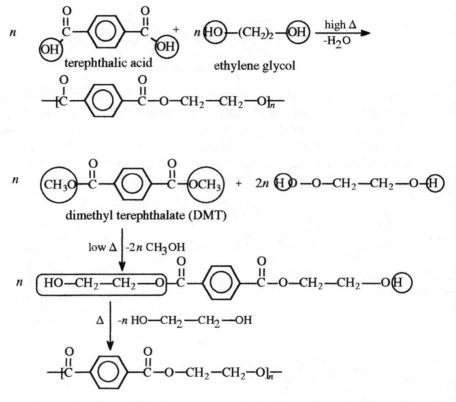

Poly (ethylene terephthalate) is known commonly by the trademarks Dacron®, Terylene®, and Fortrel® fibers and Mylar® film. The polymer melts at 270°C and has very high strength and elasticity. It is three times as strong as cellulose. It is also particularly resistant to hydrolysis (washing!) and resists creasing. Hence it has been used in clothing, especially shirts (65/35 polyester/cotton most popular). Its excellent clarity has made it useful in photographic film and overhead transparencies.

Most useful polyesters have need for the strong, rigid aromatic ring in their structure since they lack the hydrogen bonding prevalent in polyamides.

Other examples:

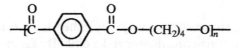

poly (tetramethylene terephthalate)

C_4 chain weaker but more flexible than C_2 link

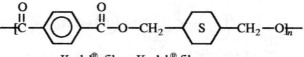

Kodel® fiber, Kodak® film

Although we will not be discussing the mechanism of each type of step growth polymer because these reactions are very similar to the corresponding monomer chemistry, we should be aware of this analogy. For instance, an acid reacts with an alcohol under acid-catalyzed conditions by a certain well-studied and proven mechanism. This same mechanism is followed each time an ester linkage of a polyester is formed. One such transformation is outlined here.

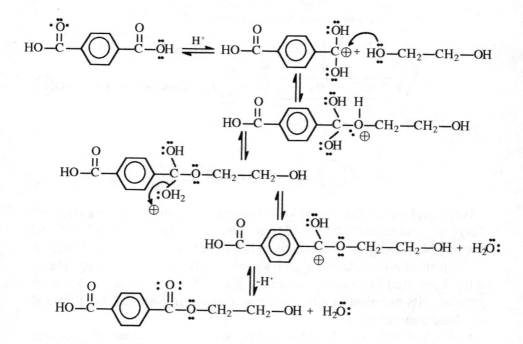

The equilibrium is shifted in the direction of the product by distillation of the water from the reaction mixture (and condensing it in a separate container—hence the name condensation polymers for this type).

Polycarbonates

The chemistry of polycarbonates is similar to the chemical behavior of polyesters. We can think of a carbonate as being a diester of carbonic acid, H_2CO_3, which is unstable itself. Polycarbonate is a strong, clear plastic used in automobiles, sheeting, electronics, appliances, sports equipment, and compact disks.

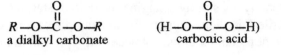

$$R-O-\overset{\overset{\textstyle O}{\|}}{C}-O-R$$
a dialkyl carbonate

$$(H-O-\overset{\overset{\textstyle O}{\|}}{C}-O-H)$$
carbonic acid

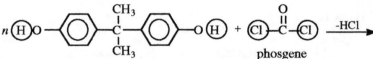

2,2-bis (4-hydroxyphenyl)-
propane ("bisphenol A")

phosgene

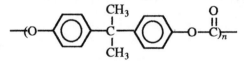

Lexan® polycarbonate
(good electrical resistance)

COPOLYMERIZATION

Copolymers are polymers made from two or more monomers. In *regular (or alternating) polymers* the monomer units alternate. Many step growth polymers are regular. An example of a regular chain growth copolymer is one based on maleic anhydride and styrene. The reaction rate between these two monomers is greater than the reaction of either of them with themselves. This is what causes the regularity.

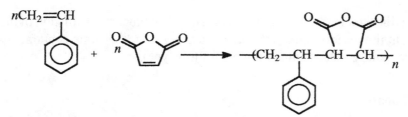

To form a *random polymer* the two monomers must react with themselves at a rate comparable to that at which they react with each other. In random polymers they need not be present in equal amounts either. The most important synthetic elastomer, styrene-butadiene rubber (SBR), is a copolymer of approximately 6 mol of butadiene to 1 mol of styrene. The properties of the final polymer are changed considerably by simply changing the ratio of starting monomers. ABS resin (acrylonitrile-butadiene-styrene) is an example of a random copolymer with three different monomer units, not necessarily present in the same amount.

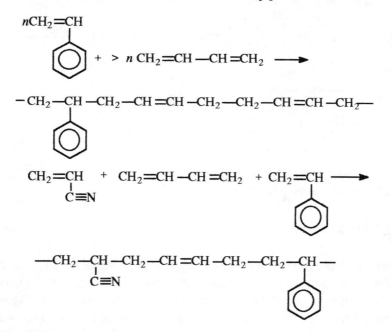

Another type of copolymer is a *block copolymer*. Here a low molecular weight polymer may be extended by reaction with a new monomer. Recall that we talked about "living" polymers previously. If, for example, we polymerized styrene alone first, then added some butadiene and polymerized it further, we would have a number of styrene units bundled together and a number of butadienes also together.

—S—S—S—S—S—B—B—B—S—S—S—S—S—B—B—

This *block copolymer* has substantially different physical properties as compared to a random styrene-butadiene copolymer.

Lastly, there are *graft copolymers* that result when a polymer chain of one monomer is grafted on to an existing polymer backbone by creation of a free radical site along the backbone which initiates growth of a polymer chain. The concept is similar to the grafting of plants in botany. To form a styrene-butadiene graft polymer, butadiene which is already polymerized is dissolved in monomeric styrene and an initiator is added. Because polybutadiene readily undergoes chain transfer at the allylic sites, polystyrene chains grow on the polybutadiene backbone. This forms high impact polystyrene, a low cost plastic that is otherwise too brittle without the grafting.

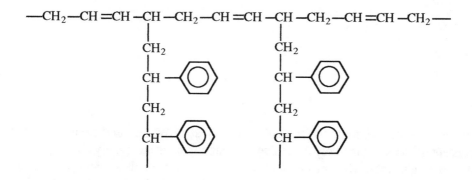

POLYMERIZATION PROCEDURES

Polymers may be made by four different experimental techniques: bulk, solution, suspension, and emulsion processes. We will not discuss these in detail because they are somewhat self-explanatory. In *bulk polymerization* only the monomers and a small amount of catalyst is present. No separation processes are necessary and the only impurity in the final product is monomer. But heat transfer is a problem as the polymer becomes viscous. In *solution polymerization* the solvent dissipates the heat better, but it must be removed later and care must be used in choosing the proper solvent so it does not act as a chain transfer agent. In *suspension polymerization* the monomer and catalyst are suspended as droplets in a continuous phase such as water by continuous agitation. Finally, *emulsion polymerization* uses an emulsifying agent such as soap, which forms micelles where the polymerization takes place.

15

Basic Polymer Chemistry: Thermoset Polymers

In the previous chapter we talked about linear polymers and mentioned the concept of cross-linking only in passing. Linear polymers are usually *thermoplastic*: they soften or melt when heated and will dissolve in suitable solvents. They can be remelted and shaped into their finished product with no further chemical reactions. *Thermoset resins*, those having elaborately cross-linked three-dimensional structures set or harden by undergoing a chemical reaction during the manufacture of finished products. They decompose on heating and are infusible and insoluble. Their chemistry and physical properties are quite different from thermoplastic polymers. The important ones are now discussed briefly.

PHENOL-FORMALDEHYDE POLYMERS (PHENOLIC RESINS)

These copolymers of phenol and formaldehyde were the first fully synthetic polymers made. They were discovered in 1910 by Leo Baekeland and given the tradename Bakelite®. They may be prepared in two ways, both involving step growth polymerization. A "one-stage" resin may be obtained using an alkaline catalyst and excess formaldehyde to form linear, low molecular weight resol resins. Slight acidification and further heating causes the curing process to give a highly cross-linked thermoset. This complex mechanism is summarized here.

The *o*- and *p*-methylolphenols are more reactive toward formaldehyde than the original phenol and rapidly undergo further reaction to give di- and trimethylol derivatives.

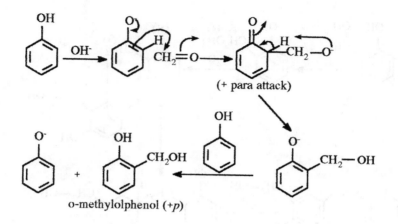

(+ para attack)

o-methylolphenol (+*p*)

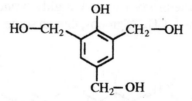

trimethylolphenol

The methylolphenols will react to form di- and trinuclear phenols at still free ortho and para positions.

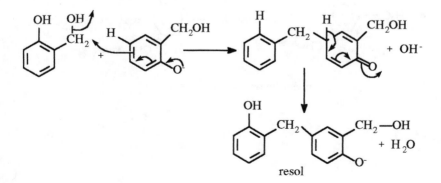

resol

Although these resols can be cross-linked under basic conditions, acidification and further heating is preferred. The mechanism of polymerization under acidic conditions involves carbocation chemistry.

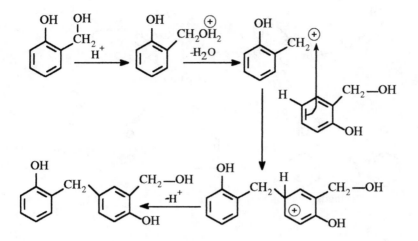

The final structure of the product is very highly branched. Most linkages between aromatic rings are —CH_2— groups, though some —CH_2—O—CH_2— linkages are present.

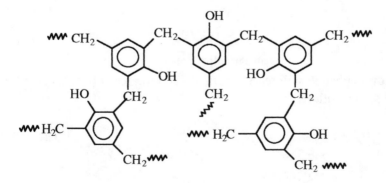

a phenolic or Bakelite® resin

The second method uses an acid catalyst and excess phenol to give a linear polymer that may be stored or sold. These are called novolacs and have no free methylol groups for cross-linking.

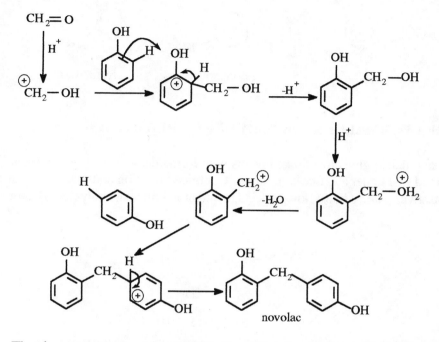

novolac

Thus in a separate second part of this "two-stage" process a cross-linking agent is added and further reaction occurs. Although formaldehyde may be added, quite often hexamethylenetetramine is used, which decomposes to formaldehyde and ammonia. Occasional nitrogen bridges occur in the final structure of some phenolics made by this method.

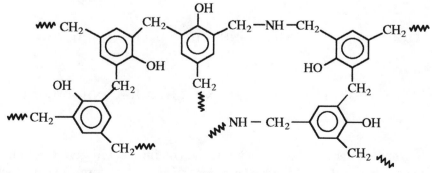

a phenolic

Other modifications in making phenolics are the incorporation of cresols or resorcinol as the phenol and acetaldehyde or furfural as the phenol.

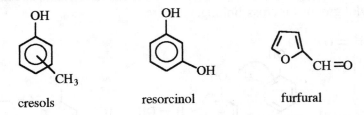

| cresols | resorcinol | furfural |

UREA-FORMALDEHYDE POLYMERS (UREA RESINS)

Urea will also give cross-linked resins with formaldehyde. Methylolureas are formed first under alkaline conditions. Continued reaction under acidic conditions gives a fairly linear, low molecular weight intermediate polymer. Heating

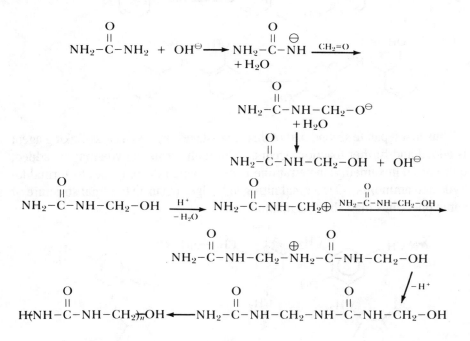

for an extended period of time under acidic conditions will give a complex thermoset polymer of poorly defined structure including ring formation of which the following may be typical.

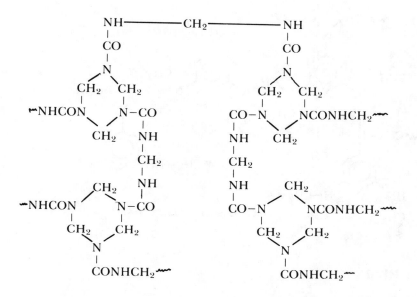

Source: W& RI. Reprinted by permission of John Wiley & Sons, Inc.

MELAMINE-FORMALDEHYDE POLYMERS (MELAMINE RESINS)

Melamine, having three amino groups and six labile hydrogens, will also form thermoset resins with formaldehyde. The chemistry is similar to that for the urea resins.

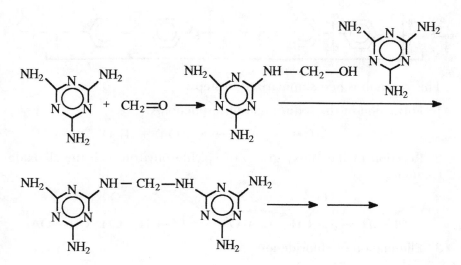

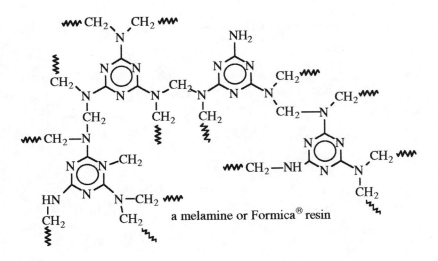

a melamine or Formica® resin

EPOXY RESIN

This type of thermoset polymer is typically made first by reaction of the sodium salt of bisphenol A and excess epichlorohydrin which forms a low molecular weight polymer with terminal epoxy groups; n is between 1 and 4.

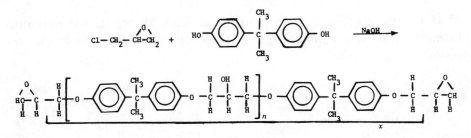

This reaction is best summarized in steps:

1. Formation of the sodium salt of bisphenol A;

$$ArOH + OH^- \longrightarrow ArO^- + H_2O$$

2. Reaction of the epoxy group of epichlorohydrin with the alkoxide anion:

$$Cl-CH_2-CH-CH_2 + ArO^- \longrightarrow Cl-CH_2-CH-CH_2-OAr$$

3. Elimination of chloride ion:

$$Cl-CH_2-CH-CH_2-OAr \longrightarrow Cl^- + CH_2-CH-CH_2-OAr$$

4. Reaction of the new epoxy group with the alkoxide ion:

$$ArO^- + \overset{\displaystyle O}{\overset{\diagup\ \diagdown}{CH_2-CH}}-CH_2-OAr \longrightarrow ArO-CH_2-\overset{\displaystyle O^-}{\overset{|}{CH}}-CH_2-OAr$$

5. Formation of a hydroxy group by protonation:

$$ArO-CH_2-\overset{\displaystyle O^-}{\overset{|}{CH}}-CH_2-OAr \xrightarrow{\text{H}_2\text{O}} ArO-CH_2-\overset{\displaystyle OH}{\overset{|}{CH}}-CH_2-OAr + OH^-$$

You may wish to go through these five steps in detail using the full structure of bisphenol A.

Cross-linking of these low molecular weight epoxy compounds is promoted by adding a curing agent such as ethylenediamine. Primary amines react with epoxides to form tertiary amines and branches. Thus a cross-linked polymeric structure is the final result.

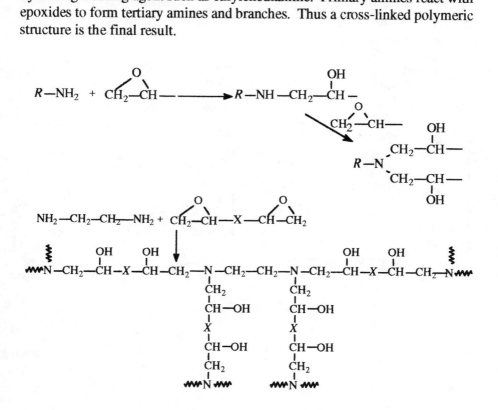

POLYURETHANE FOAMS

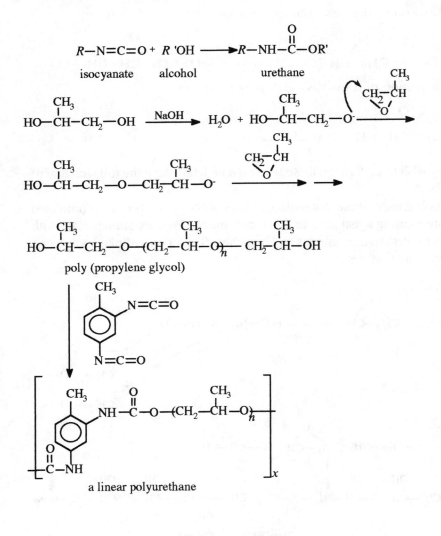

$$R-N{=}C{=}O + R\,'OH \longrightarrow R-NH-\overset{\overset{\textstyle O}{\|}}{C}-OR'$$

isocyanate   alcohol   urethane

poly (propylene glycol)

a linear polyurethane

Most useful polyurethanes are cross-linked. Those commonly used in foams start with a diisocyanate like toluene diisocyanate (TDI) and a low molecular weight polyether such as poly (propylene glycol). Recall that the basic reaction of an isocyanate plus an alcohol gives the urethane functionality.

One way of obtaining the more useful cross-linked polyurethanes is by using a trifunctional reagent. Thus either the TDI can react with a triol or the propylene oxide can be polymerized in the presence of a triol. Then the isocyanate-alcohol reaction would of course give a cross-linked urethane.

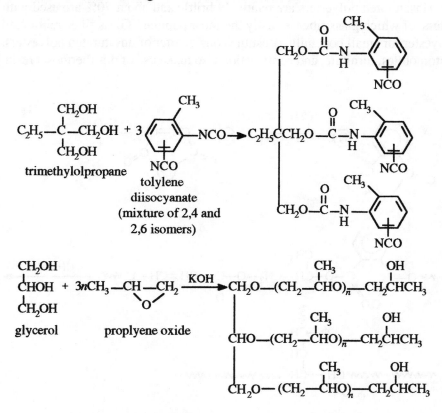

Source: W & RI. Reprinted by permission of John Wiley & Sons, Inc.

In the urethane process a small amount of water is added to convert some isocyanate functionalities into CO_2 gas and amines. The degree of foaming can be controlled by the amount of water added.

$$R-N{=}C{=}O + H_2O \longrightarrow R-NH_2 + CO_2 \uparrow$$

UNSATURATED POLYESTERS

An unsaturated polyester resin consists of a linear polyester whose chain contains double bonds and an unsaturated monomer such as styrene that copolymerizes with the polyester to provide a cross-linked product. The most common unsaturated polyester is made by step growth polymerization of propylene glycol with phthalic and maleic anhydrides. Subsequent treatment with styrene and a peroxide catalyst leads to a solid, infusible thermoset.

Unsaturated polyesters are relatively brittle and about 70% are used with fillers, of which glass fiber is easily the most popular. Glass fiber-reinforced polyester for small boat hulls consumes one quarter of unsaturated polyesters. Automobiles, furniture, and construction also make use of this thermoset resin.

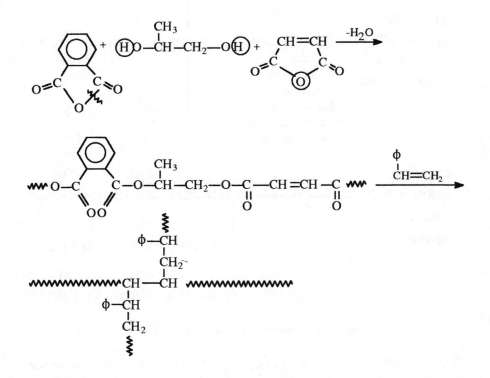

ALKYD RESINS

This is a very broad class of compounds commonly used in coatings. Over 400-500 different alkyd resins are commercially available. They also are polyesters containing unsaturation that can be cross-linked in the presence of an initiator known traditionally as a "drier". A common example is the alkyd formed from phthalic anhydride and a glyceride of linolenic acid obtained from various plants. Cross-linking of the multiple bonds in the long unsaturated chain produces the thermoset polymer.

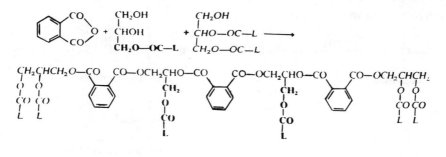

$$L{=}CH_3CH_2CH{=}CH-CH_2CH{=}CHCH_2CH{=}CH(CH_2)_7-$$

Source: W & R II. Reprinted by permission of John Wiley & Sons, Inc.

NATURAL POLYMERS

Mention has already been made of two polymers that can be obtained naturally from living animals: silk (from the silkworm) and wool (from sheep). They are proteins made of various amino acids; both are used in textiles. Other biologically derived polymers are also familiar such as wood, starch, and some sugars. We will not cover these in detail here. However, certain cellulosics we will discuss briefly since they are compared to synthetic fibers later. Cellulose is the primary substance of which the walls of vegetable cells are constructed and is largely composed of glucose residues. It may be obtained from wood or derived in very high purity from cotton fibers, which are about 92% pure cellulose.

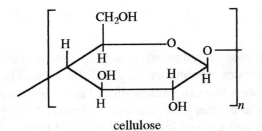

cellulose

The important fiber rayon is simply "regenerated" cellulose from wood pulp that is in a form more easily spun into fibers. Cellophane film is regenerated cellulose made into film. One method of regeneration is formation of xanthate groups from selected hydroxyl groups of cellulose, followed by hydrolysis back to hydroxy groups.

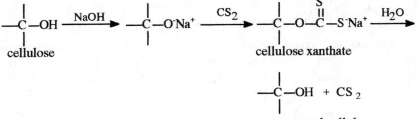

$$—C—OH + CS_2$$

regenerated cellulose

Cellulose acetate and triacetate may be used as plastics or spun into fibers for textiles. They are made by the reaction of cellulose with acetic anhydride.

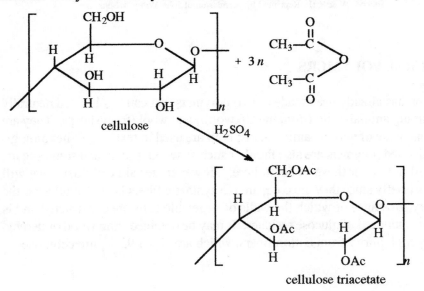

cellulose triacetate

Finally, one last type of natural polymer that will be discussed is natural rubber, obtained from the rubber tree and having the all*cis*-1,4-polyisoprene structure. This structure has been duplicated in the laboratory and is called "synthetic rubber," made with the use of Ziegler-Natta catalysis.

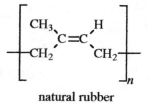

natural rubber

The biosynthesis of synthetic natural rubber has been completely worked out and appears in Fig. 15.1. Many plants and animals use this same biosynthetic pathway to make hundreds of terpenes and steroids from their common isoprenoid building blocks.

POLYMER PROPERTIES

Molecular Weight

The detailed treatment of the molecular weight analysis of polymers is left to other texts. We should be aware that there are two types of molecular weights, *number average* and *weight average*.

$$M_n = \frac{\Sigma N_i M_i}{\Sigma N}$$

M_n = number average molecular weight
N_i = number of molecules with a molecular weight of M_i
N = total number of molecules

$$M_w = \frac{\Sigma \omega_i M_i}{\Sigma \omega_i} = \frac{\Sigma N_i M_i^2}{\Sigma N_i M_i}$$

M_w = weight average molecular weight
ω_i = weight of molecules with a molecular weight of M_i

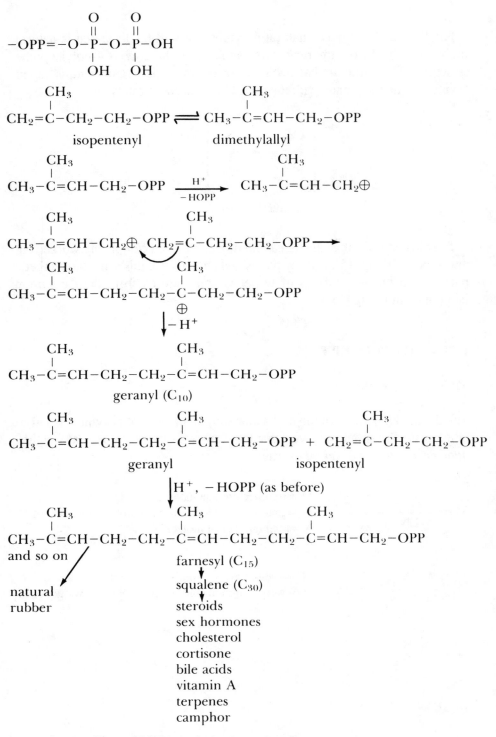

Figure 15.1 Biosynthesis of natural rubber.

Because the weight average is biased toward molecules with higher molecular weight, it is larger than the number average. Boiling point elevation, freezing point depression, osmotic pressure, and end group analysis give the number average molecular weight. Light scattering and sedimentation give the weight average molecular weight. Viscosity measurements give a value somewhere between the two. Molecular weight and mechanical strength are related since strength increases rapidly as the degree of polymerization (or the number of repeating units, *n*) increases from 50 to 500. Further increases in molecular weight have a smaller effect.

Crystallinity

This is a key factor in governing polymer properties. If the polymer molecules can align themselves with a high lateral order and the chains lie side by side, we say that the polymer is highly crystalline. Bulky groups or branching prevent the polymer from being highly crystalline. Another thing affecting crystallinity is the magnitude of attractive forces between neighboring polymer molecules. Strong intermolecular forces, such as hydrogen bonding in the nylons, promote greater crystallinity. Isotactic polymers are always more crystalline than atactic polymers because of the regularity of any large groups. Ziegler-Natta catalysts promote isotactic, crystalline polymerizations.

Crystalline polymers tend to have greater mechanical strength, higher melting points, and higher densities than amorphous polymers. On the other hand, they are usually much less transparent (more opaque) because light is reflected or scattered at the boundary between the crystalline and amorphous portions of the polymeric structure. Amorphous polymers are transparent and glasslike.

Examples of crystalline polymers are nylons, cellulose, linear polyesters, and high-density polyethylene. Amorphous polymers are exemplified by poly (methyl methacrylate), polycarbonates, and low-density polyethylene. The student should think about why these structures promote more or less crystallinity in these examples.

The crystallinity of a specific polymer may be altered by orientation or stretching the polymer mechanically in a certain direction. On stretching, the molecules align themselves and become more crystalline. Did you ever notice that a rubber band becomes more opaque on stretching? However, if a crystalline polymer is biaxially oriented, as with a drawn nylon sheet, then the whole sheet is in effect a single crystal and is very transparent! The stretching or drawing of fibers causes greater crystallinity and gives the longitudinal strength required in fibers.

Temperature Dependency of Polymers

Polymers usually do not have a single, sharp melting point like a pure chemical might. Then too each polymer is a little different in its reaction to temperature changes, and the same polymer but with different molecular weights will have different observable changes when heated or cooled. The glass transition temperature, T_g, and the crystalline melting point, T_m, are most often used to describe the rather nebulous changes of a polymer with temperature, shown here with a graph of volume versus temperature.

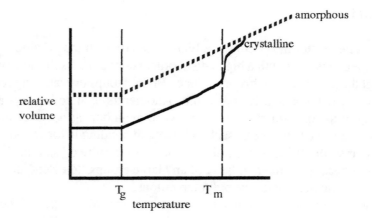

An amorphous material such as polystyrene does not solidify sharply. It goes from a viscous liquid to a rubbery solid, then to a leathery solid. Finally, it becomes a glassy solid. This last change is a sharper one and the temperature at which it occurs is called the *glass transition temperature*, T_g. Or upon heating a polymer, it is the temperature at which the polymer loses its hardness or brittleness and becomes more flexible and takes on rubbery or leathery properties. At this transition temperature noticeable changes in the specific volume, thermal conductivity, refractive index, stiffness, heat content, and dielectric loss are apparent.

More crystalline polymers have a glass transition temperature because all polymers have amorphous regions between the crystalline regions. Crystalline polymers also have a *crystalline melting point*, T_m. It is the temperature at which a molten polymer changes from a viscous liquid to a microcrystalline solid. It is accompanied by more sudden changes in density, refractive index, heat capacity, transparency, and similar properties, but it is still not so sharp as a nonpolymeric melting point. Usually T_g is about one half to two thirds of T_m for most polymers if expressed in degrees Kelvin.

Table 15.1 gives the appropriate T_g's for a few selected polymers. Note that the T_g values are low for elastomers and flexible polymers (low-density poly-

TABLE 15.1 Approximate Glass Transition Temperatures (Tg) for Selected Polymers

Polymer	Tg (°K)
Cellulose acetate butyrate	323
Cellulose triacetate	430
Polyethylene (LDPE)	148
Polypropylene (atactic)	253
Polypropylene (isotactic)	373
Polytetrafluoroethylene	160, 400[a]
Polyethyl acrylate	249
Polymethyl acrylate	279
Polybutyl methacrylate (atactic)	339
Polymethyl methacrylate (atactic)	378
Polyacrylonitrile	378
Polyvinyl acetate	301
Polyvinyl alcohol	358
Polyvinyl chloride	354
cis-Poly-1, 3-butadiene	165
trans-Poly-1, 3-butadiene	255
Polyhexamethylene adipamide (nylon 66)	330
Polyethylene adipate	223
Polyethylene terephthalate	342
Polydimethyl siloxane (silicone)	150
Polystyrene	373

[a]Two major transitions observed.
Source: S & C, p. 29. Reprinted by courtesy of Marcel Dekker, Inc.

ethylene, *cis*-poly-1,3-butadiene, silicone) and relatively high for hard amorphous plastics [polypropylene-isotactic, polyacrylonitrile, poly (vinyl alcohol), nylon 66, poly (ethylene terephthalate), polystyrene]. Notice also that T_g varies with even slight changes in structure. For instance, T_g decreases as the size of the ester groups increase in polyacrylates and polymethacrylates. T_g increases when aromatics are added [poly (ethylene adipate) versus poly (ethylene terephthalate)].

Tensile Properties

Many of the quoted physical properties of a polymer are derived from a stress-strain experiment. The polymer is cut into an appropriate shape. For instance, plastics are cut into the shape shown here (sometimes called a dogbone.) They

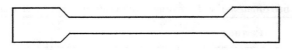

are placed in two jaws of a special instrument (Fig. 15.2). The ends are pulled apart at a certain speed and the distance pulled is plotted versus pounds per square inch of tension placed on the sample. A typical stress-strain curve for a thermoplastic is given in Fig. 15.3.

In the initial stages of the extension the graph is sometimes linear and obeys Hooke's law. The slope of this section is called Young's modulus. This portion

Figure 15.2 Tensile testing instrumentation. Polymer samples are stretched under controlled conditions and the tensile properties are evaluated. (Courtesy of Du Pont.)

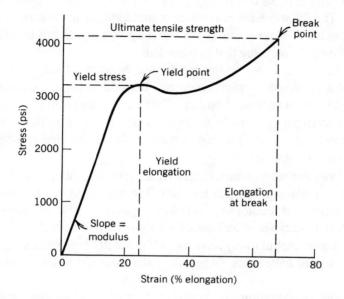

Figure 15.3 Common stress-strain curve for a thermoplastic. *(Source:* W & R I. Reprinted by permission of John Wiley & Sons, Inc.)

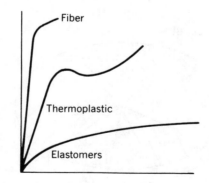

	Elastomers	*Plastics*	*Fibers*
Modulus (psi)	15–150	1,500–200,000	150,000–1,500,000
Percent elongation	100–1000	20–100	<10
Crystallinity	Low	Moderate	High
Example	Natural rubber	Polyethylene	Nylon

Figure 15.4 Stress-strain diagrams for typical polymers. *(Source:* W & R I. Reprinted by permission of John Wiley & Sons, Inc.)

of the curve is reversible. Because many polymers do not obey Hooke's law the modulus is frequently expressed as pounds per square inch at a certain elongation or extensibility. The 2% modulus is a common quotation. Some elastomers are better described as a 100% modulus. The stiffer the polymer is that is tested, the higher will be the modulus value that is recorded.

After the initial stress a yield point is reached, beyond which permanent deformation and nonreversible extension occur. Then the stress and elongation gradually increase until the plastic is broken. The stress at this point is called the ultimate tensile strength (or tensile strength) and the strain is the percent elongation (at break) where 100% would mean it could be stretched to twice its original length before breaking.

Figure 15.4 gives the stress-strain diagrams for a typical fiber, plastic, and elastomer and the average properties for each. The approximate relative area under the curve is fiber, ~1; elastomers, ~15; thermoplastics, ~150. Coatings and adhesives, the two other types of end-uses for polymers, will vary considerably in their tensile properties, but many have moduli generally between elastomers and plastics. They must have some elongation and are usually of low crystallinity.

With this brief introduction into polymer chemistry, let us now turn our attention to specific studies of the five major applications of polymers: plastics, fibers, elastomers, coatings, and adhesives, with the use percentages as shown in Fig. 15.5

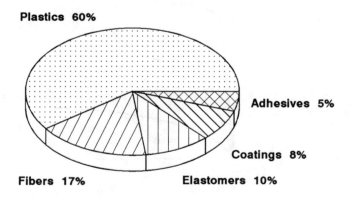

Figure 15.5 Estimated use percentage for polymers.

16

Plastics

References

W & R II, pp. 39-103
Kent, pp. 311-377
B.F. Greek, "Plastics Producers Look for Turnaround by Year's End," *C & E News,* 6-10-91, pp. 39-68.

INTRODUCTION

Having studied some of the basic chemistry and properties of polymers, we now consider in detail the major applications of these fascinating molecules. By far the most important use of polymers is in the plastics industry.

Plastics Materials and Resins (SIC 2821) makes up 12.4% of shipments for Chemicals and Allied Products (SIC 28), the highest percentage of any polymer application. Fig. 16.1 shows the growth of shipments in plastics compared to cellulosic and non-cellulosic fibers and synthetic rubber, other major uses for polymers. Note the very steep incline for plastics, now over $30 billion in commercial value. SIC 2821 includes mainly basic polymer resins and forms, including molded and extruded material. Rubber and Miscellaneous Plastics Products (SIC 30), an SIC separate from Chemicals and Allied Products, is part of our larger chemical process industries definition of "the chemical industry," as explained on page 2. This SIC deals with finished consumer products bought

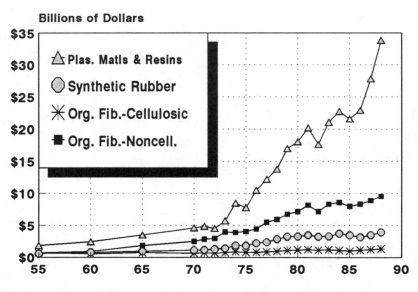

Figure 16.1 U.S. shipments of plastics, fibers, and synthetic rubber. (*Source:* AS.)

retail which contain rubber or plastic material. Note here for plastics products the large increase in recent years as shown in Fig. 16.2, where $70 billion is being approached. For these final Miscellaneous Plastics Products (SIC 308) a further breakdown is shown in Table 16.1, which lists separate subdivisions.

TABLE 16.1 U.S. Shipments of Miscellaneous Plastics Products

Industry Group	SIC	Shipments ($ billion)	%
Unsupported plastics film and sheet	3081	8.95	13.3
Unsupported plastics profile shapes	3082	2.73	4.1
Laminated plastics plate, sheet, and profile shapes	3083	2.35	3.5
Plastics pipe	3084	2.66	4.0
Plastics bottles	3085	3.30	4.9
Plastics foam products	3086	7.51	11.2
Custom compounding of purchased plastic resins	3087	2.95	4.4
Plastics plumbing fixtures	3088	0.9	1.3
Other misc. plastics products	3089	35.9	53.3
Total miscellaneous plastics products	308	67.3	100

Source: AS.

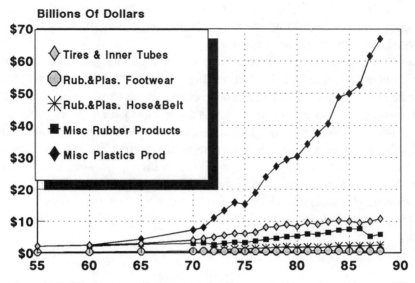

Figure 16.2 U.S. shipments of rubber and miscellaneous plastics products. (*Source*: AS.)

It cannot be denied that our modern standard of living would be changed drastically without the plastics industry. Although many criticisms of "cheap" plastic materials are sometimes justified, no one would willingly return to the

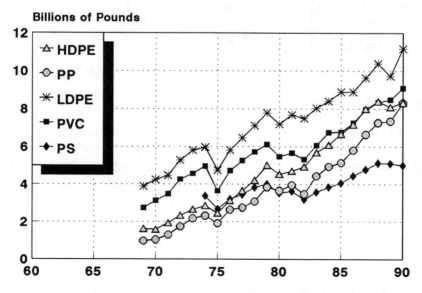

Figure 16.3 U.S. production of polymers. (*Source*: *C & E News.*) HDPE = high-density polyethylene, PP = polypropylene, LDPE = low-density polyethylene, PVC = poly (vinyl chloride), PS = polystyrene.

preplastic age, and especially have to pay for the difference. Indeed, many consumer products would not be possible without the availability of plastic materials. It is a high-growth industry.

If we look at pounds instead of dollars, we see the more gradual increases of the last twenty years in U.S. production (Fig. 16.3) for the five major polymers. continue to be the leader for some time to come. Price trends in polymers (Fig. 16.4) are more up and down depending on the economy for a given year. All of these major use polymers in the plastics industry are 40-70 ¢/lb to be competitive. PVC is the most expensive but it is also the most versatile. The reader might also wish to refer back to Table 1.18, page 27, where the top polymer production is given. For the last ten years polypropylene has been the fastest growing plastic, with an average annual growth of 8.6% per year, compared to plastics in general at 5.4% per year, close to the 6.2% per year from 1970-80 for plastics, still healthy.

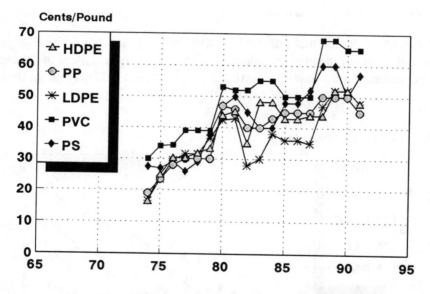

Figure 16.4 U.S. prices of polymers. (*Source*: CMR.)

TABLE 16.2 Per Capita Use of Plastics in the U.S.

Year	lb Plastics/Person
1930	0.25
1940	1.5
1950	12
1960	31
1970	90
1980	209
1990	210 est.

Table 16.2 shows the amount of plastics produced in the U.S. per person for selected years. The very large growth rate is apparent until 1980. It is amazing that each of us uses 200 lb per year.

What was the first synthetic plastic? Although some nineteenth-century experiments should be mentioned, such as the 1869 molding process for cellulose nitrate discovered by John and Isaiah Hyatt, probably the first major break-through came in 1910 with Leo Baekeland's discovery of phenol-formaldehyde resins (Bakelite®). These are still the leading thermoset plastics made today. Over 2 billion lb were made in 1984. The pioneering work of Wallace Carothers at Du Pont in 1929 produced the nylons now used primarily as fibers but noted here as the beginning of thermoplastic resin technology.

GENERAL USES OF PLASTICS

Although we will be discussing plastics according to their various types and what applications each type might fill, it is good to know something general about uses of plastics. Fig. 16.5 divides the uses of thermoplastics into some general areas. Fig. 16.6 shows some general uses of thermosets. Packaging, the largest use for thermoplastics, includes containers and lids, probably one half of this use, and packaging film, another one third. Liners, gaskets, and adhesives for packaging make up the rest. Building and construction, the largest use area for the thermosets and second largest for thermoplastics, includes various types of pipes, fittings, and conduit. Plywood adhesives are also big.

DEFINITIONS AND CLASSES

Your own intuition and experience should give you a good idea of what a plastic is. It is more difficult to define precisely because there are so many types, they have such a wide variety of properties, and their methods of fabrication are so

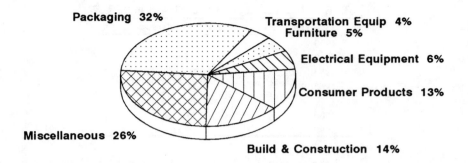

Packaging 32%

Transportation Equip 4%
Furniture 5%

Electrical Equipment 6%

Consumer Products 13%

Miscellaneous 26%

Build & Construction 14%

Figure 16.5 Uses of thermoplastics. (*Source*: *C & E News.*)

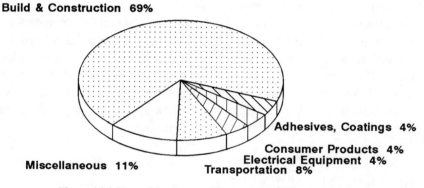

Build & Construction 69%

Adhesives, Coatings 4%

Consumer Products 4%
Electrical Equipment 4%
Transportation 8%

Miscellaneous 11%

Figure 16.6 Uses of thermosets. (*Source*: *C & E News.*)

diverse. Not all polymers are good plastics, although many polymers serve important plastic applications. Probably the best, simplest, and all inclusive definition is that plastics are polymers that have been converted into shapes by processes involving a flow of the liquid phase before solidification. In short, a polymer must be easily shaped if it is to serve in any important plastic application.

The best type of chemical classification of plastics is that same division that we use for all polymers: they are thermoplastic (linear) and thermoset (cross-linked). Unfortunately, there is an overlap in the all important mechanical properties when you use this chemical division. In 1944 Carswell and Nason categorized plastics by the shape of their stress-strain curves, one of the important properties of any plastic. These curves are pictured in Fig. 16.7, pages

318-319. The five major types are: (1) hard, tough, (2) hard, strong, (3) hard, brittle, (4) soft, weak and (5) soft, tough. General characteristics of these classes and some representative plastics for each type are given. The words hard, soft, tough, strong, brittle, and weak are not chosen lightly. A hard plastic is one that has a high tensile strength and modulus; a soft plastic has a relatively low strength and modulus. Tough refers to a high elongation; brittle refers to a very low elongation. Strong and weak are applied to plastics of moderate elongations and depend on their overall tensile strength as well.

Figure 16.8, page 320, is a graph of the range of tensile properties for each specific plastic, plotting tensile strength versus elongation. Note that the hard, tough plastics such as the nylons are in the upper right (high tensile strength, high elongation), the hard, brittle plastics such as the thermosets and polystyrene are in the upper left (high tensile strength, low elongation), and the soft, tough plastics such as low-density polyethylene are in the lower right (low tensile strength, high elongation). There are no common uses for soft, weak plastics. Specific examples and details for each of these important categories of plastic will be given in a later section.

FABRICATION OF PLASTICS

An important step in the manufacture of any plastic product is the fabrication or the shaping of the article. Most polymers used as plastics when manufactured are prepared in pellet form as they are expelled from the reactor. These are small pieces of material a couple of millimeters in size. This resin can then be heated and shaped by one of several methods. Thermoset materials are usually compression molded, cast, or laminated. Thermoplastic resins can be injection molded, extruded, or blow molded most commonly, with vacuum forming and calendering also used but to a lesser extent. Figures 16.9 to 16.14, pages 321-324, give diagrams and a concise description of each fabrication method taken from Wittcoff and Reuben, Part II, pp. 98-102, reprinted by permission of John Wiley & Sons, Inc. Study especially compression molding, injection molding, extrusion, and blow molding.

The pictures in Fig. 16.15 to 16.18, pages 326-329, show the type of equipment used in some of the processes. Following is a brief review of plastics fabrication techniques adapted with permission from Reuben & Burstall, reprinted by permission of Longmans.

Class of Plastic	Modulus	Yield Stress	Ultimate Tensile Strength	Elongation at Break	Examples
Hard and tough	High	High	High	High	ABS High density polyethylene Cellulosics Polyamides Polypropylene Fluoroplastics Engineering plastics Polyacetal, polycarbonate Polyimide, polyphenylene sulfide, polyphenylene oxide Polysulfone Poly(vinylidene chloride)
Hard and strong	High	High	High	Moderate	Rigid PVC Impact polystyrene Styrene-acrylonitrile

Hard and brittle	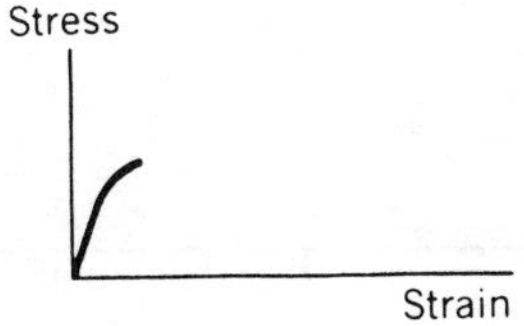 High	No well-defined yield point	High	Low	P/F, U/F and M/F resins Polystyrene Polymethyl methacrylate Unsaturated polyester resins Epoxy resins
Soft and weak	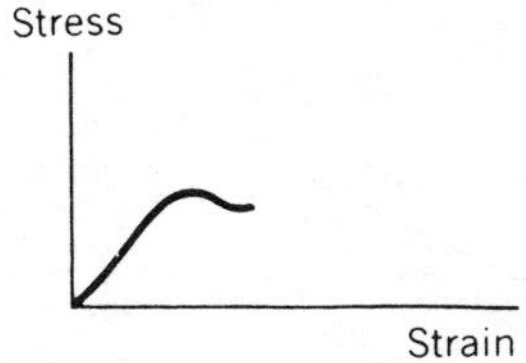 Low	Low	Low	Moderate	Polyethylene waxes
Soft and tough	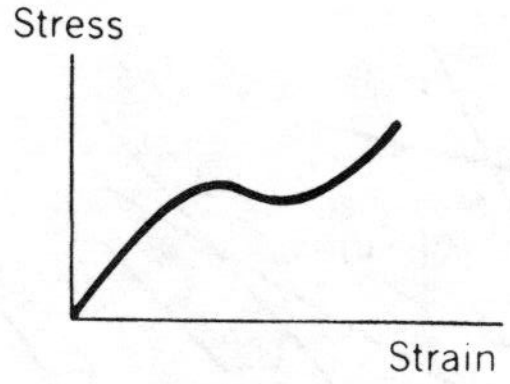 Low	Low yield point	Low	High	Low density polyethylene Plasticized PVC Ionomer

Stress strain diagrams of plastics

Figure 16.7 Classes of plastics by shape of stress-strain curve. (*Source:* W & R II. Reprinted by permission of John Wiley & Sons, Inc.)

319

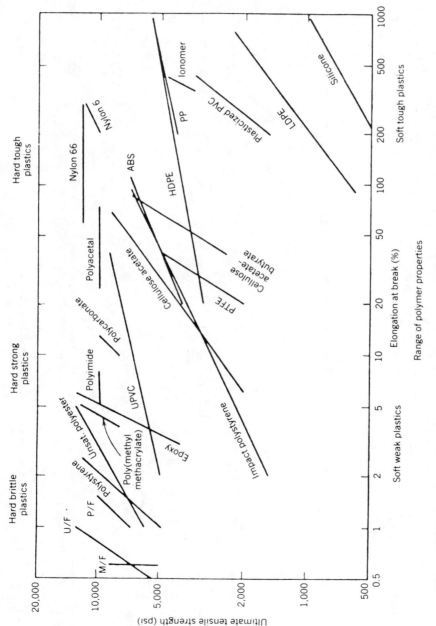

Figure 16.8 Range of tensile properties for various plastics. (*Source:* W & R II. Reprinted by permission of John Wiley & Sons, Inc.)

Compression molding (Fig 16.9). Compression molding is practically the oldest method of fabricating polymers and is still widely used. The polymer is placed in one half (the "female" half) of a mold and the second or "male" half compresses it to a pressure of about 1 ton/in^2. The powder is simultaneously heated, which causes the resin to cross-link. Transfer molding is a cross between this technique and injection molding.

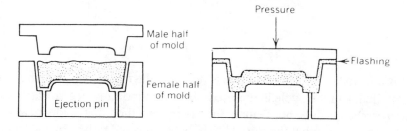

Figure 16.9 Compression molding.

Casting. In the sheet casting of poly (methyl methacrylate), monomer is partly polymerized and the viscous liquid is then poured into a cell made up of sheets of glass separated by a flexible gasket that allows the cell to contract as the casting shrinks.

Injection molding (Fig. 16.10). In injection molding the polymer is softened in a heated volume and then forced under high pressure into a cooled mold where

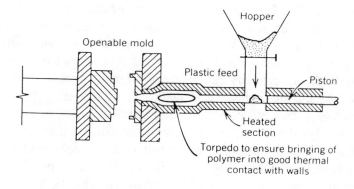

Figure 16.10 Injection molding.

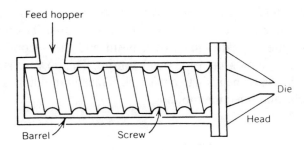

Figure 16.11 Extrusion.

it is allowed to harden. Pressure is released, the mold is opened, the molding is expelled, and the cycle is repeated. Injection molding is a versatile technique and can be used for bottle manufacture by a method that is identical with blow molding except that the initial "bubble" is injected rather than extruded.

Extrusion (Fig. 16.11). Extrusion is a method of producing lengths of plastics materials of uniform cross section. The extruder is similar to a domestic mincing machine with the added facility that it can be heated and cooled. The pellets enter the screw section via the hopper, are melted, and then pass through the breaker plate into the die. The plastic material is forced out of the die with its cross section determined by the shape of the die, but not identical with it because of stresses induced by the extrusion process. Extrusion can be used for coating electrical wiring by means of a cross-head die.

Blow Molding (Figs. 16.12 and 16.13) and Vacuum Forming (Fig. 16.14). Blow extrusion, in which the initial lump of polymer is formed by an extrusion process, is the most common form of blow molding and is shown diagrammatically in Fig. 16.12. A short length of plastic tubing is extruded through a crossed die and the end is scaled by the closing of the mold. Compressed air is passed into the tube and the "bubble" is blown out to fill the mold.

A variation of this process is used for the manufacture of thin film. A tube of plastic is continuously extruded and expanded by being blown to a large volume and consequently a small wall thickness. The enormous bubble of plastic is cooled by air jets and is continuously taken up on rollers. This material can either be slit down the side to give plastic film or turned into plastic bags by the making of a single seal across the bottom of the tube. Blown diameters of up to 7 ft. have been achieved leading to flat film widths of 24 ft. The process may be thought of as biaxial orientation of the polymer as opposed to drawing in which the polymer molecules are oriented in one dimension.

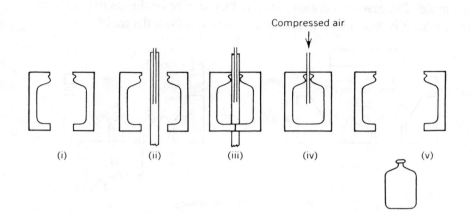

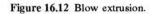

Figure 16.12 Blow extrusion.

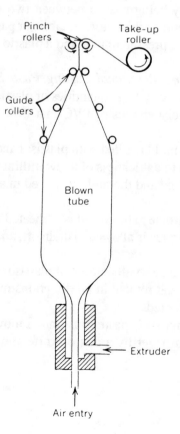

Figure 16.13 Film by extrusion blowing.

Vacuum forming (Fig. 16.14), in a sense, is the opposite of blow molding. A sheet of heat-softened plastic is placed over a mold and the air is sucked from the mold. The plastic is drawn down and its surface conforms to the shape of the mold. It is then allowed to cool and removed from the mold.

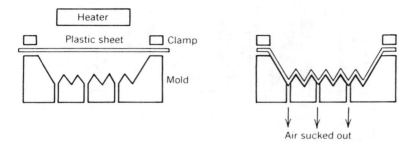

Figure 16.14 Vacuum forming.

Calendering. In the calendering process a preheated polymer mix is turned into a continuous sheet by being passed between two or more heated rolls that squeeze it to the appropriate thickness. If fabric or paper is fed through the final rolls, the plastic can be pressed onto it and a plastic coated material results.

Slush Molding, Dipping, Rotational Casting, Paste Spreading. The mixing of PVC powder with relatively large quantities of plasticizers leads to PVC pastes or plastisols, that is, dispersions of PVC that flow in a liquid or quasiliquid fashion.

In slush molding, a mold is filled with plastisol and placed briefly in an oven where the plastic gets to a thickness of a few millimeters. The excess plastisol is poured out and reused and the mold is placed in another oven where curing is completed.

In "dipping" the article to be coated is preheated and dipped into a plastisol. Some of it gels, the surplus is allowed to drain off, and the coated object is cured in an oven.

In rotational casting, a small amount of plastisol is placed in a heated mold, which is then closed and rotated in all directions in an oven so that its inside becomes uniformly coated.

In paste spreading, PVC paste runs onto a moving belt of fabric. This is spread into a uniform layer by a doctor knife and is then cured by passage through an oven.

Laminating. The best known laminates are the Formica®-type decorative laminates for household use. Brown paper is impregnated with an alcoholic solution of a resol and is then cut up and arranged in piles with a suitably printed melamine formaldehyde-impregnated decorative sheet on top. The whole is then pressed at about 150-180°C and 3 tons/in² to give the finished laminate.

IMPORTANT PLASTICS

The diversity in properties and uses of plastics is greater than any other area of polymer chemistry. It is best simply to select a few of the most important plastics and become acquainted with them individually. On pages 325-340 there is some important information on certain polymers having wide application as plastics. We will use our general categories of hard-tough, hard-strong, hard-brittle, and soft-tough to determine their order of treatment and also to emphasize which plastics compete with each other. Although some plastics are similar to others they all have their own set of advantages and disadvantages for a given application. Indeed, it is the job of the plastics companies to fit the best polymer to a particular use.

Hard-Tough

High-density Polyethylene, HDPE, $(CH_2—CH_2)_n$

Reference

CP, 5-30-88, 6-10-91

1. Manufacture
   Introduced in the 1950s
   Moderate to low pressures
   Metal oxide catalysis (usually)
2. Properties
   No branches, 90% crystalline, $T_m = 135°C$, $T_g = -70$ to $-20°C$
   More opaque than LDPE
   Stiffer, harder, higher tensile strength than LDPE
   Specific gravity = 0.96

3. Uses (Fig. 16.19, page 330)

    Blow molding: containers and lids, esp. food bottles, auto gas tanks, motor oil bottles

    Injection molding: pails, refrigerator food containers, toys, mixing bowls

4. Economics

    1990 production = 8.33 billion lb

    1990 commercial value = $4.2 billion

    1991 price = 47-55 ¢/lb

    U.S. demand for HDPE in 1995 is expected to be 10 billion lb.

    Blow molding and injection molding is expected to grow 3-4%/yr.

    Film and sheet is expected to grow 9-10%/yr.

    By the year 2000, half of some HDPE molded items may be recycled.

Figure 16.15 Research injection molding equipment. (Courtesy of Du Pont.)

$$\text{CH}_3$$
$$|$$
Polypropylene, $(\text{CH}_2\text{-CH})_n$

Reference

CP, 5-13-88, 6-17-91

1. Manufacture
    Ziegler-Natta or metal oxide catalysis
2. Properties
    T_m = 170°C, higher than HDPE, can be sterilized at 140°C for
      hospital applications
    T_g = -10°C, more brittle at low temperatures than HDPE
    Stiffer, harder, higher tensile strength than HDPE
    Specific gravity = 0.91, lightest of the major plastics
    More degraded than HDPE by heat, light, O_2 because of tertiary
      hydrogens. Therefore antioxidants and UV stabilizers can be added.
    Shiny surfaces, resists marring
3. Uses (Fig 16.20, page 330)
    Fibers: carpet backing, indoor-outdoor carpeting, rope
    Injection molding: containers, lids, bottles, toys, plastic chairs,
      luggage, steering wheels, battery cases, fan shrouds, air cleaner
      ducts
4. Economics
    1990 production = 8.32 billion lb
    1990 commercial value = $4.6 billion
    1991 price = 50-59 ¢/lb
    Average growth in the early 1990s is expected to be 5% annually.
    Recycled polypropylene may become important.

Acrylonitrile-butadiene-styrene terpolymer, ABS

$$-\text{CH}_2-\text{CH}-\text{CH}_2-\text{CH}=\text{CH}-\text{CH}_2-\text{CH}_2-\text{CH}-$$
$$|$$
$$\text{C}\equiv\text{N}$$

Reference

CP, 4-22-91

1. Manufacture
    Graft polymerization of acrylonitrile and styrene on a preformed
    polybutadiene elastomer

Figure 16.16 Polymer ribbon in a molten state at high temperatures ready to enter an extruder. (Courtesy of Amoco Chemicals Corporation, Alvin, TX.)

2. Properties
   Specific gravity = 1.06
   Opaque
   Higher tensile strength, lower elongation than HDPE or PP
3. Uses (Fig. 16.21, page 331)
   Examples of ABS products: radio housings, telephones, pocket calculators, high-quality plastic pipe and fittings, lawn mower housings, luggage
4. Economics
   1990 production = 1.16 billion lb
   1990 commercial value = $1.4 billion
   1991 price = $1.00-1.35/lb, much more expensive than HDPE or PP

Figure 16.17 Extruding equipment. The extruder is opened in the middle of this picture to show the die for making over 1000 simultaneous strands. (Courtesy of Amoco Chemicals Corporation, Alvin, TX.)

Figure 16.18 Research size blow molding equipment. (Courtesy of Du Pont.)

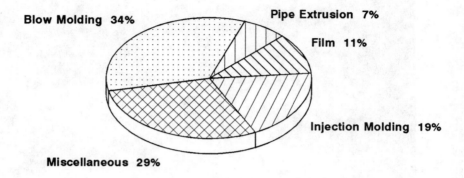

Figure 16.19 Uses of high-density polyethylene. (*Source*: CP.)

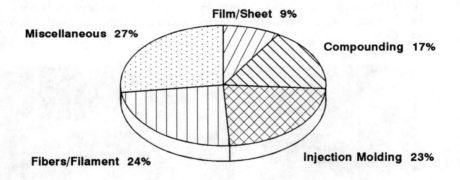

Figure 16.20 Uses of polypropylene. (*Source*: CP.)

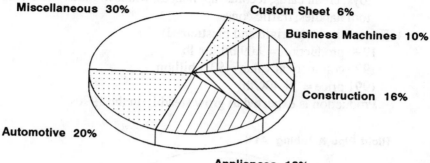

Figure 16.21 Uses of ABS resins. (*Source*: CP.)

Hard-Strong

Rigid PVC or Unplasticized PVC, Poly (vinyl chloride), $(CH_2\text{-}CH)_n$
$\quad\quad\quad\quad\quad\quad\quad\quad\quad\quad\quad\quad\quad\quad\quad\quad\quad\quad\quad | $
$\quad\quad\quad\quad\quad\quad\quad\quad\quad\quad\quad\quad\quad\quad\quad\quad\quad\quad\quad Cl$

Reference

CP, 6-5-89

1. Manufacture
   Peroxide free radical initiation
   Suspension, emulsion, or bulk procedure
2. Properties
   $T_m = 140°C$
   $T_g = 70\text{-}85°C$, higher than polyolefins because polar C-Cl bond gives
      dipole-dipole intermolecular attractions
   Low crystallinity
   Good impact strength
   Good chemical resistance
   Resistant to insects and fungi
   Nonflammable
   Easily degraded by heat and light via weak C-Cl bond
   Brittle at low temperatures
   Toxicity problem?

3. Uses (Fig. 16.22)

   Examples of rigid PVC: water pipes, gutters, roofing, siding, bottles, toys, credit cards, records, appliances, window blinds and awnings, tool handles, traffic cones

4. Economics (total rigid and plasticized)

   1990 production = 9.09 billion lb

   1990 commerical value = $3.5 billion

   1991 price = 38-40¢/lb

   Production is expected to grow steadily in the 1990s.

Rigid Pipe & Tubing 41%

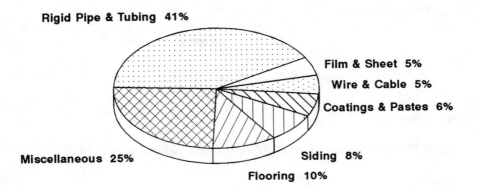

Film & Sheet 5%

Wire & Cable 5%

Coatings & Pastes 6%

Miscellaneous 25%

Siding 8%

Flooring 10%

Figure 16.22 Uses of poly (vinyl chloride). (*Source:* CP.)

High-impact polystyrene,   (CH$_2$—CH)$_n$

Butadiene polymerized first; next, styrene is grafted onto it.

More opaque, less brittle, more resistant to impact than is polystyrene.

About one third of all polystyrene is high-impact.

Hard-Brittle (polystyrene + thermosets)

Polystyrene   (CH$_2$—CH)$_n$

Reference

CP, 6-20-88, 6-24-91

1. Manufacture
    Peroxide initiation
    Suspension or bulk
2. Properties $T_m = 227°C$, $T_g = 94°C$. The wide spread is good for
    processing
    Amorphous and transparent—bulky phenyls inhibit crystallization
    Easily dyed
    Very flammable—can add flame retardants
    Not chemically resistant
    Weathers poorly
    Yellows in light—can add UV absorbers

3. Uses (Fig. 16.23)

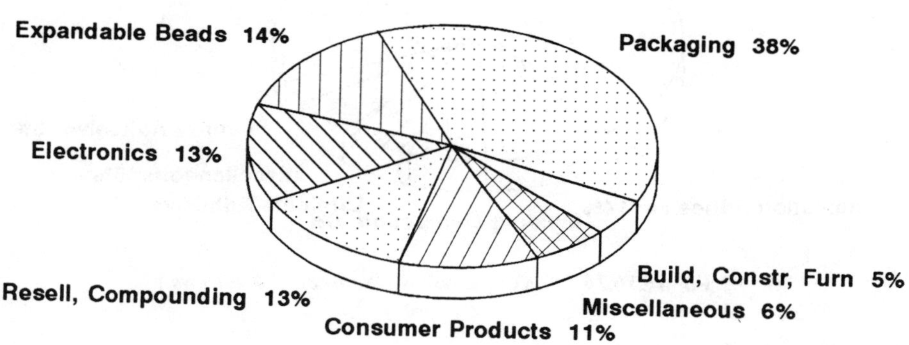

Figure 16.23 Uses of polystyrene. (*Source*: CP.)

4. Economics
    1990 production = 5.01 billion lb
    1990 commercial value = $3.5 billion
    1991 price = 57-80 ¢/lb
    Polystyrene growth in the 1990s will be less than other plastics,
    perhaps 2-3% annually.

Phenol-formaldehyde, P/F Resins, Phenolics

1. Manufacture
    One-stage cured by heat
    Two-stage cured by heat and hexamethylenetetramine
2. Properties
    Heat resistance
    H_2O and chemical resistance
    Dark color
    Problems with formaldehyde's possible carcinogenicity
3. Uses (Fig. 16.24)

Plywood Adhesive 66%

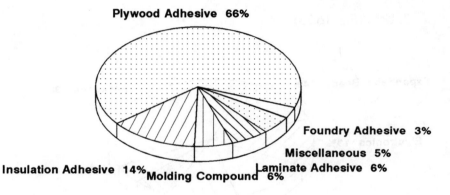

Foundry Adhesive 3%

Miscellaneous 5%

Insulation Adhesive 14% **Molding Compound 6%** **Laminate Adhesive 6%**

Figure 16.24 Uses of phenolics. (*Source*: *C & E News*.)

Urea-formaldehyde, U/F resins

1. Manufacture
    Methylolurea formation under alkaline conditions, followed by heating under acidic conditions
2. Properties
    White translucent (nearly transparent)
    Can be pigmented to wide variety of colors
    Light stable
    Less heat and water resistant than phenolics
    Good electrical properties

Figure 16.27 Polypropylene carpet backing (Amoco Fabrics) is one important application of this versatile plastic. (Courtesy of Amoco Chemicals Corporation, Alvin, TX.)

3. Uses

Electrical plugs and switches

Insulation foam (may be hazardous to health)

Melamine-formaldehyde, M/F resins

Combine good properties of P/F and U/F

H_2O and heat resistant

Pastel colors

Main use: unbreakable dinnerware, resistant to dishwasher use and can be decorated

More expensive than P/F or U/F

Unsaturated polyesters (maleic and phthalic anhydrides with propylene glycol)

1. Manufacture

Cross-linked with free radical initiator and styrene

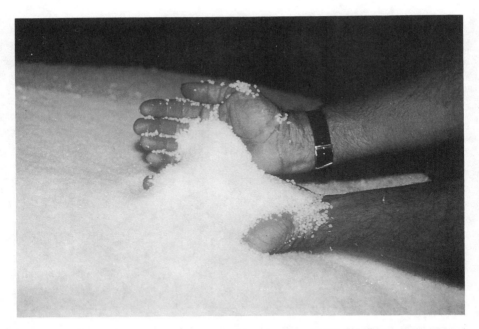

Figure 16.28 Polymers are often pelletized during the extrusion process, a convenient state in which to store and ship plastics before final fabrication. (Courtesy of Amoco Chemicals Corporation, Alvin, Tx.)

2. Properties

Variation of monomers and their percentage gives wide range of properties

3. Uses

Glass fiber reinforced polyester boat hulls (no rot or saltwater corrosion)

Automobile bodies

General construction

Casting:fireplaces, vanities, plaques, shower stalls, playground equipment

Misc.: bowling balls, sewer pipe, pistol grips, corrosion resistant tanks

Epoxies (Epichlorohydrin, Bisphenol A, and Ethylenediamine)

1. Uses
   Electrical printed circuits
   Structural parts for jets
   Rocket motor casings
   Insulators for power transmission lines
   Coatings and adhesives

Soft-Tough

Low-density polyethylene, LDPE, $(CH_2-CH_2)_n$

Reference

CP, 6-6-88, 6-3-91

1. Manufacture
   Free radical initiators
   High pressure required
   Discovered by ICI in the U.K. in 1933, commercialized in 1938
2. Properties
   Much C_4 branching, only 50-60% crystalline, 30 branches per 100
      carbons
   $T_m = 115°C$, lower than HDPE
   Specific gravity = 0.91-0.94, lower than HDPE
   Easily processed
   Flexible without plasticizers
   Resists moisture and chemicals—but porous to O_2
   Translucent—clear enough for packaging
   Cheap price
3. Uses (Fig. 16.25)
   Film and sheet: packaging, trash bags, household wrap, drapes,
      tablecloths
   Injection molding: squeeze bottles, toys, kitchen utilityware
   Extrusion: wire and cable insulation
4. Economics
   1990 production = 7.2 billion lb (not counting LLDPE)
   1990 commercial value = $3.7 billion
   1991 price = 48-55 ¢/lb
   An increase of more than 2% annually is expected.

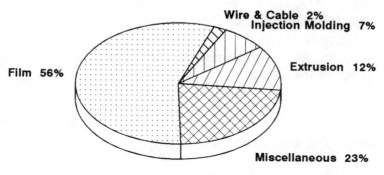

Figure 16.25 Uses of LDPE. (*Source*: CP.)

Linear low-density polyethylene, LLDPE. It is a copolymer of ethylene with small amounts of 1-butene (or other alkenes) added purposely to give some branching. Thus it is made by a HDPE process with metal catalysts but resembles LDPE because of occasional ethyl branches due to the 1-butene being copolmerized. Whereas conventional LDPE requires pressures of 15,000-40,000 psi and temperatures of 300-500°C, LLDPE is made at 1,000 psi and 100-200°C. Less energy and less costly process equipment is the end result. It may take a substantial market away from LDPE in the years to come because of this lower cost and also because it is stronger and more rigid. The properties can be tailor-made by varying the percentage of the comonomer.

    1. Manufacture

        Low pressure, low density

        Introduced in the late 1970s

    2. Uses (Fig. 16.26)

    3. Economics

        1990 production = 4.7 billion lb (not counting LDPE)

        1990 commercial value = $2.4 billion

        1991 price = 48-55 ¢/lb

        A growth rate of 20% annually for 1980-90 will slow to 7-15% for 1990-2000.

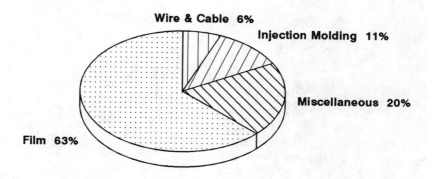

Figure 16.26 Uses of linear low-density polyethylene. (*Source*: CP.)

Plasticized PVC, Poly(vinyl chloride), Vinyl, $(CH_2\text{-}CH)_n$
$$\underset{Cl}{\big|}$$

1-2% plasticizer such as dioctyl phthalate

1. Manufacture
   Suspension or bulk
2. Uses;
   Wire insulation, hose, floor tiles, auto tops, raincoats, footwear, baby pants, baggage, furniture, car seats—leather replacement, wallpaper, handbags, blood and intravenous bags, tablecloths

Figure 16.29 Polymer storage in pellet form can be done in large silos, each of which can hold 185,000 lb (7-10 hr of production). The silos can be mixed to ensure uniformity before the pellets are added to a tank car holding nearly the same amount as one silo. (Courtesy of Amoco Chemicals Corporation, Alvin, TX.)

17

Fibers

References

W & R II, pp. 104-125
Kent, pp. 378-427
S & C, selected sections

HISTORY AND TYPES OF FIBERS

Fibers have been used for thousands of years to make various textiles. For centuries certain natural products have been known to make excellent fibers. Probably the first synthetic fiber experiment came with the work of Christian Schönbein, who made cellulose trinitrate in 1846. After breaking a flask of nitric and sulfuric acids, he wiped up the mess with a cotton apron and hung it in front of the stove to dry. Cellulose trinitrate was developed by Hilaire de Chardonnet as a substitute for silk in 1891, but it was very flammable and was soon nicknamed "mother-in-law silk," being an appropriate gift for only disliked persons. When rayon came along in 1892, "Chardonnet silk" was soon forgotten. Then the entirely synthetic fibers came, with the pioneering work of W. Carothers at Du Pont synthesizing the nylons in 1929-30. Commercialization occurred in 1938. Polyesters were made by Whinfield and Dixon in the U.K. in 1941. They were commercialized in 1950.

It is important to understand the different types of fibers. Subclasses are best differentiated based on both the origin of the fiber and its structure. The structure and chemistry of many of these polymers was discussed earlier in Chapter 14.

Table 17.1 contains a list of the three important types of fibers—natural, cellulosic, and noncellulosic —as well as a list of specific polymers as examples of each type. The ones marked with an asterisk are the most important.

Referring back to Fig. 16.1, page 312, we see that the value of U.S. shipments for cellulosic and noncellulosic fibers, though quite small compared to plastics, is still a big industry. While Plastics Materials and Resins (SIC 2821) in 1988 was $33.8 billion, Noncellulosic Fibers (SIC 2824) was $9.5 billion and Cellulosic Fibers (SIC 2822) was $1.3 billion. These two fibers together have a $10.8 billion value, which is 4.7% of Chemicals and Allied Products. We must also remember that many of these fibers are sold outside the chemical industry, such as in Textile Mill Products, Apparel, and Furniture, all large segments of the economy. The importance of fibers is obvious. In 1920 U.S. per capita use was 30 lb/yr, whereas in 1990 it was 66 lb/yr. From 1920 to 1970 the most important

TABLE 17.1 Types of Fibers

Natural
1. From animal sources—all are proteins
    a. Silk (from the silkworm)—mostly glycine and alanine
    b. *Wool (from sheep)—complex mixture of amino acids
    c. Mohair (from the Angora goat)
2. From plant sources—all are cellulose polymers
    a. *Cotton (from the cotton plant)
    b. Flax (from blueflowers)—used to make linen
    c. Jute (from an east Indian plant)—used for burlap and twine
3. From inorganic sources
    a. Asbestos—mostly calcium and magnesium silicates
    b. Glass—silicon dioxide

Cellulosic
These fibers are also called semisynthetic since the natural cellulose is modified in some way chemically.
1. *Rayon (regenerated cellulose)
2. *Cellulose acetate
3. Cellulose triacetate

Noncellulosic
These are entirely synthetic, made from polymerization of small molecules.
1. *Nylons 6 6 and 6
2. *Polyester (linear)—poly (ethylene terephthalate) mostly
3. Acrylic—polyacrylonitrile
4. Polyurethane
5. Polypropylene

Billions of Pounds

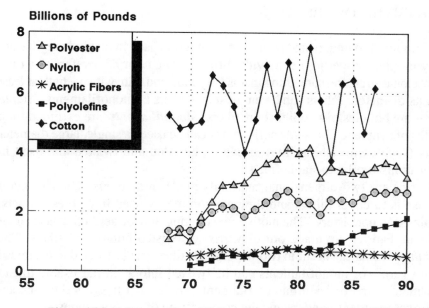

Figure 17.1 U.S. production of fibers. (*Source: C & E News* and CEH.)

fiber by far was cotton. However, synthetic fibers (cellulosic and noncellulosic) increased much more rapidly in importance, with cellulosics booming between World Wars I and II and noncellulosics dominating after World War II, while all that time cotton showed only a steady pace in production. The more recent competition between the various fibers in the United States is given in Fig. 17.1. Nylon was originally the most important synthetic (1950-1971) but polyester now dominates the market (1971-present). For a few years (1970-1980) acrylics were third in production, but since 1980 polyolefins have been rapidly increasing. Cotton, being an agricultural crop, certainly demonstrates its variable production with factors such as weather and the economy. It is an up-and-down industry much more so than the synthetics.

The student should also review Table 1.18, page 27, where the top polymer production is given numerically. Overall no growth was recorded for 1980-90 in synthetic fibers. A net decrease in the cellulosics of 4.6% shows their diminishing importance. Polyester also decreased 2.2% annually in this period. The rising star is polyolefins, which increased 9.3% per year in the past decade.

U.S. price trends are given in Fig. 17.2, page 345. Cotton and polyester are the most popular for good reason: they are the cheapest. Nylon and acrylic have had price increases over the last few years, part of which may be due to improvements and safeguards needed to manufacture their precursors acrylonitrile, butadiene, and benzene, which are on the suspect carcinogen list.

PROPERTIES OF FIBERS

It is important to understand some common terms used in this industry before studying fiber properties or individual fibers. The term *fiber* refers to a one-dimensional structure, something that is very long and thin, with a length at least 100x its diameter. Fibers can be either staple fibers or monofilaments. *Staple fibers* are bundles of parallel short fibers. *Monofilaments* are extruded long lengths of synthetic fibers. Monofilaments can be used as single, large diameter fibers (such as in fishing line) or can be bundled and twisted and used in applications similar to staple fibers.

Fabrics are two-dimensional materials made from fibers. Their primary purpose is to cover things and they are commonly used in clothes, carpets, curtains, and upholstery. The motive for covering may be aesthetic, thermal, or acoustic. Fabrics are made out of *yarns* or twisted bundles of fibers. The *spinning of yarns* can occur in two ways: staple fibers can be twisted into a thread ("spun yarn") or monofilaments can be twisted into a similar usable thread ("filament yarn" or "continuous filament yarn"). All these definitions are important in order to understand the conversation of the fiber industry.

The tensile properties of fibers are not usually expressed in terms of tensile strength (lb/in^2 or kg/cm^2). The strength of a fiber is more often denoted by tenacity. *Tenacity* (or breaking tenacity) is the breaking strength of a fiber or yarn in force per unit denier (lb/denier or g/denier). A *denier* is the weight in grams of 9,000 m of fiber at 70°F and 65% relative humidity. A denier defines the linear density of a fiber since it depends on the density and the diameter of a fiber. To give you an idea of common values of deniers for fibers, most commercially useful fibers are 1-15 denier, yarns are 15-1600 denier, and monofilament (when used singly) can by anywhere from 15 denier on up. Table 17.2 lists common values of tenacity for various fibers and compares these values to tensile strength. Tenacities can be converted into tensile strength in pounds per inch square by the following formula:

tensile strength (lb/in^2) = tenacity (g/denier) x density (g/cm^2) x 12,791

Note that tenacity values for most fibers range from 1-8 g/denier.

In general, you should recall that the tensile strength and modulus of fibers must be much higher than that for plastics and their elongation must be much lower. Synthetic fibers are usually stretched and oriented uniaxially to increase their degree of parallel chains and increase their strength and modulus.

TABLE 17.2 Physical Properties of Typical Fibers

Polymer	Tenacity (g/denier)	Tensile Strength (kg/cm²)	Elongation (%)
Cellulose			
Cotton	2.1-6.3	3000-9000	3-10
Rayon	1.5-2.4	2000-3000	15-30
High-tenacity rayon	3.0-5.0	5000-6000	9-20
Cellulose diacetate	1.1-1.4	1000-1500	25-45
Cellulose triacetate	1.2-1.4	1000-1500	25-40
Proteins			
Silk	2.8-5.2	3000-6000	13-31
Wool	1.0-1.7	1000-2000	20-50
Vicara	1.1-1.2	1000	30-35
Nylon-66	4.5-6.0	4000-6000	26
Polyester	4.4-5.0	5000-6000	19-23
Polyacrylonitrile	2.3-2.6	2000-3000	20-28
Saran	1.1-2.9	1500-4000	20-35
Polyurethane (Spandex)	0.7	630	575
Polypropylene	7.0	5600	25
Asbestos	1.3	2100	25
Glass	7.7	2100	3

Source: S & C, p. 441. Reprinted by courtesy of Marcel Dekker, Inc.

What makes a polymer a good fiber? This is not an easy question to answer. If one fact can be singled out as being important, it would be that all fibrous polymers must have strong intermolecular forces of one type or another.

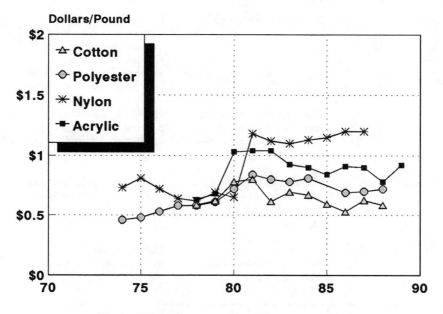

Figure 17.2 U.S. prices of fibers. (*Source*: CEH.)

TABLE 17.4 Properties and Uses of Important Fibers

Fiber Name	Definition	Properties
Acrylic	Acrylonitrile units, 85% or more by weight	Warm, lightweight; shape retention; resilient; quick drying; resistant to sunlight
Modacrylic	Acrylonitrile units, 35-85% by weight	Resilient; softenable at low temperatures; easy to dye; abrasion resistant; quick drying; shape retentive; resistant to acids, bases
Polyester	Dihydric acid-terephthalic acid ester, 85% or more by weight	Strong, resistant to stretching and shrinking; easy to dye; quick drying; resistant to most chemicals; easily washed, wrinkle resistant; abrasion resistant; retains heat-set pleats and creases (permanent press)
Spandex	Segmented polyurethane, 85% or more by weight	Light, soft, smooth, resistant to body oils; stronger and more durable than rubber; can be stretched repeatedly and 500% without breaking; can retain original form, abrasion resistant; no deterioration from perspirants, lotions, detergents
Nylon	Recurring amide groups	Exceptionally strong elastic; lustrous; easy to wash; abrasion resistant; smooth, resilient, low moisture absorbency; recovers quickly from extensions
Rayon	Regenerated cellulose with substitutes no more than 15% of the hydroxyl groups' hydrogens	Highly absorbent; soft; comfortable; easy to dye; good drapability
Acetate	Not less than 92% of the hydroxyl groups are acetylated, includes some triacetates	Fast drying; supple, wide range of dyability; shrink resistant
Triacetate	Derived from cellulose by combining cellulose with acetic acid and/or acetic anhydride	Resistant to shrinking, wrinkling, and fading; easily washed

TABLE 17.4 (Continued)

Typical Uses	Patent Names (assignees)
Carpeting, sweaters, skirts, baby clothes, socks, slacks, blankets, draperies	Orlon (Du Pont), Acrilon (Monsanto), CHEMSTRAND (Monsanto)
Simulated fur, scatter rugs, stuffed toys, paint rollers, carpets, hairpieces and wigs, fleece fabrics	Verel (Eastman), Dynel (Union Carbide)
Permanent press wear: skirts, shirts, slacks, underwear, blouses; rope, fish nets, tire cord, sails, thread	Vycron (Beaunit), Dacron (Du Pont), Kodel (Eastman) Fortrel (Fiber Ind., Celanese), CHEMSTRAND (polyester, Monsanto)
Girdles, bras, slacks, bathing suits, pillows	Lycra (Du Pont)
Carpeting, upholstery, blouses, tents, sails, hosiery, suits, stretch fabrics, tire cord, curtains, rope, nets, parachutes	Caprolan (Allied Chemical) CHEMSTRAN nylon (Monsanto), Astroturf (Monsanto), Celanese polyester (Fiber Ind., Celanese), Cantrece (Du Pont)
Dresses, suits, slacks, blouses, coats, tire cord, ties, curtains, blankets	Avril (FMC Corp.), Cuprel (Beaunit), Zantrel (American Enka)
Dresses, shirts, slacks, draperies, upholstery, cigarette filters	Estron (Eastman), Celanese acetate (Celanese)
Skirts, dresses, sportswear (pleat retention important)	Arnel (Celanese)

Source: S & C, pp. 160-161. Reprinted by courtesy of Marcel Dekker, Inc.

TABLE 17.3 Ideal Properties of Polymers for Fiber Applications

1. High-tensile strength, tenacity, and modulus
2. Low elongation
3. Proper T_g and T_m. A low T_g aids in easy orientation of the fiber. The T_m should be above 200°C to accept ironing (as a textile) but below 300°C to be spinnable.
4. Stable to chemicals, sunlight, and heat
5. Nonflammable
6. Dyeable
7. Resilient (elastic) with a high flex life (flexible lifetime)
8. No static electrical buildup
9. Hydrophilic (adsorbs water and sweat easily)
10. Warm or cool to the touch as desired
11. Good "drape" (flexible on the body) and "hand" (feel on the body)
12. No shrinking or creasing except where intentional
13. Resistance to wear after repeated washing and ironing

Usually this means hydrogen bonding or dipole-dipole interactions, shown here for various types of fibers. Besides the high tenacity, a number of other properties are considered necessary for most fiber applications. Although no one polymer is superior in all of these categories, the list in Table 17.3 represents ideals for polymers being screened as fibers. Table 17.4 gives some direct comparisons among the various fibers.

H-bonding: dipole-dipole:

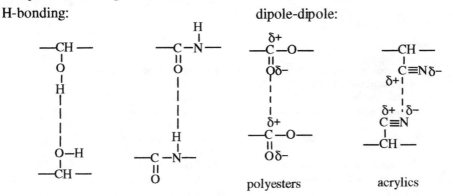

cellulosics   proteins and nylons   polyesters   acrylics

IMPORTANT FIBERS

A study of the fiber industry is not complete without some knowledge of the characteristics of individual fibers. Since each is so different it is difficult to generalize or compare directly as in Table 17.4. This section presents a summary of each important fiber, including pertinent information on their manufacture, properties, uses, and current economics in a brief, informal but concise manner.

Natural

Wool
(protein, mostly keratin, complex mixture of amino acids)

1. Processing requires 20 stages, therefore very expensive
2. Good resilience (elasticity) because orientation changes α-keratin (helical protein) into ß-keratin (zigzag). Example: woolen carpet recovers even after years of heavy furniture.
3. Good warmth due to natural crimp (many folds and waves) which retains air, therefore a good insulator
4. Hydrophilic—absorbs perspiration away from body and is comfortable, not sweaty
5. Dyed easily since it has acidic and basic groups in the amino acids
6. Disadvantages: (1) expensive; (2) retains water by H-bonding with washing, causing shrinkage; (3) many people allergic to protein

Cotton
(pure form of cellulose)

1. Fewer processing steps, cheaper than wool
2. Good hand
3. Wears hard and long
4. Easily dyed via free hydroxyls
5. Absorbs water but dries easily. If preshrunk, it is stable to washing and ironing more than other fibers.
6. Hydrophilic—cool and comfortable. Comfort never matched by synthetics. Used especially in towels and drying cloths.
7. Disadvantages: (1) creases easily, requiring frequent ironing; (2) agricultural variables in growing the plant; (3) brown lung disease in workers in mills; (4) waste about 10% in harvest; (5) variable strength; (6) unpredictable price

Cellulosic

Rayon
(regenerated cellulose from wood pulp, especially higher molecular weight "alpha" fraction not soluble in 18% caustic)

1. Manufacture
   a. Steeping 1 hr in 18% caustic gives "soda" cellulose,

$$(C_6H_{10}O_5)_n + NaOH \longrightarrow [(C_6H_{10}O_5)_2 \cdot NaOH]_n$$

where some $-\overset{\displaystyle |}{\underset{\displaystyle |}{C}}-OH$ are converted into $-\overset{\displaystyle |}{\underset{\displaystyle |}{C}}-O^-Na^+$

b. Reaction with CS_2,

$$-\overset{|}{\underset{|}{C}}-O^-Na^+ \; + \; \underset{\underset{S}{\parallel}}{C}=S \longrightarrow -\overset{|}{\underset{|}{C}}-O-\underset{\underset{S}{\parallel}}{C}-S^-Na^+$$

about 1 xanthate for each two glucose units, soluble in 6% NaOH for spinning, called "viscose rayon"

c. Ripening—slow hydroysis

$$-\overset{|}{\underset{|}{C}}-O-\underset{\underset{S}{\parallel}}{C}-S^-Na^+ \xrightarrow{H_2O} -\overset{|}{\underset{|}{C}}-O-\underset{\underset{S}{\parallel}}{C}-SH \; + \; NaOH \xrightarrow{H_2O}$$

$$-\overset{|}{\underset{|}{C}}-OH \; + \; HO-\underset{\underset{S}{\parallel}}{C}-SH \longrightarrow CS_2 + H_2O$$

"viscose"—different molecular weight than starting cellulose, has some xanthate groups

d. Spinning—$ZnSO_4$, H_2SO_4 bath. H_2SO_4 neutralizes NaOH of viscose solution. $ZnSO_4$ gives intermediate which decomposes more slowly.

$$2-\overset{|}{\underset{|}{C}}-O-\underset{\underset{S}{\parallel}}{C}-S^-Na^+ + ZnSO_4 \longrightarrow -(\overset{|}{\underset{|}{C}}-O-\underset{\underset{S}{\parallel}}{C}-S)_2Zn \; + \; Na_2SO_4$$

$$\Big\downarrow H_2SO_4$$

$$2-\overset{|}{\underset{|}{C}}-OH \; + \; ZnSO_4 \; + \; 2CS_2$$

2. Properties

Dyeable (free hydroxyls), hydrophilic (comfortable), stable, low price, poor wash and wear characteristics

3. Uses are given in Fig. 17.3.

4. Economics

Peak production in 1950s, 1960s. Declined by 5.6% per year in 1970-1980, 4.9% annually in 1980-1990.

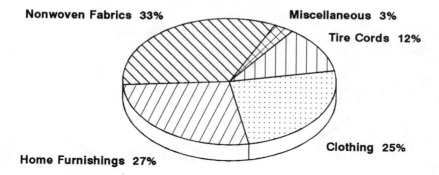

Nonwoven Fabrics 33%

Miscellaneous 3%

Tire Cords 12%

Home Furnishings 27%

Clothing 25%

Figure 17.3 Uses of rayon.

Acetate

1. Manufacture
    a. Esterification

$$[C_6H_7O_2(OH)_3]_n + H_2SO_4 + CH_3-\overset{\overset{O}{\|}}{C}-O-\overset{\overset{O}{\|}}{C}-CH_3$$

$$\downarrow$$

$$[C_6H_7O_2(OSO_3H)_{0.2}(AcO)_{2.8}]_n$$

    b. "Ripening" with $Mg(OAc)_2$, H_2O

cellulose sulfoacetate $\xrightarrow[\text{H}_2\text{O}]{\text{Mg(OAc)}_2}$ $[C_6H_7O_2(OH)_{0.65}(AcO)_{2.35}]_n$ + $MgSO_4$

   Usually the primary carbon has the OH, secondary carbons the OAc. Cellulose triacetate has three OAc groups.
2. Properties
    a. Acetate has lower tenacity than any other fiber because bulky OAc group keeps molecules far apart.
    b. Acetate has free OH so more easily dyed and more hydrophilic than triacetate.
    c. Random acetate groups make it less crystalline than triacetate. So triacetate is better for ironing. But triacetate gives stiffer fabrics with inferior drape and hand.

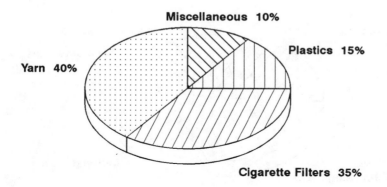

Figure 17.4 Uses of acetate. (*Source:* CP.)

    d. Both are softer than rayon but not so strong, both have poor crease re
       sistance, both are not colorfast.
3. Uses are shown in Fig. 17.4.
4. Economics
    Down 4.5%/yr from 1970-1980, 4.2% annually from 1980-1990

Noncellulosic

Nylon 6 and 66

1. Manufacture
    a. Nylon 66 developed by W. H. Carothers of Du Pont in 1930s. Adipic acid,
       HMDA, 280-300°C, 2-3 hr, vacuum. Trace of acetic acid terminates
       chains with acid groups and controls molecular weight.
    b. Nylon 6 developed by Paul Schlak in Germany in 1940. Caprolactam +
       heat + water as a catalyst.
    c. Fibers are melt spun (no solvent) while still above T_m (= 265-270°C
       for 66 and 215-220°C for 6). Extruded through spinerette.
    d. Fibers oriented by stretching to 4 x original length by cold drawing (two
       pullies at different speeds) or by spin drawing as it is being cooled.

Figure 17.5 Pilling chambers to test, by rapid rotation, the tendency of a fabric to form pills as in repeated washings and use. (Courtesy of Du Pont.)

2. Properties
   Strongest and hardest wearing of *all* fibers
   Heat stable
   Dyeable
   Disadvantages: hydrophobic ("cold and clammy"), degraded by UV, yellows with age, poor hand, tends to "pill" (gentle rubbing forms small nodules or pills, where surface fibers are raised, see Fig. 17.5).

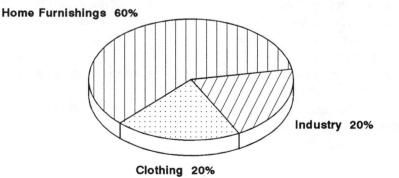

Figure 17.6. Uses of nylon. (*Source: C & E News.*)

3. Uses are given in Fig.17.6.

The clothes market includes especially stockings. Home furnishings are dominated by carpet and rope.

4. Economics

Production increased 5.7%/yr from 1970-1980, but only 1.2% annually from 1980-1990.

Nylon 66 is two thirds of the U.S. market, nylon 6 one third

1990 production = 2.7 billion lb

1990 commercial value = $3.2 billion

Polyester—poly (ethylene terephthalate)

1. Manufacture

a. Developed by Whinfield and Dixon in the U.K. Originally made by trans-esterification of DMT and ethylene glycol in a 1:2.4 ratio, distillation of the methanol, then polymerization at 200-290°C *in vacuo* with SbO_3 as catalyst. In early 1960s pure TA began to be used with excess ethylene glycol at 250°C, 60 psi to form an oligomer with n = 1-6, followed by polymerization as in the DMT method. In the 1970s three times as much DMT as TA was used. In the 1980s DMT =TA.

b. Melt spin like nylon, T_m = 250-265°C.

c. Orientation above T_g of 80°C to 300-400%.

2. Properties

Stable in repeated laundering

Complete wrinkle resistance

Blends compatibly with other fibers, especially cotton!

Can vary from low tenacity, high elongation to high tenacity, low elongation by orientation

Disadvantages: stiff fibers (aromatic rings), poor drape except with cotton blends, hydrophobic, pilling, static charge buildup, absorbs oils and greases easily (stains)

3. Uses are given in Fig. 17.7.

Clothing: suits, pants, shirts, dresses, and underwear either nonblended or blended with other fibers such as cotton

Home furnishings: carpets, pillows, bedspreads, hose, sewing thread, draperies, sheets, pillowcases

Industrial: tire cords

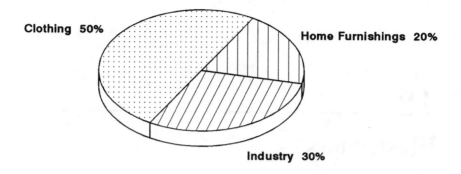

Clothing 50%

Home Furnishings 20%

Industry 30%

Figure 17.7 Uses of polyester. (*Source: C & E News.*)

4. Economics

Production increased 10.5%/yr from 1970-1980, but decreased 2.2% annually from 1980-1990.

1990 production = 3.2 billion lb

1990 commercial value = $2.3 billion

Figure 17.8 Spinning of fibers for use in carpets. (Courtesy of Du Pont.)

18

Elastomers

References

W & R II, pp. 126-142
S. Stinson, "Rubber Chemicals Industry Strong, Slowly Growing Despite Changes," *C & E News*, 5-21-90, pp. 45-66.
B. F. Greek, "Rubber Demand Is Expected to Grow After 1991," *C & E News*, 5-13-91, pp. 37-54.
G. B. Kauffman and R. B. Seymour, "Elastomers," Part I, *J. Chem. Educ.* **1990**, *67*, 422-425; Part II, **1991**, *68*, 217-220.

INTRODUCTION

Although naturally occurring rubber from the tropical tree has been known for ages, the Spanish navigator and historian Gonzalo Valdez (1478-1557) was the first to describe the rubber balls used by Indians. Natural rubber was brought back to Europe from the Amazon in 1735 by Charles Condamine, a French mathematical geographer, but it remained only a curiosity. Michael Faraday made a rubber hose from it in 1824. But it was not until Charles Goodyear discovered vulcanization in 1839 that natural rubber got its first wave of interest. As the story goes, Charles became so involved with his job that he set up a laboratory at home to study the chemistry of rubber. Because his wife hated the odor of his experiments, he could only continue his work at home when she was not around. While studying the effect of sulfur and other additives on the properties of rubber he was interrupted unexpectedly by his wife one day when she returned home early from shopping. He quickly shoved his latest mixture into the oven to hide it. As fate had it, the oven was lit, the rubber was vulcanized, and the modern era of elastomer research was born. His first patent covering this process was issued in 1844.

Today both natural rubber, an agricultural crop, and synthetic elastomers are multi-billion dollar businesses. In Fig. 16.1, page 312, we see that Synthetic Rubber (SIC 2812) totals $3.9 billion. It is a large area of polymer use and is 1.5% of Chemicals and Allied Products. But in the related industry covering final end products called Rubber and Miscellaneous Plastics Products (SIC 30, see Fig. 16.2, page 313), the subdivisions of Tires and Inner Tubes, Rubber and Plastics Footwear, Rubber and Plastics Hose and Belting, and Miscellaneous Rubber Products total $19.7 billion. The top polymer production summary in Table 1.18, pages 27-28, gives a numerical list of important synthetic elastomers. Styrene-butadiene rubber (SBR) dominates the list at 1.88 billion lb for U.S. production in 1990. All other synthetic elastomers are much smaller. While elastomers had a slight increase in production from 1980-1990, only 0.5% annually, SBR was down 2.3%/yr. The fastest growing elastomer is ethylene-propylene, up 5.9% annually. Fig. 18.1 gives a breakdown in consumption of elastomers including natural rubber.

NATURAL RUBBER

Natural rubber can be found as a colloidal emulsion in a white, milky fluid called latex and is widely distributed in the plant kingdom. The Indians called it "wood tears." It was not until 1770 that Joseph Priestly suggested the word *rubber* for the substance, since by rubbing on paper it could be used to erase pencil marks, instead of the previously used bread crumbs. At one time 98% of the world's natural rubber came from a tree, *Hevea brasiliensis*, native to the Amazon Basin of Brazil which grows to the height of 120 ft. Today most natural rubber is produced on plantations in Malaysia, Indonesia, Singapore, Thailand, and Sri Lanka. Other rubber-bearing plants can be cultivated, especially from a guayule shrub, which is now more important than the tree.

VULCANIZATION

The process that makes the chemistry, properties, and applications of elastomers so different from other polymers is cross-linking with sulfur, commonly called vulcanization.

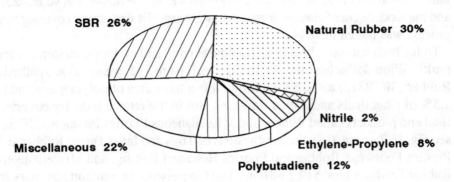

Figure 18.1 U.S. consumption of elastomers. (*Source*: *C & E News.*)

The modern method of cross-linking elastomers involves using a mixture of sulfur and some vulcanization accelerator. Those derived from benzothiazole account for a large part of the market today. Temperatures of 100-160°C are typical.

   2-mercaptobenzothiazole

Zinc oxide and certain fatty acids (*R*-COOH) are also added. Although this mechanism is by no means completely understood, it is proposed that the benzothiazole and zinc oxide give a zinc mercaptide, and this forms a soluble complex with the fatty acid:

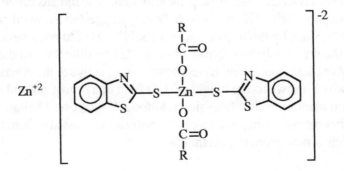

Reaction of this with S_8 molecules gives a persulfidic complex (X = benzothiazole).

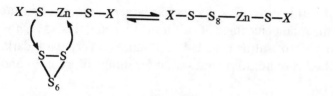

Interchange with the original complex leads to the formation of a mixture of polysulfidic complexes, which are considered to be the active sulfurating species:

$$X-S-S_8-Zn-S-X \quad + \quad X-S-Zn-S-X$$

$$\Updownarrow$$

$$X-S-S_n-Zn-S-X \quad + \quad X-S-S_{8\text{-}n}-Zn-S-X$$

$$\Updownarrow$$

$$X-S-S_x-Zn-S_y-S-X$$

These complexes then react with the allyl carbons of rubber, the most reactive sites in the polymer:

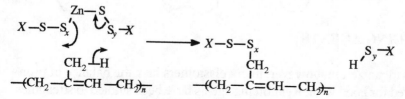

The cross-linking occurs by reactions of the following type, where

$$R- \; = \; -(CH_2-\underset{\underset{CH_2-}{|}}{C}=CH-CH_2)_n-$$

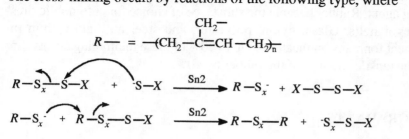

Usually a mono- or disulfide cross-link occurs but larger numbers of sulfur atoms are possible. If the total percentage of sulfur in the material is <5%, it is usually very elastic. If >5% of sulfur is added, it produces a very hard, dark, nonelastic material called ebonite, sometimes used for things like combs and buttons.

ACCELERATORS

In 1906 Oenslager and Marks at Diamond Rubber Co. (later B. F. Goodrich Co.) began working on accelerators for cross-linking with sulfur. These substances not only increase the rate of vulcanization but create a final product that is more stable and less susceptible to aging. Benzothiazoles now own 22% of the accelerator market, which is about 83 million lb/yr. Other types of accelerators are sulfenamides (50%), dithiocarbamates (5%), thiurams (4%), and others (19%). The reason for the differences is that some cause very fast vulcanization rates like sulfenamides, and some slower like benzothiazoles. Sulfenamides such as N-cyclohexyl-2-benzothiazolsulfenamide can be made from benzothiazoles by reaction of an amine and an oxidizing agent such as NaOCl, HNO$_2$, or H$_2$O$_2$.

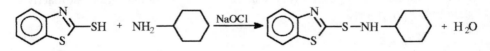

REINFORCING AGENTS

Even with vulcanization, however, many elastomers lack the balance of properties required for good wear. Reinforcing agents have been studied to strengthen the rubber mechanically. In 1912 the Diamond Rubber Co. found that addition of carbon to rubber tires caused them to last ten times longer than without this reinforcing agent. Rubber became the substance of choice for automobile tires and conveyor belts. Glass, nylon, polyester, and steel now aid carbon in reinforcement for many applications. Up to 20% of these reinforcing agents can increase the tensile strength of the rubber by 40%.

ANTIDEGRADANTS

Most polymers are attacked by oxygen, ozone, and ultraviolet light. Rubber is one such polymer that is rapidly degraded in molecular weight and mechanical strength. Over 100 chemicals, collectively called age resistors or antidegradants, are added to elastomers to keep them from becoming brittle, turning sticky, developing cracks, etc. The present market for these resistors is quite large, over 151 million lb/yr.

Most oxidation inhibitors today are either amines, phenols, or phosphites. Phenols were suggested as early as 1870 to combat aging. Amines are now used more than phenols in elastomers. Combinations are often used for heat, oxygen, ozone, UV, and moisture resistance. Two examples of amine age resistors are given here. The market is about 150 million lb, of which amines are about 60%. The market breakdown for antidegradants is phenylenediamines (50%), phenolics (13%), phosphites (13%), quinolines (10%), diphenylamines (6%), and others (8%).

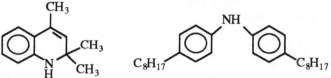

Other kinds of rubber chemicals are blowing agents, peptizers, and retarders. The total market for all chemical additives for rubber is nearly 250 million lb.

DEVELOPMENT OF SYNTHETIC RUBBER

Until the 1930s natural rubber from *Havea brasiliensis* was the only available elastomer. The United States had to, and still does, import every pound. Although research on synthetic substitutes began before 1940 in this country, World War II influenced speedy development of substitutes when our supply of natural rubber from the Far East was stopped. Gasoline had to be rationed not because of its shortage, but because of the automobile tire shortage.

In 1910 scientists concluded that natural rubber was *cis*-1,4-polyisoprene.

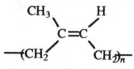

In 1931 Du Pont introduced the first synthetic elastomer, polychloroprene (Neoprene®, Duprene®), and Thiokol Corporation introduced a polysulfide rubber called Thiokol®. Polychloroprene, although very expensive compared to polyisoprene, has superior age resistance and chemical inertness. It is also nonflammable.

The Government Rubber Reserve Company in the 1940s pioneered the development of styrene-butadiene copolymers, by far the largest volume of

Billions of Pounds

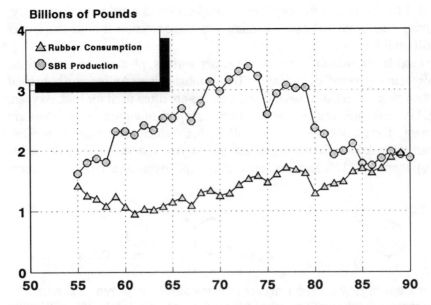

Figure 18.2 U.S. consumption of natural rubber vs. SBR production. (*Source:* CEH and *C & E News.*)

synthetic rubber used today. Now usually known as SBR, it has also been called Buna-S, Butadiene with a sodium (Na) catalyst and copolymerized with styrene, or GR-S ,Government Rubber Styrene. Although it took many years to develop, it is now the rubber of choice for most applications today, especially automobile tires.

Polyisobutylene, commonly called butyl, was first developed in 1937 by Esso Research and Engineering Co. Its main repeating unit is isobutylene but it contains some isoprene for cross-linking. Originally butyl was used for automobile tire inner tubes, where it was replaced in 1955 by tubeless tires. However, most inner tubes still employed today are butyl rubber. In addition to being used for engine mounts and suspension bumpers, it has found large volume uses as liners in reservoirs and in irrigation projects.

Hypalon® was introduced by Du Pont in 1952. Although not a high volume rubber it has found use in coatings and hoses.

CATALYSTS AND MECHANISMS

The mid-1950s saw the first commercial production by Goodrich, Firestone, and Goodyear of polymers with stereochemistry which is consistent or regular.

In the early 1950s Karl Ziegler in Germany and Giulio Natta in Italy found catalysts that polymerized olefins with regular configurations. The Ziegler-Natta catalysts were primarily a combination of a transition metal salt ($TiCl_3$ or $TiCl_4$) and an organometallic compound (Et_3Al). By proper manipulation of the ratio of these two substances either *cis*-1,4- or *trans*-1,4-polyisoprene from isoprene can now be prepared. The mechanism of Ziegler-Natta polymerization was given for polypropylene in Chapter 14, page 280. Review this and work through the mechanism with an elastomer monomer such as butadiene.

Many of the synthetic rubbers now made are still polymerized by a free radical mechanism. Polychloroprene, polybutadiene, polyisoprene, and styrene-butadiene copolymer are made this way. Initiation by peroxides is common. Many propagation steps create high molecular weight products. Review the mechanism of free radical polymerization of dienes given in Chapter 14, page 277.

Butyl rubber, polyisobutylene, is an example of cationic polymerization with an acid. Review Chapter 14, page 279. A small amount of isoprene is addedto enable cross-linking during vulcanization through the allylic sites.

$$R \oplus \; + \; CH_2{=}\overset{\overset{\displaystyle CH_3}{|}}{C}{-}CH{=}CH_2 \longrightarrow R{-}CH_2{-}\overset{\overset{\displaystyle CH_3}{|}}{C}{=}CH{-}CH_2\oplus$$

The more complex structure of this polymer must therefore be

$$(CH_2{-}\underset{\underset{\displaystyle CH_3}{|}}{\overset{\overset{\displaystyle CH_3}{|}}{C}})_{\overline{n}}{-}(CH_2{-}\overset{\overset{\displaystyle CH_3}{|}}{C}{=}CH\;{-}CH_2)_{\overline{1}}$$

SBR VERSUS NATURAL RUBBER

By far the largest selling elastomers are SBR and natural rubber. SBR accounts for about 40% of the U.S. synthetic rubber market and 26% of the total rubber market at 1.88 billion lb in 1990. The United States imported about 2 billion lb of natural rubber. A distant third is polybutadiene at 0.89 billion lb in 1990. In 1940 natural rubber had 99.6% of the U.S. market. Today it has only 30%. It was in 1950 that synthetic elastomer consumption passed natural rubber use in the United States. Since then it has been a battle between the leading synthetic, SBR, and the natural product. It is apparent that these two polymers are very important. Table 18.1 summarizes and compares them by their properties.

The balance between natural rubber and SBR is a delicate one. Natural rubber has made a comeback and reversed its downward trend. Developments of rubber

TABLE 18.1 SBR and Natural Rubber

Property	Natural	SBR
Tensile strength, psi	4,500	3,800
Percent elongation	600	550
300-400% modulus, psi	2,500	2,500
Temperature range for use, °C	-60-100	-55-100
Degree of elasticity	Excellent	Good
Tear resistance	Good	Moderate
Abrasion resistance	Moderate	Good
Age resistance	Poor	Moderate
Solvent resistance	Poor	Poor
Gas impermeability	Good	Moderate
Uniformity	Variable	Constant
Versatility	Lower	Higher
Processibility	Easier	Harder
Tolerance for oil & carbon additives	Lower	Higher
Price stability	Bad	Good
Percent of U.S. market	30	26

farming have raised the yield from 500 lb/acre/yr to 2000-3000. Petrochemical shortages and price increases have hurt SBR. Finally, the trend toward radial-ply tires, which contain a higher proportion of natural rubber, favors this comeback. Fig 18.2 shows the U.S. natural rubber consumption trends vs. U.S. SBR production, where this "bounceback" of the natural rubber market is very evident from 1980 to the present. Fig. 18.3 demonstrates the competitive price structure for these two elastomers through the years.

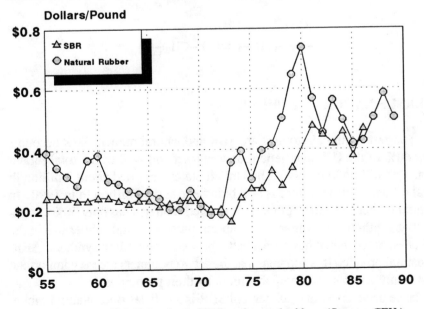

Figure 18.3 U.S. prices of SBR and natural rubber. (*Source*: CEH.)

TIRES

No discussion of elastomers is complete without a mention of tire technology. About 70% of all synthetic elastomers in the United States are used in tires. About 264 million tires are produced in the United States annually, 217 million for cars and the rest for trucks and busses. A typical tire is made up of four parts: (1) the tread, which grips the road; (2) the sidewall, which protects the sides of the tire; (3) the liner, which prevents air loss; and (4) the carcass, which holds the layers together (Fig. 18.4).

The tire is about 50% rubber by weight. Carbon black (as a reinforcing agent), extender oil, and the tire cord in the carcass make up the rest. The cord was rayon for many years. Glass fiber has also been popular. But now nylon, polyester, and steel are the major cord components. Steel is most popular in radial tires of the 1980s and is growing in importance as the primary reinforcing agent. About 75% of car radial tires and 92% of truck radials are steel belted.

The tread must have the best possible "grip" to the road. Grip is inversely related to elasticity, and natural rubber has good elasticity but poor grip, so no natural rubber is used in automobile tire treads. Treads are blended of SBR and polybutadiene in an approximate ratio of 3:1. Truck tire treads do have natural rubber, between 65-100%, to avoid heat buildup and because grip is not so necessary in heavy trucks. Aircraft tires consist of 100% natural rubber.

The carcass requires better flexing properties than the tread and is a blend of natural rubber and SBR, but at least 60% of natural rubber. The sidewalls have a lower percentage of natural rubber, from 0-50%. The liner is made of butyl rubber because of its extreme impermeability to air.

The most important single trend in the U.S. tire market is the switch from cross-ply and belted bias-ply to radial-ply tires. Radials held only 8% of the U.S. car tire market in 1972, but by 1977 it had grown to 50% and it is now 89%. The difference in the three is shown in Fig. 18.5.

A tire carcass contains plies of rubberized fabric. In the cross-ply the cords cross the tire at an angle. In the belted bias-ply the cords cross at an angle and an additional belt of fabric is placed between the plies and the tread. In the radial-ply the cords run straight across the tire and an extra belt of fabric is included. Radial tires have better tread wear average (66,000 miles radial, 40,000 miles bias-ply) and better road-holding ability. However, they are more easily damaged on the sidewall and they give a less comfortable ride. They also require a higher proportion (80% vs. 50%) of the more expensive natural rubber. It seems likely that the popularity of radial-ply tires will continue, and natural rubber consumption may continue its comeback. (See Fig. 18.6.)

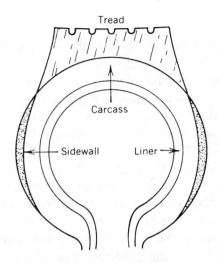

Figure 18.4 Parts of a typical tire. (*Source:* W & R II. Reprinted with permission of John Wiley & Sons, Inc.)

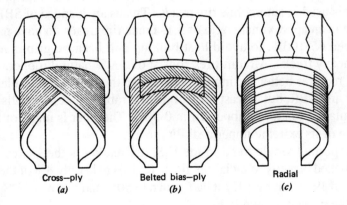

Figure 18.5 Types of plies in tires. (*Source:* W & R II. Reprinted with permission of John Wiley & Sons, Inc.)

Today the elastomer can be reclaimed from discarded tires. Over 2 billion are available for recycling. Most reclaiming of the elastomer is done by an alkali process with 5-8% caustic soda and heating. Reclaiming is not profitable unless it costs no more than half as much as pure elastomer, since reclaimed material contains only 50% elastomer hydrocarbon. Approximately 0.66 billion lb of elastomer is reclaimed each year in the U.S., only about 10% of the total elastomers used. Efforts are also being made to burn discarded tires for fuel to generate electricity, since each tire contains energy equivalent to 2.5 gallons of oil as fuel, enough to heat an average house for a day.

We will finish this chapter with Table 18.2 which gives many of the details for important elastomers including chemical structure, properties compared to natural rubber, uses, amount of production, and manufacturing process. Some familiarity with these elastomers is important.

Figure 18.6 Research size equipment for ply building in tires to test the usefulness of various fibers as the plies in tires. (Courtesy of Du Pont.)

TABLE 18.2 Important Elastomers

Name (s)	Trade name or Abbrev.	Structure and Average M.W.	Companies	Date Intro.	Uses	Properties	1990 U.S. Prod. billion lb	Approx. % in Tires	Manufacturing Process
natural rubber cis-1, 4-polyisoprene	NR	CH_3 H $C=C$ $-CH_2$ CH_2- 98% cis 350,000-500,000	—	1840	see Table 18.1	Automobile tires, conveyer belts	None (about 2.0 imported)	76	Biological polymerization
Poly(styrene-butadiene)	SBR Buna-S GR-S Ameripol Hycar	$(CH_2-CH=CH-CH_2)(CH_2-CH)\phi$ 75% butadiene by weight, 85% by moles butadiene units are mixed, 1,2 and 1,4	Copolymer Dow, Firestone, Goodyear, Reichhold	1933	Good overall properties, see Table 18.1	Automobile tires, carpet backing, paints, paper coatings, brake linings, chewing gum, footware, cement additive, floortile	1.88	57	Emulsion and solution polymerization, free radical catalyst at low temperature
Polybutadiene	BR Buna CB Ameripol CB Cariflex BR Budene	H H $C=C$ $-CH_2$ CH_2- mostly cis	Goodyear, Firestone, American, Synthetic, Polysar	1955	Excellent, abrasion resistance, low-temperature flexibility, poor traction, cheap price	Automobile tires, high impact polystyrene modifiers, belts, hoses, seals	0.89	68	Ziegler cat, solution & emulsion polymerization
Polychloroprene	CR neoprene Duprene GR-M Souprene	Cl $-CH_2-C=CH-CH_2-$ Mostly trans	Du Pont Mobay	1931	Flame, solvent, age, and heat resistance	Wire insulation, cable jackets, conveyer belts, safety clothing, hoses, building seals, highway joints, adhesives, shoe soles	0.19	1	Free radical emulsion polymerization

Type	Symbol/Trade names	Structure	Year	Properties	Applications			Method
Ethylene-propylene (5% diene for cross-linking)	EPDM EPM	$(CH-CH_2)(CH-CH)$ CH_3 55% E 40% P 5% D	1963	Low-temperature, flexibility, age, heat and abrasion resistance	Wire coatings, automobile tires (walls), auto bumpers, auto window seals, radiator & heater hose, footwear, roofing membrane, oil additives, polymer modifier	0.56	3	Ziegler catalyst
Polyisobutylene (0.6%-3.5% isoprene for cross-linking), Butyl rubber	IIR GR-I	$(CH_2-C_n\,(CH_2-C=CH-CH_2)_1$ with CH_3 groups 350,000	1937	Low permability to air & water, weather resistance, noise & vibration absorbance	Inner tubes, engine mounts, bumpers, reservoir liners, coatings, automobile tires (liner), adhesives	0.44	63	Low-temperature solution polymerization, cationic initiation
Poly (butadiene-acrylonitrile), Nitrile rubber	Buna-N NBR CHemigun Butaprene Hycar OR, G-R-N	$(CH_2-CH=CH-CH_2)(CH_2-CH)$ CN 10-40% acrylonitrile	1937	Solvent, fat & oil resistance, wide temperature performance, low coefficient of friction	Adhesives, hoses, footwear, coated fabrics, sealants, oil seals, flexible fuel tanks, O-rings, molded products, sponges	0.12	1	Emulsion polymerization, free radical catalyst
Synthetic natural rubber, cis-1,4-Polyisoprene	IR Ameripol SN Coral Natsyn Cariflex IR	CH_3, H ... $C=C$ ... $-CH_2$, CH_2- 94% cis	1955	Nearly identical to natural rubber, but not as flexible	Automobile tires, mechanical goods, footwear	0.1	65	Ziegler catalyzed

TABLE 18.2 Important Elastomers (continued)

Name (s)	Trade name or Abbrev.	Structure and Average M.W.	Date Intro.	Properties	Uses	1990 U.S. Prod. billion lb	Approx. % in Tires	Manufacturing Process
Polysulfide	Thiokol (in U.S.), Ethanite (in Belgium)	$-CH_2-CH-S_x-$ $x = 2-4$ 500-10,000	1963	Solvent resistance, low-temperature flexibility	Paint thinner, flexible gasoline hoses, sealants	Small	—	Condensation polymerization
Polysiloxane (silicone)	—	$(Si-O-Si-O)$ with R groups $R = CH_3\phi, CH_2=CH, CF_3CH_2CH_2$	1945	Chemical resistance, soluble, non-toxic, low flammability, heat stability, not biologically rejected.	Hydraulic fluids, lubricants, sealants, artificial tubes for human body, contact lenses, GC column packing, stopcock grease	Small	—	Hydrolysis of R_2SiCl_2 and dehydration
Chlorosulfonated polyethylene	Hypalon	$(CH_2-CH_2)(CH_2-CH)(CH_2-CH)$ Cl SO_2Cl 20,000 1 SO_2Cl for 90C's 1 Cl for 7C's	1952	Age resistance, electrical insulation	Protective coatings, wire coverings, chemical hoses	Small	—	Solution polymerization of polyethylene followed by chlorination and sulfonation

19

Coatings and Adhesives

References

Kent, pp. 787-808
W & R II, pp. 143-181
M. S. Reisch, "High-Tech Uses Drive Demand for High-Performance Adhesives," *C & E News*, 3-4-91, pp. 23-43.
M. S. Reisch, "Demand Puts Paint Sales at Record Levels," *C & E News*, 10-30-89, pp. 29-60.
P. Coombes, "Adhesives Find New Strength in Diversity," *Chemical Week*, 3-13-91, pp. 32-34.

INTRODUCTION TO COATINGS

Having treated three major end uses of polymers in the last three chapters, we now present the last two general areas of polymer use, coatings and adhesives. These are quite large areas of the chemical industry. The size of the coatings industry is best estimated by SIC 2851, Paints and Allied Products, which had 1988 U.S. shipments of $12.9 billion. The increase in this segment is graphed in Fig. 19.1 along with the three main subdivisions of the industry, Architectural Coatings (SIC 28511), Product Finishes for Original Equipment Manufacturers (SIC 28512), and Special Purpose Coatings (SIC 28513). The industry usually grows 1-2% annually, although in some years it is closer to 4%. It is 4.9% of Chemicals

371

Billions Of Dollars

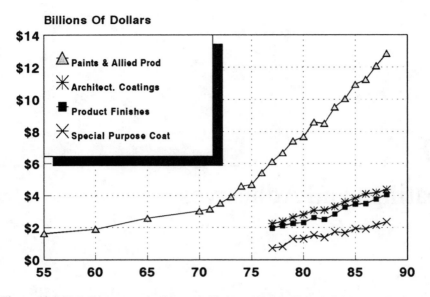

Legend:
- △ Paints & Allied Prod
- ✳ Architect. Coatings
- ■ Product Finishes
- ✕ Special Purpose Coat

Figure 19.1 U.S. shipments of paints and allied products and its subdivisions. (*Source*: AS.)

and Allied Products, which rivals that of fibers in size.

Coatings are sold in terms of volume rather than weight because they are usually solids dissolved in solvents. Approximately 1.1 billion gal/year are sold in the U.S. with 49% of this going to architectural coatings, 36% to product coatings, and 15% to special-purpose coatings. Table 19.1 gives a listing of the types of uses falling into these categories. Architectural coatings are used on houses and buildings and are applied by the personal consumer or by a professional painter. Product coatings are used by the manufacturer for many types of products. The automobile industry is the largest single user, applying around 10% of all coatings. Special-purpose coatings usually are specifically designed for one of the purposes listed in the table. The average price of coatings is $8-9/gal for architectural, $11-12/gal for product, and $13-14/gal for special-purpose.

Coating is a general term for a thin film covering something. It may be decorative, protective, or functional. Adhesives have a separate use because they are applied between two things. Adhesives will be discussed later in this chapter.

Coatings must have both adhesion and cohesion. *Adhesion* is the attraction of the coating to the substrate. It is the "outward force" of a coating. For good adhesion the coating must have weaker, stickier mechanical properties; it must be soluble, soft, and permeable to gases. *Cohesion* is the attraction of the coating to itself, the "inward force" of the coating. Good cohesion requires stronger mechanical properties; the coating must be only sparingly soluble, hard, and nonpermeable to gases. As you can see adhesion and cohesion are diametrically opposed. Coatings are a happy medium. They must have balanced properties.

TYPES OF COATINGS

There are hundreds of different types of coatings, many being complex formulations with numerous components. Indeed, to be successful a coatings chemist

TABLE 19.1 Coatings Fall into Three Major End-Use Groups

Architectural Coatings
    Exterior house paints
    Interior house paints
    Undercoaters, primers, and sealers
    Varnishes
    Stains
Product coatings-OEM[a]
    Wood furniture and fixtures
    Automotive Metal containers
    Machinery and equipment
    Factory-finished wood
    Metal furniture and fixtures
    Sheet, strip, and coil
    Transportation (nonautomotive)
    Appliances
    Electrical insulation
    Marine
    Film, paper, and foil
    Pipe
    Toys and sporting goods
    Miscellaneous consumer and industrial products
Special-Purpose Coatings
    Automotive and machinery refinishing
    High-performance
    Traffic paint
    Roof
    Bridge
    Aerosol
    Swimming pool
    Arts and crafts
    Metallic
    Multicolored

Source: C & E News, and SRI International.

[a] Original equipment manufacturers.

and company must be able to change a formulation to fit the needs of a given specific coating application, problem, and use. Close cooperation and communication between supplier and user is necessary.

But in general there are four broad classes of coatings: paints, varnishes, lacquers, and shellac. A definition of each class follows.

Paint: A coating with a colored pigment included. It is always opaque.

Varnish: A clear, transparent coating that dries by evaporation of solvent
(Enamel)   and oxidation or polymerization of a resin. It has no pigment.

Lacquer: A rapid-drying coating by evaporation of solvent only. It will redissolve in the solvent, and it may be lightly pigmented.

Shellac: A natural product from insect secretions. It is a hard tough coating like a varnish, which contains mostly aliphatic polyhydroxy acids about C_{60}.

BASIC COMPOSITION OF COATINGS

Although coatings are complex formulations, they can be divided into four main types of materials: a pigment, a binder, a thinner, and additives.

Pigment: An opaque coloring; some dyes used (transparent coloring) but not so common in the coatings industry

Binder: A polymer resin or resin-forming material

Thinner: A volatile solvent that deposits the film on evaporation

Additives: Defoamers, thickeners, coalescing agents, flowing agents, driers, biocides, etc.

The binder plus the thinner is often called the *vehicle* for the coating.

PIGMENTS

There many reasons why pigments are added to coatings to make paints. The pigment imparts opaqueness and color to the coating. It can adjust the gloss or shininess of the coating, improve the anticorrosive properties of the coating, and does help to reinforce the binder.

Pigments must have certain key properties. They are rated in *hiding power*, the ability to obscure an underlying color. Hiding power is proportional to the the difference between the index of refraction of the pigment and the vehicle. It is expressed in square meters of surface that can be covered per kilogram of paint. Dark pigments are usually higher in hiding power than are light ones. Smaller particle size is better than the larger size pigment particles.

Pigments also have a *tinting strength*, the relative capacity of a pigment to impart color to a white base. In addition to these properties, pigments must be inert, insoluble in the vehicle but easily dispersed, unaffected by temperature changes, and nontoxic. We will briefly discuss two important inorganic and organic pigments as examples. The inorganic pigments are known for their superior hiding power, the organic pigments for their tinting strength. Dyes and pigments are a vast industry in themselves. We will cover just a couple of examples.

Inorganic Pigments

Titanium Dioxide. This is by far the most important white pigment. It has dominated the market since 1939 and is in the top 50 chemicals. About 2.16 billion lb were sold in 1990, and total commerical value was $2.1 billion in 1990. Two types of titanium dioxide are made, anatase and rutile. Anatase is made by taking ilmenite ore ($FeO \cdot TiO_2$) and treating it with sulfuric acid to purify and isolate TiO_2. Rutile is made by taking rutile ore (mostly TiO_2), chlorinating and distilling $TiCl_4$, and oxygenating back to pure TiO_2. The hiding power and index of refraction of the two types of titanium dioxide, along with two other common inorganic pigments, is given here. Note the superiority of rutile and anatase in hiding power. Most vehicles have an index of refraction about 1.5. Although the hiding power increases as the difference in index of refraction between pigment and vehicle increases, it jumps very high for titanium dioxide. Rutile is slightly more expensive than anatase. They are about $1.00/lb in 1991.

Pigment	Hiding Power (m^2/kg)	Index of Refraction	Tinting Strength
Rutile	30.1	2.76	1800
Anatase	23.6	2.55	1250
Lithopone ($BaSO_4 \cdot ZnS$)	5.5	2.37	280
Zinc oxide (ZnO)	4.1	2.02	210

Carbon Black. This most widely used black pigment is also in the top 50 chemicals. About 2.87 billion lb of carbon black were made in 1990. Commercial value in 1990 was $0.6 billion at 22¢/lb, but 94% of this is used in the tire industry for reinforcement of elastomers. Only 1% is used in paints. Carbon black is made by the partial oxidation of residual hydrocarbons from crude oil. The hydrocar-

$$CH_2 \longrightarrow C + H_2$$
$$CH_2 + {}^1\!/_2O_2 \longrightarrow CO_2 + H_2O$$

bons are usually the heavy by-product residues from petroleum cracking, ideally high in aromatic content and low in sulfur and ash, bp around 260°C.

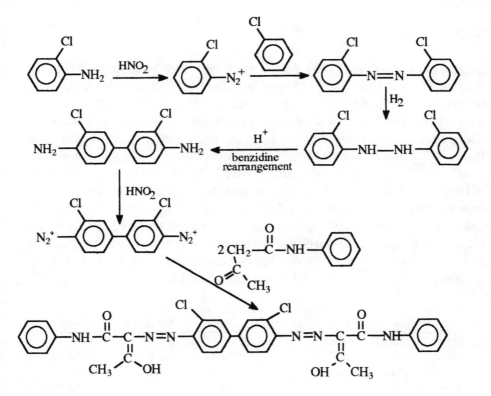

Figure 19.2 Synthesis of benzidine yellow.

Organic Pigments

Benzidine yellow (pigment yellow 12) is the highest volume organic pigment; over 8 million lb are made yearly. It is an example of a large class of organic pigments that contain an azo linkage, -N=N-. Its synthesis relies very heavily on diazonium salt coupling reactions and the benzidine rearrangement. Although benzidine is banned in the United States because of suspected carcinogenicity, 3,3'-dichlorobenzidine is only a "controlled substance" and its manufacture is at present permitted under rigorous safeguards. The synthesis is outlined in Fig. 19.2.

Pigment blue 15:3 is a copper phthalocyanine that is easily made from phthalic anhydride, urea, and a copper salt. About 7 million lb are made each year and it is the second largest volume organic pigment.

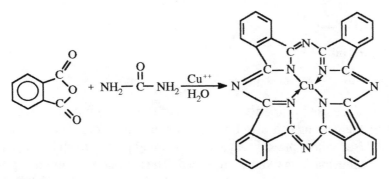

BINDERS

Perhaps the most important part of a coating is the binder or resin. Binders can be in the form of solutions, where the resin is dissolved in a solvent, or a dispersion, where the resin is suspended in water or an organic liquid with a particle size of 10 microns or less. We will subdivide our discussion of binders into four primary types and summarize their important properties and uses.

Natural Oils

The oldest type of coating is that which has a binder made from natural oil, especially linseed oil. Linseed oil is the triglyceride of linolenic acid, a natural fatty acid. The double bonds polymerize and cross-link when exposed to atmospheric oxygen catalyzed by lead, cobalt and manganese salts of fatty acids.

The binders are good for coating "receptive" (porous) surfaces such as wood, paper, and cellulose but not metal, glass, or plastic. They must be highly pigmented for good protection from ultraviolet aging. Linseed oil based coatings make very flexible films, especially good for easily swelled wood substrates. They are porous films that allow moisture to escape. They are easily applied with a brush because of their low molecular weight but are not very resistant to abrasion or chemicals. These coatings have been partially replaced by alkyd resins for exterior paints and by waterborne emulsions for interior use.

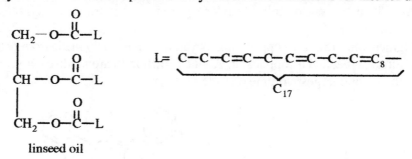

linseed oil

Vinyls

Vinyl coatings are used primarily on metal surfaces. They provide excellent protection by their strong cohesive forces, although their adhesion to the metal is not good. Used with a phosphoric acid-containing primer to etch the metal surface, this adhesion is markedly improved. The primer also contains poly (vinyl butyral) and is approximately 0.2-0.3 mil thick (1 mil = 1/1000th inch). Poly (vinyl butyral) is made from polymerized vinyl acetate by hydrolysis and reaction with butyraldehyde.

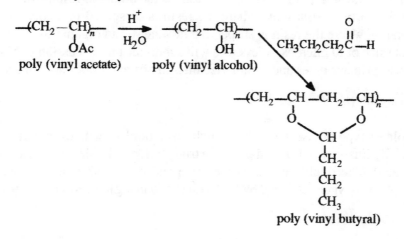

poly (vinyl butyral)

After this primer is applied vinyl chloride-vinyl acetate copolymer is added in a series of thin films. The total thickness is usually 5 mils. Pigments like iron oxide, lead, or zinc chromate prevent corrosion of the metal substrate in acid environments and may also be included in the coating. The final coated metal has good resistance to water and many chemicals with about a ten year lifetime.

Alkyd Resins

The structural chemistry of alkyds has already been covered in Chapter 15, page 301. Although there are over 400-500 varieties of such resins, they all are polyesters with carbon-carbon double bonds that can be cross-linked. They are very versatile in coatings, and their diverse properties can be matched for particular uses. They are the most widely used resins for protective coatings. Their best points can be summarized as follows: (1) easy to apply; (2) can have flat, semigloss, or high-gloss finish; (3) useful for most surfaces except concrete or plaster (alkaline); (4) good color retention; and (5) odorless (some of them).

There are important modifications of alkyds that help in specific applications. Phenolics can be added to improve film hardness and water resistance, but these confer increased yellowing tendencies on the final coating. Silicones impart heat resistance and exterior durability. Styrene increases the drying speed. Methyl methacrylate when added also gives faster drying properties and improves the color and durability of the coating.

Latex or Emulsion (Water Dispersion)

About 90% of all interior wall paint is now water dispersed. These coatings have become very popular because of easy cleanup, nonflammability, and minimization of air pollution. There are three principal types.

1. Styrene-butadiene binders started the "water revolution" in coatings after World War II. Most formulations used about a 2:1 ratio of styrene to butadiene (just the reverse of SBR elastomer). The coatings are low cost and provide excellent alkali resistance.
2. Poly (vinyl acetate) dispersions are more sensitive to hydrolysis but have good color retention and resistance to grease and oil.
3. "Acrylics" is a general term referring especially to polymers of methyl methacrylate and acrylic acid esters as binders. They are tough, flexible films with excellent durability and color retention, along with alkali, water, grease, and oil resistance. But they are more expensive. The acrylics became popular when

Figure 19.3 Full-scale coating line for industrial application. This line processes water-based coatings. (Courtesy of W. H. Brady Co., Milwaukee, Wl.)

Du Pont found that a molecular weight of 100,000 gives the desired properties. Today most automobiles are finished with acrylics.

SOLVENTS

Although we usually think of solvents as being just something to dissolve a solute, the chemistry and formulation technology of these materials in the coatings industry is very complex and critical to the success of the finished product. There is a large market for solvents, too, since 20% of all industrial solvents are used in coatings. Aliphatic hydrocarbons are the preferred solvents because they are inexpensive and lack environmental effects, but they are not so good at dissolving most binders. Alcohols are generally well-liked and accepted in the industry. Although aromatic hydrocarbons and ketones or esters dissolve well they cause much more profound pollution problems in the form of smog. Their emissions are restrictive. Water, of course, is ideal for those binders that form good dispersions (see Fig. 19.3). Finally, chlorinated solvents are non-flammable but have for the most part found limited applications lately due to their high toxicity. Today in the coatings industry there is a big drive to replace solvents which cause air pollution problems. Certain ones, such as toluene and xylene, are categorized as VOCs (volatile organic compounds) and have restrictions on their use. One chemical, although it is chlorinated, which is replacing some VOCs in coatings is 1,1,1-trichloroethane, or methyl chloroform. It does get the VOCs down, but it is more expensive. Methyl chloroform is however only a temporary solution, since it is known to cause some stratospheric ozone depletion, though much less than CFCs. Methyl chloroform is also somewhat toxic, though less than other chlorinated solvents. Other newer solvents are methyl amyl ketone (2-heptanone) and methyl propyl ketone (2-pentanone), which have high solvency power and low weight per gallon.

INTRODUCTION TO ADHESIVES

An adhesive or bonding agent is any substance that produces a bond between two or more similar or dissimilar substrates. In other words, it holds two things together. The term *adhesive* has become generic and includes more popular terms such as cement, glue, and paste. *Sealants* or caulks, having a different type of use, fill gaps or joints between two surfaces. They must remain flexible and also prevent the passage of liquid or gas between surfaces. Some products serve both adhesive and sealant purposes. The adhesives and sealants business is particularly difficult to define since many substances with adhesive and sealant properties are used in borderline applications where bonding and sealing are secondary to such primary functions as coating.

The importance of adhesives lies in the fact that they allow for a combination of the properties of dissimilar materials. For example, a laminate of polyethylene, with its heat sealability and water resistance, is ideally combined with cellophane, a grease resistant material that accepts ink printing, for packaging applications.

MARKET FOR ADHESIVES

Fig. 19.4 gives the trend in U.S. shipments of Adhesives and Sealants (SIC 2891) and its major subdivisions of Natural Base Glues and Adhesives (SIC 28913), Synthetic Resin and Rubber Adhesives (SIC 28914), and Caulking Compounds and Sealants (SIC 28916 plus 28917). Adhesives and sealants are a fast growing business. U.S. shipments for 1988 totalled $4.8 billion, which is 1.9% of Chemicals and Allied Products. From 1970-1980 the average annual growth was 14%. In 1980-1988 the growth was 16% annually. The predictions for the early 1990s is a growth rate of 8%/yr.

The total weight of adhesives and sealants is difficult to estimate, but exceeds 15 billion lb containing probably 3 billion lb of polymers. The business is very diffuse and is one of the few areas of the chemical industry where a small

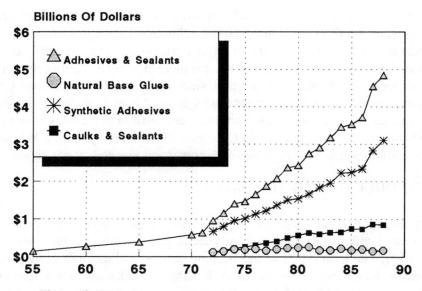

Figure 19.4 U.S. shipments of adhesives and sealants. (*Source:* AS.)

company can thrive. Over 500 U.S. companies manufacture adhesives and sealants. Henkel has the most sales. Industrial use is high (60%), but consumers (20%) and craftspersons (20%) also use them.

THE ADHESION PROCESS AND ADHESION PROPERTIES

There is still much art as well as science in the adhesion field, but a few generalizations can be made. Adhesion occurs by forming electrovalent or covalent bonds or by the use of weaker secondary attractions such as hydrogen bonding, London, dispersion, induced dipole, or van der Waals forces. Molecular "nearness" and good "wetting" properties are important. The adhesive must be flexible for good nearness. The substrate (to which the bond is made) must have a clean surface. A thin layer of adhesive is better than a thick layer. The thinner the layer is, the greater will be the adhesive force and the less will be the cohesive force. Putting on an adhesive in a series of these layers is ideal.

Finally, the solubility parameter of the adhesive and the substrate must be close. Without getting too technical, the solubility parameter is a rough estimate of polarity. The old saying "like dissolves like" can be extended to "like bonds like." More accurately, the solubility parameter is the calculated potential

TABLE 19.2 Solubility Parameters

Solvent	S.P.	Polymer	S.P.
n-Hexane	7.3	Poly (tetrafluoroethylene)	6.2
Cyclohexane	8.2	Poly (chlorotrifluoroethylene)	7.2
1,1,1-Trichloroethane	8.3	Poly (dimethyl siloxane)	7.3-7.6
Carbon tetrachloride	8.6	Ethylene-propylene rubber	7.9
Toluene	8.9	Polyethylene	7.9-8.1
Ethyl acetate	9.1	Natural rubber	7.9-8.3
Trichloroethylene	9.2	Polystyrene	8.6-9.1
Methyl ethyl ketone	9.3	Poly (methyl methacrylate)	9.3
Methyl acetate	9.6	Butadiene-acrylonitrite rubber	9.5
Cyclohexanone	9.9	Poly (vinyl chloride)	9.5-9.7
Dioxane	10.0	Epoxy resin	9.7-10.9
Acetone	10.0	Polyurethane resin	10.0
Carbon disulfide	10.0	Ethyl cellulose	10.3
Nitrobenzene	10.0	Poly (vinyl chloride-acetate)	10.4
Dimethylformamide	12.1	Poly (ethylene terephthalate)	10.7
Nitromethane	12.6	Cellulose acetate	10.4-11.3
Ethanol	12.7	Cellulose nitrate	9.7-11.5
Dimethyl sulfoxide	13.4	Phenol-formaldehyde resin	11.5
Ethylene carbonate	14.5	Poly (vinylidene chloride)	12.2
Phenol	14.5	Nylon 66	13.6
Methanol	14.5		
Water	23.2		

Source: W & R II. Reprinted by permission of John Wiley & Sons, Inc.

energy of 1 cm^3 of material for common solvents. Polymers are assigned solubility parameters of solvents in which they are soluble. Table 19.2 lists solubility parameters for various solvents and polymers. As an example of how to use this table, butadiene-acrylonitrile rubber with $\delta = 9.5$ bonds natural rubber ($\delta = 7.9$-8.3) to phenolic plastics ($\delta = 11.5$). Note that its solubility parameter is between that of the two substrates.

FORMS OF ADHESIVES

Adhesives are available in various physical states and are applied by many different techniques: water dispersions (43%), organic soluble (16%), two-component (11%), hot melt (21%), and miscellaneous (9%). Water dispersions deposit a film of adhesive after the water evaporates. Organic soluble adhesives form films when the organic solvent evaporates. Some adhesives are two-component solventless systems that react and form a strong cross-linked thermoset upon mixing. Examples would be epoxy resins (epoxy plus amine) and unsaturated polyesters (polyester plus styrene). Hot melts are thermoplastic resins that can be melted to a freely flowing material, be applied to the substrate, and form a good bond upon cooling. The advantage of hot melts is that they can be applied in an automated high-speed process. We are all familiar with common pressure-sensitive adhesives (PSAs), such as Scotch® tape, which have a permanently sticky adhesive on the tape that is applied to a second substrate with minimum pressure, good for temporary bonding uses. Finally, newer methods of bonding are being researched, such as those involving ultrasonic energy, magnetic fields, dielectric sealing, or ultraviolet light.

CHEMICAL TYPES OF ADHESIVES AND SAMPLE USES

Adhesives are so numerous and versatile that it is difficult to generalize by chemical type. For purposes of organization we divide them into four general areas and list a few representative examples of these types. Review the structure and chemistry of these materials as discussed in Chapters 14 and 15.

Thermoplastic

1. Poly (vinyl acetate) emulsions are used in bookbinding, milk cartons, envelopes, and automobile upholstery.
2. Polyethylene and polypropylene are used as carpet backing and as hot melts in packaging.
3. Poly (vinyl chloride) is used as a plastisol (dispersion with plasticizers), solution, or water dispersion. It is a good cement for pipes and is used extensively as an adhesive in automobiles.
4. Poly (vinyl butyral) has clarity and a refractive index similar to glass. It has good flexibility and adhesion to glass, even at low temperatures. It is used as safety glass interlining.
5. Nylon 6,36 has good adhesion to metal. It is a common beer can adhesive.

Thermoset

1. Phenolics are used in bonding wood and plywood. They are also good adhesives for automobile brake linings. A phenolic plus poly (vinyl butyral) is used to bond copper to paper or glass fiber for printed circuits.
2. Urea-formaldehyde and melamine-formaldehyde adhesives are resins in particleboard.

3. Epoxy adhesives are common two-part consumer glues that bond concrete blocks together and keep glass reflectors on highways in place.
4. Unsaturated polyesters are auto body fillers. They are commonly used in place of solder for many applications.
5. Urethanes (alkyd resin plus diisocyanate) bind sand to form a temporary mold in many foundry operations.
6. Cyanoacrylate adhesives, the famous consumer Super glue®, is a monomer that polymerizes when it comes in contact with moisture, even with atmospheric moisture (general structure $-CH_2-\underset{\underset{\displaystyle COOR}{|}}{\overset{\overset{\displaystyle C \equiv N}{|}}{C}}-$).

Elastomeric

1. Both natural rubber and SBR are used in Scotch® tape, masking tape, and adhesive-backed floor tiles. They are used in automobiles to bond fabric, carpets, and tire cord.
2. Polychloroprene plus a phenolic is a "contact adhesive" to bond flooring to concrete or wood, to attach soles to footwear, and to bond vinyl seats and roofs to automobile bodies.

Natural Products

1. Starch glues are water dispersions for "library paste" and wallpaper paste. They coat paper for better receptivity of inks and to keep the inks "held out" on the paper surface. They are also used in corrugated cardboard and paper laminating.
2. Protein glues are good for bonding rubber to steel, cork to plywood.
3. Asphalt makes a good adhesive for roofing of homes.

TABLE 19.3 Adhesives and Their Applications

	Thermoplastics						
	Poly(vinyl Acetate)	Polyolefins and Copolymers	Poly(vinyl Chloride) and Copolymers	Acrylics	Poly(vinyl Alcohol)	Poly(vinyl Butyral)	Polyamides
Construction	A		C		B		
Wood bonding and furniture	C	A		C			
Packaging	A	A			A		C
Textile fabrication	C	B	C	A			C
Automotive manufacture			A	A		B	C
Foundry binders							
Pressure sensitive adhesives				B			
Aircraft and aerospace					C		C
Electrical/electronics		C			C	C	C
Shoe construction		C				C	C
Bookbinding	C	C					
Abrasives							
Product assembly		B					C
Consumer	B						

Source: W & R II. Reprinted by permission of John Wiley & Sons, Inc.
  AA, very large usage (over 200 million lb/yr); A, large usage (50-200 million lb/yr); B, medium usage (10-50 million lb/yr); C, small usage (<10 million lb/yr).

TABLE 19.3 (Continued)

	Thermosets									Elastomers					Natural Products		
Phenolics	Urea and Melamine Resins	Epoxy Resins	Furan Resins	Polyurethane	Polyesters	Silicones	Cyanoacrylates	Anaerobics	SBR	Natural Rubber	Neoprene	Butadiene-Acrylonitrile Copolymer	Polybutenes-Polyisobutenes	Starch and Dextrins	Bituminous Resins	Animal Glue
A		B			C	C			A	B				A	A	
AA	AA										A					
			C						C					AA	A	A
			C						AA							
				A						A						
A			A	B												
									B	A			C			
		C							C		C	C				
		C			C		C	C							C	
			C							C	C					
																C
B	C	C								C						C
	C	C			C		C	C					C		C	C
	C	C						C		C	C	C				

387

USE SUMMARY

The single largest industrial application for adhesives is now in paper and packaging, accounting for over 35% of total consumption. About 21% of adhesives are utilized in woodworking and various wood products, especially plywood and particleboard. The construction industry accounts for 24% of the physical volume of adhesives, especially with increased numbers of prefinished products and factory-built home and modular units. Automotives is fourth with about 10% of the adhesive market. Table 21.2 gives a good summary of various adhesives and their areas of application, along with the approximate market volume. Note those labeled as AA, the large-use applications over 200 million lb/yr.

20

Pesticides

References

Kent, pp. 747-786
W & R II, pp. 339-362
W. J. Storck, "Pesticides Growth Slows," *C & E News,* 11-16-87, pp. 35-42.

WHAT NEXT?

Thus far in our study of industrial chemistry we have covered in some detail the top 100 chemicals produced as well as the important polymers made by the chemical industry. We have especially studied their manufacture and end uses, but have also looked at some history, economics, and toxicological and environmental problems associated with some of these products. In terms of net worth of shipments coming from these sectors, this is over half of the chemical industry. What should we study next? Table 20.1 lists the 1988 U.S. shipments of the most important sectors of Chemicals and Allied Products along with other sectors of the chemical process industries. If we add up Industrial Inorganic Chemicals, Industrial Organic Chemicals, Plastics, Cellulosic and Noncellulosic Fibers, Synthetic Rubber, Paints and Allied Products, and Adhesives and Sealants, we have 57% of Chemicals and Allied Products. We have also covered some of the chemistry associated with sectors outside Chemicals and Allied Products, such as Petroleum and Coal Products and Rubber and Miscellaneous Products, also big industries. We will now present in the following chapters some specific selected technologies which will allow us to learn about other important areas of the

TABLE 20.1 Shipments of Selected Sectors of Chemicals and Allied Products and Other Chemical Process Industries

SIC Code	Name	Shipments $ Billion	%
281	Industrial Inorganic Chemicals	22.110	8.5
286	Industrial Organic Chemicals	59.972	23.1
2821	Plastics	33.823	13.0
2823	Cellulosic Fibers	1.346 ⎫	4.2
2824	Organic Fibers, Noncellulosic	9.512 ⎭	
2822	Synthetic Rubber	3.916	1.5
2851	Paints and Allied Products	12.851	4.9
2891	Adhesives and Sealants	4.845	1.9
2873	Nitrogenous Fertilizers	3.281 ⎫	
2874	Phosphatic Fertilizers	4.150 ⎬	3.4
2875	Fertilizers, Mixed Only	1.506 ⎭	
2879	Agricultural Chemicals (Pesticides)	6.360	2.4
283	Drugs	43.987	16.9
284	Soaps, Cleaners, and Toilet Goods	37.856	14.6
	Other	14.184	5.6
28	**Chemicals and Allied Products**	**259.699**	**100.0**
26	Paper and Allied Products	122.560	
29	Petroleum and Coal Products	131.415	
30	Rubber and Miscellaneous Plastic Products	94.200	

Source: AS.

chemical industry. After we complete our study of Agricultural Chemicals (Pesticides), Fertilizers, Drugs, and Soaps, Cleaners, and Toilet Goods, we will have increased our coverage of SIC 28 to 94%. Other miscellaneous areas are suggested for self-study in the Appendix with some leading references. We have also included a chapter on Paper and Allied Products, a fascinating industry with some interesting chemistry.

INTRODUCTION TO PESTICIDES

Although pesticides have been criticized for many years as having many side effects, it should be remembered that chemicals have been a prime factor in agriculture's ability to keep pace with the hunger problem in the world. Production of food crops would decrease by 30% without pesticides. Production of livestock would drop 25%. Food prices would increase by 50-75%. Because of pesticides, farmers can conserve wildland since they need only half the land

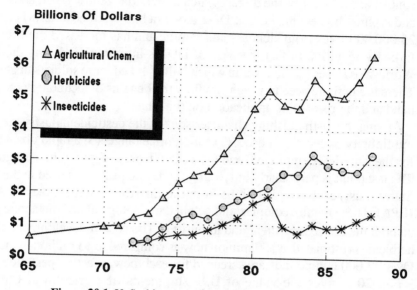

Figure 20.1 U. S. shipments of pesticides. (*Source*: AS.)

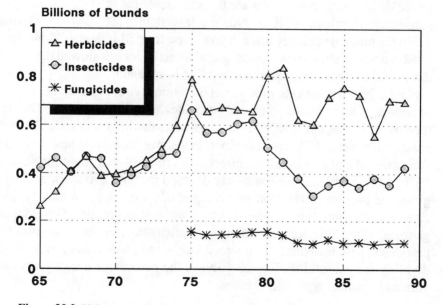

Figure 20.2 U.S. production of pesticides. (*Source*: *C & E News* and CEH.)

they used previously for the same amount of crops. Also, pesticides have helped control many insect-borne diseases such as malaria, yellow fever, encephalitis, and typhus. It is estimated that DDT alone (a bad word nowadays) has saved 25,000,000 lives from sickness and famine and has increased the lifespan in India by 15 years. In fact, because of DDT's introduction during the 1940s, World War II was the first war in which bullets killed more soldiers than insects. There is a definite need to weigh carefully the benefits of technology on the one hand and the risks in its application on the other.

Among the distinguishing characteristics of the pesticide industry are (1) the multiplicity of chemicals used, (2) a high price range, (3) a rapid obsolescence for the chemicals employed, and (4) a high degree of government regulation. The last point is particularly interesting. Today all pesticides used in the United States must receive registrations from the Environmental Protection Agency (EPA). These require complex and detailed toxicological and metabolic studies on both the active ingredient and impurities. The cost of development of a new herbicide is estimated at $50 million now as compared to $3 million in the 1950s. Over 22,000 compounds are screened for each new effective pesticide.

Fig. 20.1 gives the value of U.S. shipments of Agricultural Chemicals excluding fertilizers (SIC 2879), the best government estimate of the size of the pesticide industry. Also given are the shipments for the two most important subsectors, Herbicides (SIC 28796) and Insecticides (SIC 28795). During the 1970s the pesticide industry had a dramatic rise from $1 billion to $5 billion. The 1982 recession and the decreasing use of insecticides has slowed the growth in the 1980s, though some recent rebounding is apparent.

Fig. 20.2 shows the U.S. production of various types of pesticides in billions of pounds. Over a billion lb of total pesticides are made each year. Throughout the 1940s, 1950s, and 1960s insecticides were the largest branch. Herbicides passed them up in 1970 and have increased since then while insecticides have decreased. Total production of insecticides in 1989 was about the same as in 1970, while herbicide use has nearly doubled in those two decades. Fig 20.3 gives the present production percentages of types of pesticides, which are categorized by the type of pest that they attempt to control. Other kinds are germicides, rodenticides, and miticides. Agriculture uses about two thirds of all pesticides, with industrial, commercial, home and garden, and government use dividing the other third. Fig. 20.4 shows the percentage use of pesticides on various important agricultural crops.

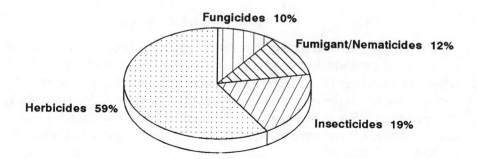

Figure 20.3 Types of pesticides. (*Source*: CEH.)

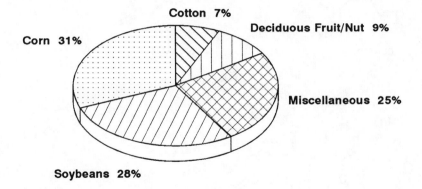

Figure 20.4 Uses of pesticides on crops. (*Source*: CEH.)

INSECTICIDES

History

Besides causing sickness, death, famine, and suffering, insects alone cause a large financial loss. People have tried to control insects since antiquity. Most early insecticides (first-generation insecticides) were inorganic compounds of arsenic, lead, copper, and sulfur. Bordeaux mixture (copper sulfate/calcium hydroxide) is still used sparingly today. Millions of lb of lead and arsenic pesticides used in the first half of this century still cause occasional problems as soil residues where concentrations were high, such as at mixing stations. Lead arsenate is still used against the potato beetle, since no residue stays in the potato itself.

$$Pb(NO_3)_2 + H_3AsO_4 \longrightarrow PbHAsO_4 + 2HNO_3$$

Early insecticides also included organic natural products such as nicotine, rotenone, and pyrethrin. Rotenone is used today as a method of killing rough fish when a lake has been taken over completely by them. In a couple of weeks after treatment the lake is then planted with fresh game fish. The pyrethrins, originally obtained from Asian or Kenyan flowers, can now also be synthesized laboriously. Nicotine is no longer used as an insecticide because it is not safe for humans (smokers note).

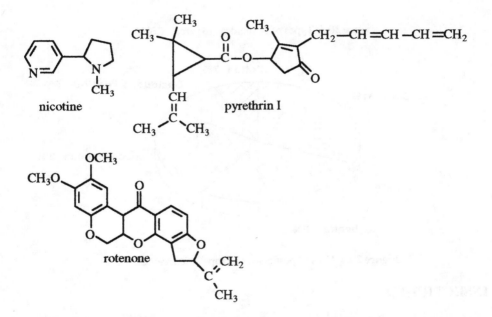

nicotine

pyrethrin I

rotenone

As late as 1945 inorganic chemicals accounted for almost 75% of all pesticide sales, with oil sprays and natural products being most of the remainder. None of these first-generation insecticides are used much now. During the 1950s the second generation of insecticides made an explosive growth with the development of DDT and other chlorinated hydrocarbons. Second-generation insecticides are of three major types: chlorinated hydrocarbons, organophosphorus compounds, and carbamates. Synthetic pyrethroids are a recent fourth type. Fig. 20.5 pictures the trend in production of types of insecticides through the years. A very dramatic decline of the chlorinated hydrocarbons in the late 1960s and the 1970s, while organophosphates and carbamates increased, is striking.

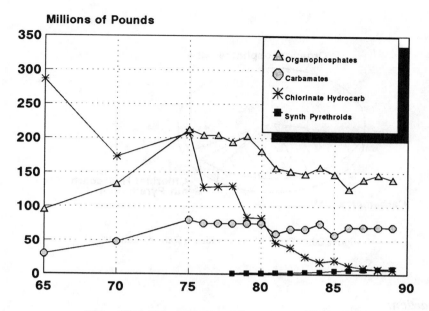

Figure 20.5 U.S. production of insecticides. (*Source*: CEH.)

Most of the chlorinated hydrocarbons are banned from use in the U.S. today, but they still are applied in other countries and historically they are very important, so we will cover them first, even though they are presently only 3% of the total production of insecticides, as shown in Fig. 20.6.

Chlorinated Hydrocarbons

DDT. DDT is no longer being used in large amounts in this country because of its persistence in the environment, although for many uses there were no good substitutes available. *D*ichloro*d*iphenyl*t*richloroethane (DDT) was first made back in 1874 by Zeidler in Germany, but its insecticidal properties were not discovered until 1939 by Dr. Paul Mueller of Geigy Chemical Company in Switzerland. He received the Nobel Prize in Medicine and Physiology in 1948 for this work. Chloral (trichloroacetaldehyde) can be made from the chlorination and oxidation of ethanol in one step.

$$CH_3-CH_2-OH + 3Cl_2 + {}^1\!/_2O_2 \longrightarrow Cl_3C-CH=O + 3HCl + H_2O$$
$$\text{chloral}$$

The synthesis of DDT is a good example of an electrophilic aromatic substitution. The chloral is protonated and attacks the aromatic ring to generate a carbocation. Loss of a proton regenerates the aromatic ring.

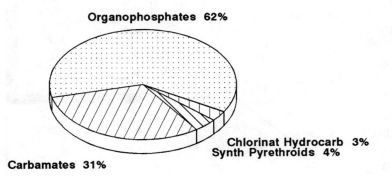

Figure 20.6 Types of insecticides. (*Source*: CEH.)

Reaction:

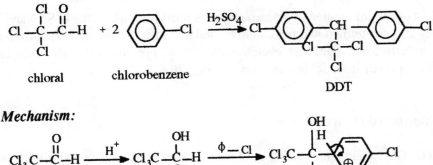

chloral   chlorobenzene   DDT

Mechanism:

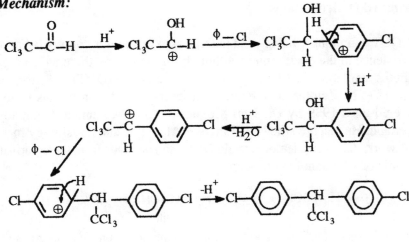

The commercial product is actually a mixture of about 80% *p,p* isomer and 20% *o,p* isomer. The *p,p* isomer has the most insecticidal properties. Structure-activity relationships have been studied in detail for DDT and its analogs. For good biological activity there must be at least one para chlorine. The *m,m*-dichloro isomer is not active. The activity increases as the para halogen is changed: I (which is inactive) < Br < Cl < F. If the para chlorines are replaced by alkyl groups of about the same size (like CH_3- or CH_3O-), then the compound is still active, but larger R groups show no activity. Methoxychlor has been used as an insecticide. Finally, as the chlorines of the Cl_3C- group are replaced by hydrogens the activity also declines ($Cl_3C \longrightarrow Cl_2CH \longrightarrow ClCH_2 \longrightarrow CH_3$—).

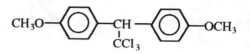

methoxychlor

DDT is still used extensively overseas out of necessity. For example, in India malaria cases went from 75 million in the early 1950s to 50 thousand in 1961 thanks to DDT. But when spraying stopped the figure went back up to 6 million in 1976, then down to 2.7 million in 1979 when other insecticides were sprayed. In December, 1972 the Environmental Protection Agency imposed a near total ban on DDT used in the United States.

The use of chlorinated hydrocarbons has declined in the United States for three main reasons: (1) concern over the buildup of residues (half-lives of 5-15 years in the environment, especially in the fat tissue of higher animals [10—20 ppm not uncommon]), (2) the increasing tendency of some insects to develop resistance to the materials, and (3) the advent of insecticides that can replace the organochlorines. Domestic consumption of DDT fell from 70 million lb in 1960 to 25.5 million in 1970 to none in 1980 (although worldwide demand is still high).

Cyclodienes. All the chlorinated hydrocarbons belonging to this second group of compounds, once used in large amounts, have been banned for use in the United States since 1974. They are made by the Diels-Alder reaction, named after two chemists who won the Nobel Prize in Chemistry in 1950 for the discovery of this important reaction. The synthesis of the important insecticides chlordane, heptachlor, aldrin, dieldrin, and endrin are summarized in Fig. 20.7.

Other Chlorinated Hydrocarbon Insecticides. Free radical addition of chlorine to benzene with ultraviolet light gives *benzene hexachloride*, $C_6H_6Cl_6$, known as BHC or 666. Although first synthesized by Faraday in 1825, its insecticidal

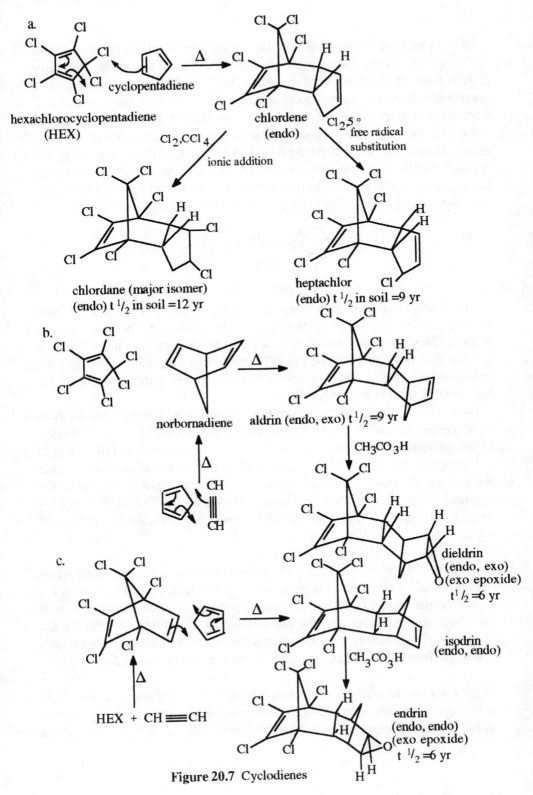

Figure 20.7 Cyclodienes

properties were not discovered until 1943. Of the nine possible stereoisomers of this compound only five are known and only one has insecticidal properties: lindane, the *a,a,a,e,e,e* isomer .

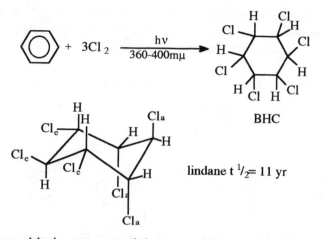

BHC

lindane t $^1/_2$= 11 yr

An organochlorine compound that was still in use in 1982 at 12 million lb because of its better biodegradability was toxaphene. Toxaphene was made by the chlorination of camphene with UV light in carbon tetrachloride until 68-69% chlorine is introduced. Over 170 different compounds are present in this mixture. Toxaphene was also used against rodents and it is not too toxic to bees. However, in 1989 consumption was down to 0.

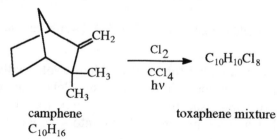

camphene
$C_{10}H_{16}$

toxaphene mixture

Polychlorinated Biphenyls. Polychlorinated biphenyls (PCBs), known by their trade marks of Arochlor® (Monsanto in the U.S.), Phenochlor® (in France), and Clophen® (in Germany) are chemically similar to the chlorinated insecticides. Although not used for this purpose, their existence and persistence in the environment is well established. They were used to make more flexible and flame retardant plastics and are still used as insulating fluids in electrical transformers since there is no substitute in the application. They have been made by Monsanto since 1930 and were first discovered as a pollutant in 1966. U.S. production peaked at 72 million lb in 1970 but in 1975 it was down to 40 million lb/yr because in 1971 Monsanto voluntarily adopted the policy of selling PCBs only for electrical systems. At least 105 PCBs are present in the environment.

PCBs are made by the chlorination of biphenyl by electrophilic aromatic substitution (know this mechanism!). A typical sample might contain some of the hexachloro derivatives shown here:

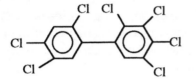

$$\xrightarrow[\substack{\text{Fe} \\ 500\,^\circ\text{C} \\ 12\text{-}36\ \text{hr}}]{\text{Cl}_2}$$ complex mixture of PCBs

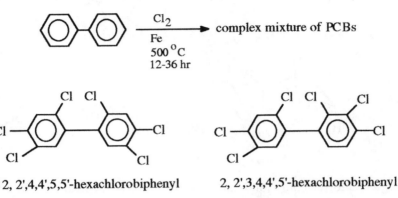

2, 2',4,4',5,5'-hexachlorobiphenyl   2, 2',3,4,4',5'-hexachlorobiphenyl

2, 2',3,4,4',5,5'-heptachlorobiphenyl   2, 2',3, 3',4,4',5-heptachlorobiphenyl

Much work has been done on the PCB problem. Potentially dangerous amounts of PCBs have been found in lake trout and Coho salmon from Lake Michigan (15 and 10 ppm, respectively). A very toxic trace contaminant in European PCBs that may be present in Monsanto's PCBs is tetrachloro-dibenzofuran, the second most toxic chemical known to humans.

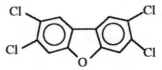

tetrachlorodibenzofuran

The search for the ideal PCB replacement continues, especially for the difficult electrical transformer application. Approximately 324 million lb of PCBs are still present in some 150,000 transformers. Possible substitutes range from mineral oil to high-temperature hydrocarbons, with silicones by far the most popular. There may be as much as a $2 billion market in replacing PCB-containing transformers, which under 1985 EPA rules cannot be used where they would present a contamination risk in human food or animal feed.

Organophosphorus Insecticides

In 1976 organophosphorus compounds became the leading type of insecticide and they still are, mainly because they are less persistent. Over 40 such compounds are registered in the United States today as insecticides. Gerhard Schrader synthesized the first organophosphorus insecticide in Germany during 1938, commonly called *tetraethyl pyrophosphate* (TEPP).

$$CH_3-CH_2-O-\overset{\overset{\displaystyle O}{\|}}{\underset{\underset{\displaystyle O}{|}}{P}}-O-\overset{\overset{\displaystyle O}{\|}}{\underset{\underset{\displaystyle O}{|}}{P}}-O-CH_2-CH_3$$

$$CH_3-CH_2 \qquad\qquad CH_2-CH_3$$

TEPP

Methyl parathion was developed around 1948 when the German technology was discovered after the war. Parathion (the ethyl analog) is not so safe and is used to a lesser extent. Both methyl parathion and parathion are synthesized by reacting the sodium salt of *p*-nitrophenol with *O,O*-dialkyl phosphorochloridothioate, which is made from phosphorus pentasulfide, the alcohol, and chlorine.

$$P_2S_5 + ROH \longrightarrow RO-\overset{\overset{\displaystyle S}{\|}}{\underset{\underset{\displaystyle RO}{}}{P}}-SH + H_2S \xrightarrow{Cl_2} RO-\overset{\overset{\displaystyle S}{\|}}{\underset{\underset{\displaystyle RO}{}}{P}}-Cl + HCl + S$$

$$Na^+ \, {}^-O -\!\!\!\bigcirc\!\!\!- NO_2$$

$$RO-\overset{\overset{\displaystyle S}{\|}}{\underset{\underset{\displaystyle RO}{}}{P}}-O-\!\!\!\bigcirc\!\!\!- NO_2 + NaCl$$

R=CH₃ , methyl parathion

R=CH₃—CH₂ , parathion

The parathions, although not persistent (half-life of one to ten weeks in the environment), are highly toxic to humans and deaths have been attributable to careless use. Operators in organophosphate plants must take blood tests once a month. The discovery of the safer malathion by American Cyanamid in the early 1950s was therefore welcome. Malathion is widely used today. It is synthesized by condensing diethyl maleate with the *O,O*-dimethyl phosphorodithioic acid obtained as previously discussed.

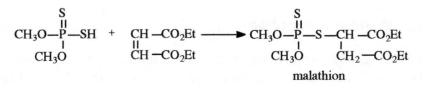

malathion

Methyl parathion is used primarily on cotton. Malathion and parathion are the broadest spectrum organophosphate insecticides. There are many other organophosphate insecticides in use in the United States today. Many are produced at the 1-5 million lb/yr level for specific applications. Two other leading organophosphates besides those mentioned already are chlorpyrifos and terbufos. Table 20.2 gives the 1989 U.S. consumption of the leading organophosphates and the total used if all 40 are counted.

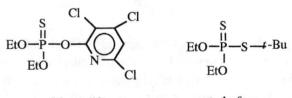

chlorpyrifos   terbufos

TABLE 20.2 U.S. Consumption of Organophosphates as Insecticides

Name	Million lb
Organophosphates, Total for 40 Compounds	93.5
methyl parathion	15.1
chlorpyrifos	13.2
malathion	13.0
terbufos	10.6
parathion	7.2

Source: CEH.

Carbamate Insecticides

Carbamates are sold at a lesser volume than are organophosphorus compounds. The first carbamate (urethane) insecticides were developed in the late 1940s at Geigy Chemical Co. in Switzerland. Research on carbamates was inspired by the known toxicity of the alkaloid physostigmine which occurs naturally in a West African bean.

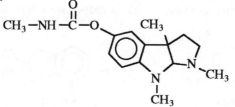

physostigmine

In the 1950s Kolbezen and Metcalf at the University of California-Riverside laid the foundation for Union Carbide's development of carbaryl (Sevin®), the first major carbamate. Still the most important carbamate insecticide, it is made by condensing l-naphthol with methyl isocyanate. The l-naphthol is made from naphthalene by hydrogenation, oxidation, and dehydrogenation. Carbaryl finds use in practically all the agricultural crop markets and is popular for home lawn and garden use.

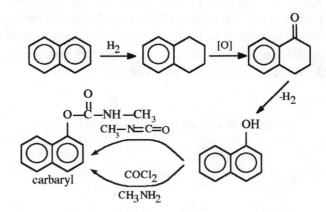

The 1-naphthol is made from naphthalene, which is obtained from coal tar distillation or from petroleum.

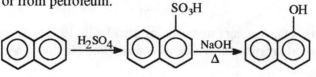

Methyl isocyanate is a very dangerous chemical. It was responsible for the deaths of over 2500 people in the worst industrial accident ever, that of the carbamate insecticide plant in Bhopal, India on December 3, 1984. It is a very toxic chemical. This tragedy is discussed in more detail in Chapter 25. Methyl isocyanate can be made from phosgene and methylamine, which would circumvent use of the isocyanate. Phosgene is made from chlorine and carbon monoxide, but it is also very toxic and dangerous.

Another important carbamate insecticide is carbofuran, whose synthesis is outline here.

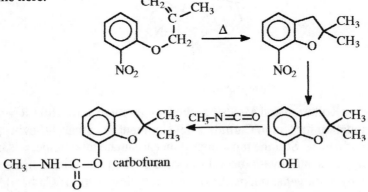

A third important carbamate is aldicarb, or Temik®, an insecticide and nematocide for potato and vegetable crops. This chemical has been found in water wells in 11 states above the 1 ppm EPA safety threshold, barring use in some locales in 1982. According to Union Carbide, one manufacturer, humans can safely ingest 500 ppb. But it is one of the most acutely toxic pesticides registered by the EPA. A fourth carbamate insecticide is methomyl.

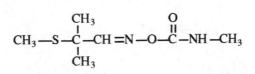

aldicarb

$$CH_3 - S - \underset{\underset{CH_3}{|}}{\overset{\overset{CH_3}{|}}{C}} = N - O - \overset{\overset{O}{\parallel}}{C} - NH - CH_3$$

methomyl

Table 20.3 lists the leading carbamate insecticides by 1989 U.S. consumption. Carbofuran and aldicarb are also used as nematocides but this is not reflected in the table.

TABLE 20.3 U.S. Consumption of Carbamates as Insecticides

Name	Million lb
Carbamates, Total for 11 Compounds	21.7
carbaryl	10.2
carbofuran	4.0
methomyl	3.2
aldicarb	2.3

Source: CEH.

Advantages of the carbamate insecticides are lower toxicity to animals and use immediately up to harvest of crops (half-life is one week). Prolonged protection against insects requires frequent sprayings. But in 1978 34 species of malarial mosquitoes were resistant to DDT, 10 to malathion, and only 4 to carbamates.

Synthetic Pyrethroids

Mention was made of the natural product pyrethrins and the structure of pyrethrin I was given on page 394. Because of the unique structures of these cyclopropane-containing natural products and their high insecticidal properties, syntheses of analogs have been studied. The isobutenyldimethylcyclopropane-carboxylic acid moiety, called chrysanthemic acid, has been modified by using different ester groups. As a result a number of synthetic pyrethroids are available for certain specific uses, though their expense is $50/lb compared to methyl parathion at $3/lb, carbaryl at $4/lb, and terbufos and carbofuran at $10/lb. Names and structures of some synthetic pyrethroids are given below. Their main advantages are (1) few side effects on plants, livestock, and humans; (2) no resistance buildup by insects; and (3) lower quantities needed. The synthetic pyrethroids can have up to 30 times the quickkill power of the natural pyrethrins and longer half-lives.

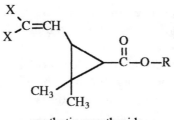

synthetic pyrethroids

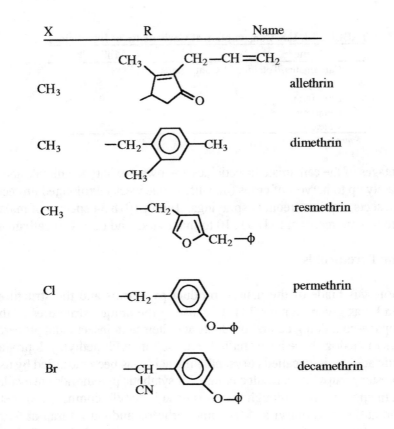

X	R	Name
CH₃		allethrin
CH₃		dimethrin
CH₃		resmethrin
Cl		permethrin
Br		decamethrin

Third-Generation Insecticides

Many people think the ultimate pesticide should be developed from research now being done on certain insect attractants and juvenile hormones. Isolation of naturally occurring sex attractants (pheromones) and juvenile hormones has been accomplished. The attractants could be used to congregate large numbers of insects in one place for extermination by the already existing insecticides. Alternatively, juvenile hormones have been found that prevent maturation or cause sterility in many pests.

The U.S. Forest Service scientists in New York have isolated and identified chemical sex attractants used by elm bark beetles that are responsible for transmitting the fungus causing Dutch elm disease. Examples of attractants are 2,4-dimethyl-5-ethyl-6,8-dioxabicyclo [3.2.1] octane (called multistriatin) and 4-methyl-3-heptanol. In field trials an artificially produced mixture of the compounds has proved attractive to the elm bark beetle.

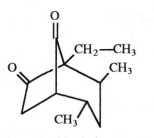

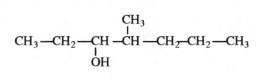

multistriatin

4-methyl-3-heptanol

The boll weevil sex attractant is a mixture of four compounds, two alcohols and two aldehydes. The epoxide disparlure has been isolated as the gypsy moth sex attractant. Both of these pheromones are in experimental use for control of these pests.

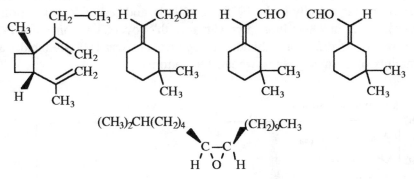

Insects may emit as little as 10^{-15} g of pheromone and a female insect contains typically only 50 mg of the material. The structure is sometimes exacting, as in the case of the pink bollworm, where the ratio of *cis* and *trans* double bonds is species-specific to avoid hybridization of insects.

In 1965 the first juvenile hormone was isolated and it was synthesized in 1967. The substance studied was methyl *trans, trans, cis*-10-epoxy-7-ethyl-3,11-dimethyl-2,6-tridecadienoate from the male silkmoth. In 1975 a juvenile hormone was okayed for commercial marketing by the Environmental Protection Agency. Approved for control of floodwater mosquitoes, Zoecon Corporation of Palo Alto, California, is selling Altosid SR-10, which is isopropyl 11-methoxy-3,7,11-trimethyldodeca-2,4-dienoate, also known as methoprene. The big advantage of this type of insecticide is its relatively rapid degradation and low toxicity to applicators, fish, birds, beneficial insects, and other wildlife. The price is competitive with more conventional pesticides. No doubt more third generation insecticides will be developed in the future.

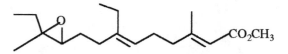

methyl *trans, trans, cis*-10-epoxy-7-ethyl-3,11-dimethyl-2,6-
tridecadienoate

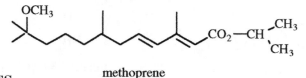

methoprene

HERBICIDES

In the 1980s the Department of Agriculture estimated that about 10% of U.S. agricultural products is lost because of weeds, probably $20 billion/yr. About 1500 species of weeds cause economic loss. As we mentioned earlier, herbicide use has risen dramatically in the 1970s and 1980s, and herbicide production is just now leveling at 700 million lb/yr with a worth of over $3 billion/yr. In 1950 there were only 15 different herbicides; today there are over 180. Herbicides are used mainly on corn, soybeans, wheat, and cotton.

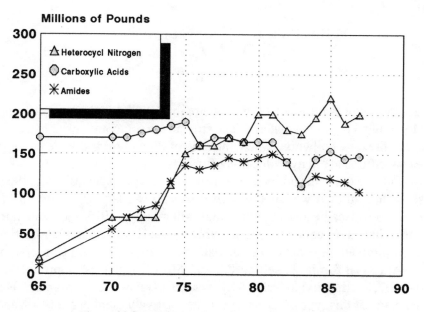

Figure 20.8 U.S. Production of Herbicides. (*Source:* CEH.)

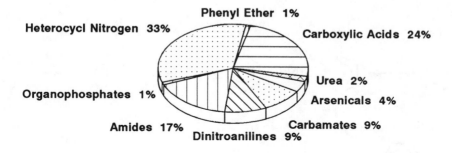

Figure 20.9 Types of Herbicides. (*Source:* CEH.)

Fig. 20.8 shows the trend in production for the three most important types of herbicides. Carboxylic acids have been the standby for many years and were the first type of herbicide. They were replaced as number one by the heterocyclic nitrogen compounds in the 1970s. A close third is the amide herbicides. The present percentage of production for herbicides is given in Fig. 20.9.

Carboxylic Acids (Phenoxy Herbicides)

Rapid growth of chemical weed control did not occur until after World War II when a herbicide was introduced by Jones in 1945 at the Imperial Chemical Industries of England: 2,4-dichlorophenoxyacetic acid (2,4-D). Its utility has come from its ability to kill selectively broadleaf weeds in cereal grains, corn, and cotton. It does not disturb the soil and is not persistent. 2,4,5-T was launched commercially by American Chemical Paint Co. in 1948 (now Union Carbide) to combat brush and weeds in forests, along highways and railroad tracks, in pastures, and on rice, wheat, and sugarcane.

2,4-Dichlorophenoxyacetic acid (2,4-D) and 2,4,5-trichlorophenoxyacetic acid (2,4,5-T) dominated the herbicide market up to the late 1960s. Phenol is the starting material for 2,4-D. Chlorination via electrophilic aromatic substitution (know the mechanism!) gives 2,4-dichlorophenol. The sodium salt of this compound can react with sodium chloroacetate (S_N2) and acidification gives 2,4-D.

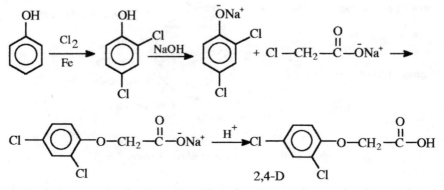

2,4,5-T can be synthesized easily. Chlorination of benzene gives 1,2,4,5-tetrachlorobenzene (why?) which reacts with caustic to give 2,4,5-trichlorophenol. A conversion similar to the preceding one yields the phenoxyacetic acid 2,4,5-T.

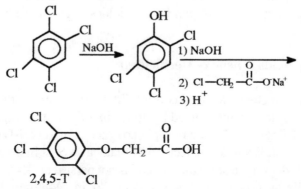

   The phenoxy herbicides' inexpensiveness, selectivity, nonpersistency and low toxicity to animals are difficult to beat. Application is usually accomplished by spraying on the leaves. The herbicides cannot themselves be applied to the soil because they are washed away or decomposed by microorganisms in a few weeks. They can be applied by this method using a sulfonic acid derivative that, after hydrolysis in the soil and oxidation by bacteria, can form 2,4-D in the plant. 2,4-D is still the main herbicide used on wheat.

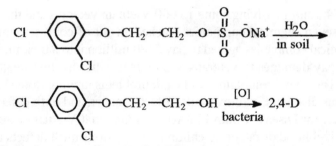

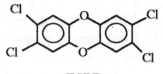

Much publicity has been given to 2,4,5-T recently. A trace impurity called 2,3,7,8-tetrachlorodibenzo-*p*-dioxin (TCDD) has been called the most toxic small molecule known to humans. TCDD kills animals and causes birth defects at lower levels than any other chemical tested in the laboratory. In 1969 Saigon newspapers claimed use of 2,4,5-T as a defoliant was causing illness, stillbirths, and fetal deformities.

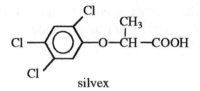

TCDD

Agent Orange (named after the color of its storage drums) contained a 50:50 mixture of the butyl esters of 2,4,5-T and 2,4-D. The U.S. Defense Department stopped using it in 1970. The National Cancer Institute funded a study in 1968 that seemed to indicate that large doses of 2,4,5-T (containing 28 ppm of dioxin) increased the incidence of birth defects in certain strains of mice. The 2,4,5-T marketed more recently contained 0.01 to 0.02 ppm. In 1971 the EPA prohibited the use of 2,4,5-T on most food crops except rice. In 1979 the EPA issued an emergency order to suspend its use on forests, pastures, home gardens, rights-of-way, aquatic weeds, and ornamental turf. It was permitted on rice fields and rangelands. In 1985 the EPA cancelled all uses of 2,4,5-T as well as another dioxin-containing herbicide, silvex.

silvex

In 1984 a suit involving some 15,000 Vietnam veterans and their dependents against seven chemical companies reached an out-of-court settlement in which the chemical companies agreed to pay $180 million into a trust fund that will be used to pay damages to veterans and their families as their health claims are proven. They have complained of health problems ranging from skin conditions to nervous disorders, cardiovascular effects, cancer, and birth defects. However, the Center for Disease Control in Atlanta found that Vietnam veterans have no greater likelihood of fathering children with serious birth defects than do other American males. The Center also believes that the 1982 evacuation of Times Beach, MO may have been unnecessary. In the last ten years researchers have suggested that TCDD is not the cancer-causing agent at first believed. Its effects are much worse in animals than in humans. The 1976 toxic cloud containing TCDD released in Seveso, Italy did not cause an increase in cancer rate, birth defects, or other diseases, and the signs of chloracne, a skin disfiguring disease, have disappeared from most of the people who suffered from it. In 1990 the National Cancer Institute reviewed 2,4-D, still the most widely used carboxylic acid herbicide. They found that wheat farmers exposed to the herbicide may face an increased risk of cancer. Studies so far have only shown increased cancer risk for heavy users of the weedkiller, not the occasional user, such as homeowners. One of the active ingredients in Ortho Weed-B-Gon® lawn weed killer is 2,4-D.

Other carboxylic acids that have become popular herbicides, though they are not phenoxyacetic acid derivatives, are dicamba and glyphosate or Roundup®.

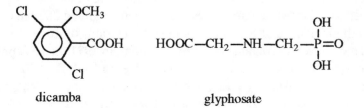

dicamba   glyphosate

Notice that dicamba does have a methoxy group and two chlorines, so it is similar to 2,4-D. Glyphosate is used on cotton and soybeans. Monsanto, the maker of Roundup®, has genetically engineered cotton and soybeans to develop tolerance to the lower selectivity of this herbicide.

Heterocyclic Nitrogen Herbicides (Triazines)

The most widely used herbicides today are triazine compounds (three nitrogens in the heterocyclic aromatic ring). Atrazine is used especially on corn but also on pineapple and sugarcane. It is synthesized by reacting cyanuryl chloride successively with one equivalent of ethylamine and one equivalent of isopropylamine. Cyanuryl chloride is made in one step from chlorine and hydrogen cyanide. Simazine is made with two equivalents of ethylamine and is used on corn. Cyanazine is another important triazine. Trazines will kill most types of plants but corn because something in this crop degrades the triazines before their toxic action can take place.

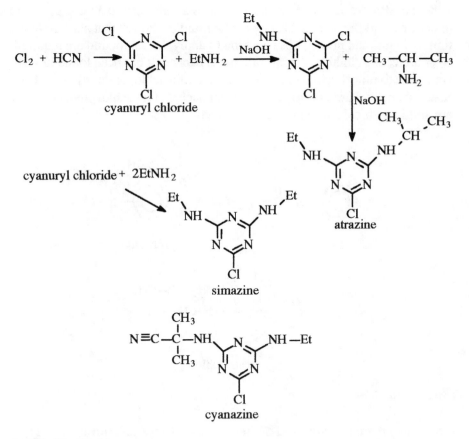

Amide Herbicides

A number of other herbicides have specific uses. The amide herbicides, of which propanil is typical, are used in large quantities. Propanil is made by the reaction of propionyl chloride and 3,4-dichloroaniline.

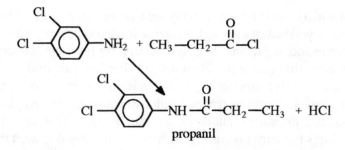

propanil

Another common amide herbicide, alachlor (Lasso®) is used on corn and soybeans in large amounts, as well as on potatoes, peanuts, and cotton. In late 1984 the EPA determined that alachlor poses a significant potential cancer risk to persons working with it. Dietary feeding studies showed that alachlor induces tumors in rats and mice. Those wishing to apply the compound are required to wear protective clothing. Aerial spraying is banned. It has been found in surface water due to runoff. More than 70 million lb of alachlor are used in the United States annually. Further testing is being conducted. Propachlor and metolachlor are amide herbicides used in large volume.

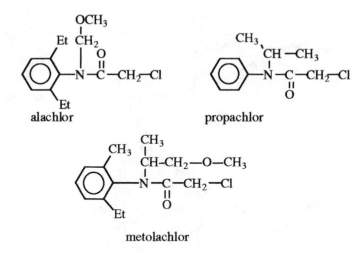

Other Herbicides

Some thiocarbamates are used as herbicides. These are sulfur analogs of carbamates, which we covered under insecticides. The leading thiocarbamates are butylate and EPTC.

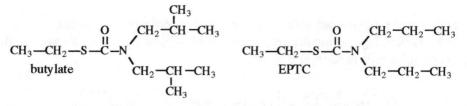

butylate

EPTC

Diuron and linuron are urea herbicides. Diuron, used on cotton, is made from dimethylamine and an isocyanate (which one?).

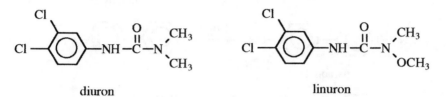

diuron

linuron

A class of compounds, discovered at Ely Lilly in 1961, are the dinitroanilines, with trifluralin (Treflan®) being the most important member. Trifluralin is used on soybeans and cotton. Pendimethalin is a second important dinitroaniline. Benefin is a common crabgrass preventer for home lawns.

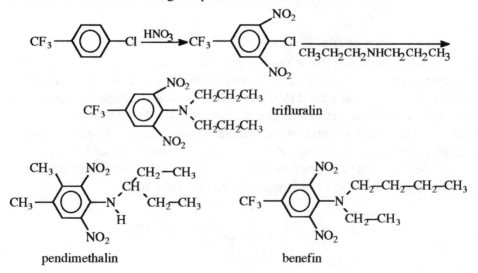

trifluralin

pendimethalin

benefin

The bipyridyl herbicides paraquat and diquat are interesting. Paraquat is made by reduction of pyridine to radical ions, which couple at the para positions. Oxidation and reaction with methyl bromide gives paraquat. Diquat is formed by dehydrogenation of pyridine and quaternization with ethylene dibromide. Both of the herbicides kill green foliage effectively but are deactivated as soon as they come into contact with the soil. Fields can be cleared of vegetation before and after sowing, calling for only a minimum of plowing. They are also used to control weeds in lakes and rivers.

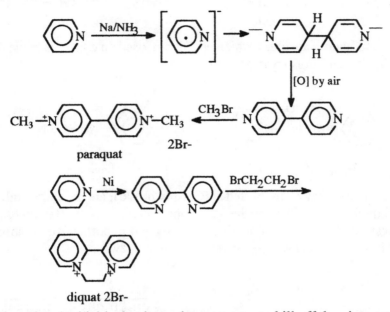

paraquat

2Br-

[O] by air

diquat 2Br-

Dinoseb, a herbicide that is used on potatoes to kill off the vines, making it easier to harvest the tubers, was outlawed by the EPA in 1986. It was used on cotton, soybeans, and peanut fields as well. This was only the third pesticide banned by the EPA completely, the first being 2,4,5-T and the second being ethylene dibromide, a fumigant that had entered into much of the nation's grain supply. Dinoseb may also seep into groundwater. Dinoseb causes a very serious risk of birth defects in pregnant women or sterility in men for the people who are applying it. It has also been found to be carcinogenic in mice.

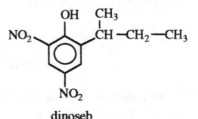

dinoseb

Table 20.4 lists the herbicides discussed here and a few others with their annual U.S. consumption, together with the total for each type of herbicide.

In this chapter we have taken up our first example of a sector of the chemical industry which involves multistep organic syntheses, very diverse organic chemical structures, and final compounds which are unique in their biological action and selectivity. We will see this type of industrial sector again when we study the pharmaceutical industry in Chapter 23. Despite this complexity of chemistry we see that most of the major pesticides fall into one chemical class or another that has been shown to give the desired biological response. Then slight modifications of structures are used for specific applications to maximize the desired effect and minimize the side effects. Thus the insecticide market is now dominated by the organophosphates and the carbamates. The herbicide market is a little more diverse, but heterocyclic nitrogens, carboxylic acids, amides, dinitroanilines, and carbamates/thiocarbamates are the main materials.

TABLE 20.4 U.S. Consumption of Herbicides

Name	Million lb	
Carboxylic Acids, Total	100.0	
2,4-D		36.5
glyphosate		18.6
dicamba		11.5
diuron		5.2
linuron		
Heterocyclic Nitrogens, Total	142.6	
atrazine		73.2
cyanazine		27.2
bentazon		8.0
metribuzin		6.6
simazine		5.2
Amides, Total	110.0	
alachlor		72.0
metolachlor		15.0
propanil		10.0
Carbamates/Thiocarbamates, Total	50.5	
butylate		20.0
EPTC		15.0
Ureas, Total	16.5	
diuron		5.2
linuron		4.4
Dinitroanilines, Total	45.9	
trifluralin		27.4
pendimethaline		11.4
benefin		1.7
Bipyridyls, Total	5.0	
paraquat		3.4
diquat		1.0

Source: CEH

21

Fertilizers

References

KO, Vol. 10, pp. 31-125
Kent, pp. 251-280
B.F. Greek, "Fertilizer Production Up Modestly for Second Straight Year," *C & E News,*
7-8-91, pp. 21-22.

INTRODUCTION

Besides the 3 basic elements of carbon, hydrogen, and oxygen that are common
to all plants, there are 16 other elements known to be essential to good plant
growth. Their percentages are given here. This chapter is concerned with the
three primary nutrients making up most fertilizers: nitrogen, phosphorus, and
potassium. The usual sources of nitrogen are ammonia, ammonium nitrate, urea,
and ammonium sulfate. Phosphorus is obtained from phosphoric acid or
phosphate rock. Potassium chloride is mined or obtained from brine and the
sulfate is mined in small amounts. Potassium nitrate is made synthetically. These
chemicals have already been described under inorganic chemicals of the top 50.
Sources for the three primary nutrients are given in Fig. 21.1.

> 95% basic elements—44% C, 6% H, 45% O
> 3.5% primary nutrients—2.0% N, 0.5% P, 1.0% K
> 1.3% secondary nutrients—Ca, Mg, S
> 0.1% micronutrients—B, Cl, Cu, Fe, Mn, Mo, Zn (Co, F, I in animals also)

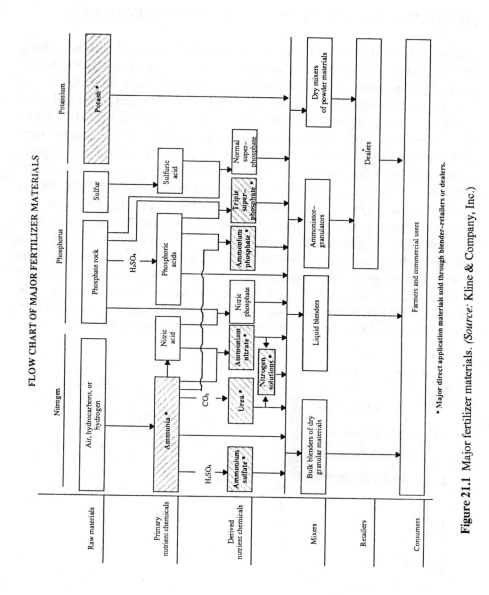

Figure 21.1 Major fertilizer materials. (*Source:* Kline & Company, Inc.)

HISTORY AND ECONOMICS OF FERTILIZER PRODUCTION

Although the modern era of fertilizers began with the work of Justus von Liebig in 1840 and the first U.S. patent for a mixed fertilizer was granted in 1849, the use of large amounts of synthetic fertilizers was popularized only after World War II. Fertilizer consumption increased eight times between 1950 and 1980 worldwide. U.S. shipments of fertilizers are summarized in Fig. 21.2. Phosphatic Fertilizers (SIC 2874) had a very fast increase from $1 billion in the early 1970s to $4.4 billion in 1980. A recent comeback from poorer early 1980 years now puts 1988 shipments back at $4.1 billion. Nitrogenous Fertilizers (SIC 2873) have also had a similar trend and were back to $3.3 billion in 1988. Mixed Fertilizers (SIC 2875) have had a more constant value in the last few years. Fig. 21.3 gives trends in nitrogen, phosphate, and potash fertilizer production for the last few years. Nitrogen and phosphorus production in billions of pounds has paralleled shipments in billions of dollars, with a recovery occurring in the late 1980s. Potash production is always much less and has been steady or decreasing at best in the last 20 years. Fig. 21.4 shows the uses of fertilizers on various types of crops. Note that nearly half of all fertilizers is used on one crop: corn. Wheat, hay, soybeans, and cotton consume most of the rest of fertilizers used on crops.

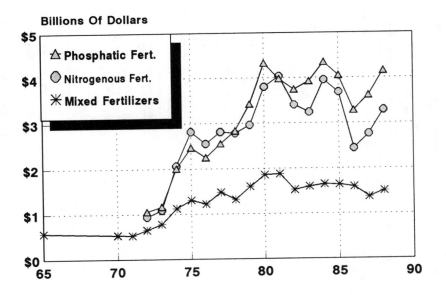

Figure 21.2 U.S. shipments of fertilizer. (*Source*: AS.)

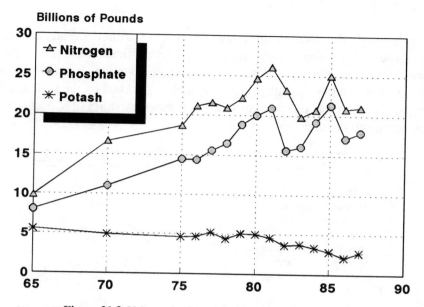

Figure 21.3 U.S. production of fertilizers. (*Source*: CEH.)

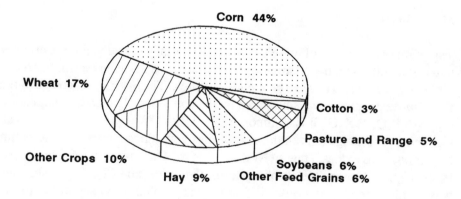

Figure 21.4 Uses of fertilizers on crops. (*Source*: CEH.)

FERTILIZER MATERIALS

Fertilizers may contain all three primary nutrients, in which case they are called mixed fertilizers, or they may contain only one active ingredient, called direct application fertilizers. Recently the ratio of direct application to mixed fertilizers was approximately 1:1 in the United States.

Direct Application Fertilizers

The following table lists all important direct application materials and their percentage of direct application fertilizers. Direct application use is increasing mainly because of anhydrous ammonia gas becoming popular. It can be pumped in 3-6 in. beneath the soil during plowing and is absorbed by the soil rapidly. Nitrogen solutions can also be applied in this manner (mixtures of free ammonia, ammonium nitrate, urea, and water.)

Fertilizer	Percent of Direct Application Materials
Nitrogen solutions	24
Anhydrous and aqueous ammonia	22
Potassium chloride	20
Ammonium nitrate	10
Urea	8
Superphosphates (enriched in P_2O_5)	5
Ammonium sulfate	3
Ammonium phosphate	2
Other	6
Total	100%

Mixed Fertilizers

The primary advantage of these fertilizers is that they contain all three primary nutrients—nitrogen, phosphorus, and potassium—and require a smaller number of applications. They can be liquids or solids. The overall percentage of the three nutrients must always be stated on the container. The grade designation is %N-%P_2O_5-%K_2O. It is commonly called the *NPK value*. Note that it is an elemental percentage only in the case of nitrogen. Phosphorus and potassium are expressed as oxides. Thus an NPK value of 6-24-12 means that 6% by weight is elemental nitrogen, 24% is phosphorus pentoxide, and 12% is potash. One way of remembering the order is that they are alphabetical according to the English name (*n*itrogen, *p*hosphorus, *p*otassium). A changeover to a grade designation by the three elemental bases is being resisted by the industry.

Nitrogen Sources. The nitrogenous chemicals ammonia, urea, ammonium nitrate, and ammonium sulfate are used as sources of nitrogen in mixed fertilizers. A mixture is also quite popular and is relatively cheap, since the mixed nitrogen solution from which pure urea is made can be used as fertilizer. Nitrogen solutions have their own code number. An example would be 414 (19-

Billions of Pounds

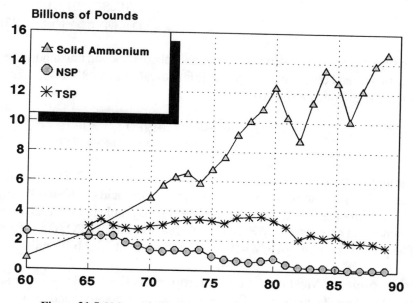

Figure 21.5 U.S. production of phosphate fertilizers. (*Source*: CEH.)

66-6), meaning 41.4% total nitrogen, 19% free ammonia, 66% ammonium nitrate, and 6% urea with the rest being water. Over 100 nitrogen solutions are marketed. Although the solutions are cheap, the solids do not have a vapor pressure problem and are more easily transported. The present breakdown of nitrogen fertilizer production is ammonia, 50%; ammonium nitrate, 9%; ammonium sulfate, 15%; and nitrogen solutions, 17%.

Phosphorus Sources. The production of different phosphate fertilizers is given in Fig. 21.5 All phosphorus fertilizers come from wet process phosphoric acid or directly from phosphate rock. Normal superphosphate, triple or concentrated superphospate, and ammonium phosphate are the three common types used. Normal or ordinary superphosphate (NSP or OSP) is mostly monocalcium phosphate and calcium sulfate. It is made from phosphate rock and sulfuric acid and is equated to a 20% P_2O_5 content. It led the market until 1964. The production of normal superphosphate is similar to that for the manufacture of wet process phosphoric acid (Chapter 2, page 51) except that there is only partial neutralization. The following reaction is one example of an equation which represents this process.

$$CaF_2 \cdot 3Ca_3(PO_4)_2 + 17H_2O + 7H_2SO_4 \longrightarrow 3[CaH_4(PO_4)_2 \cdot H_2O] + 2HF + 7(CaSO_4 \cdot 2H_2O)$$

normal superphosphate (NSP)

Triple superphosphate (TSP), made from phosphate rock and phosphoric acid, is mostly mono- and dicalcium phosphate. It is equivalent to a 48% P_2O_5 content. It led the market from 1965-1967.

$$CaF_2 \cdot 3Ca_3(PO_4)_2 + 14H_3PO_4 \longrightarrow 10CaH_4(PO_4)_2 + 2HF$$

triple superphosphate (TSP)

The ammonium phosphates took over the lead in about 1967. Diammonium phosphate (DAP) is made from wet process phosphoric acid of about 40% P_2O_5 content and ammonia. The usual finishing $NH_3:H_3PO_4$ mole ratio is 1.85-1.94:1. Monoammonium phosphate (MAP) is made with a final $NH_3:H_3PO_4$ ratio of about 1:1. The present ratio of DAP:MAP production is about 13:2 billion lb.

Potassium Sources. Most potassium in fertilizers is the simple chloride salt, having a 60-62% K_2O equivalent. Certain crops such as potatoes and tobacco do not like high amounts of chloride. For these crops KNO_3, K_2SO_4, or $K_2Mg(SO_4)_2$ may be used.

Ammoniation. When an ammonia fertilizer is mixed with a superphosphate a chemical reaction occurs which changes the active ingredient's structure. The following equations illustrate this chemistry:

$$H_3PO_4 + NH_3 \longrightarrow NH_4H_2PO_4 \tag{1}$$
$$Ca(H_2PO_4)_2 \cdot H_2O + NH_3 \longrightarrow CaHPO_4 + NH_4H_2PO_4 + H_2O \tag{2}$$
$$NH_4H_2PO_4 + NH_3 \longrightarrow (NH_4)_2HPO_4 \tag{3}$$
$$2CaHPO_4 + CaSO_4 + 2NH_3 \longrightarrow Ca(PO_4)_2 + (NH_4)_2SO_4 \tag{4}$$
$$NH_4H_2PO_4 + CaSO_4 + NH_3 \longrightarrow CaHPO_4 + (NH_4)_2SO_4 \tag{5}$$

Reactions (1) and (2) are common for both normal and triple superphosphate. Reaction (3) is important in triple superphosphate because of the lack of large amounts of calcium sulfate. Reaction (5) is important with normal superphosphate because of the large surplus of calcium sulfate in this formulation.

LIQUIDS VERSUS SOLIDS

There are many different types of liquid and solid fertilizers but we give only some generalizations about advantages of each. By *liquid* is meant a clear solution, a suspension of a solid in a liquid (aided by a suspending agent), or a simple slurry of a solid in a liquid. Solid fertilizers contain no liquid. The following table summarizes the advantages of liquids and solids.

Advantages of Liquids	*Advantages of Solids*
Lower capital investment by the company	Less corrosion of equipment
Less labor, handling and conditioning costs	Better economies of costs of storing smaller volumes
More uniform composition	Solubility restrictions are not present
More uniform distribution on land	No crystallization problems in cold weather

Mixed solid fertilizers can be made by either direct granulation methods (40%) or bulk blending (40%). Bulk blending is made by mechanical mixing of the separate granular intermediate materials. It is usually done in small plants near the point of use. This technique is employed because the fertilizer can be "tailor-made" to fit the exact requirements of the user. Fluid or liquid fertilizers (clear, suspension, and slurry) account for 20% of all NPK mixed fertilizers.

CONTROLLED-RELEASE FERTILIZERS

Much recent research has centered on developing long-lasting slow-release fertilizers to make application requirements less. Urea-formaldehyde resins in combination with nitrogen fertilizers tie up the nitrogen for a longer time, whereas degradation of the polymer occurs slowly by sunlight. This type of fertilizer is especially popular for the high nitrogen content of home lawn fertilizers. Sulfur-coated urea (SCU) is also becoming a popular slow-release nitrogen formulation. *sym*-Tetrahydrotriazone, made by reacting urea, formaldehyde, and ammonia, can be added to urea fertilizers. Triazones form ammonium ions much more slowly than urea. Slow-release potassium is also being developed. A coating of sulfur seems to delay its release. For phosphorus $Mg(NH_4)PO_4$ is becoming popular because it has a slower dissolution rate in the soil. Despite the simple chemicals used in most fertilizers, some interesting research and formulation work will keep chemists involved in the industry for some time to come.

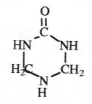

sym-tetrahydrotriazone

22

Pulp, Paper, and Other Wood Products

References

Austin, pp. 602-632
Kent, pp. 519-576

INTRODUCTION

Although the pulp and paper industry is not part of Chemicals and Allied Products, it is one of the major divisions of the chemical process industries. Containing some interesting chemistry, this industry employs many chemists and chemicals. It takes over 400 lb of chemicals to make 1 ton of paper.

Writing paper was first used in Egypt as far back as 2500-2000 BC, made from the papyrus reed. Paper manufacture began in China about AD 105. In 1690 the first American paper mill began its operation. Two recent dates of importance to modern paper technology are 1867, when Tilghman in the United States developed the sulfite process, and 1884, when Dahl in Germany discovered the kraft or sulfate process.

The student should review Fig. 7.1, p. 115 , to see the relative size of the Paper and Allied Products industry (SIC 26) compared to other chemical process industries. Its 1988 U.S. shipments totalled $123 billion, about half the size of Chemicals and Allied Products (SIC 28) at $260 billion. It is about the same size as Petroleum and Coal Products (SIC 29) at $131 billion. It has undergone a steady increase over the years, even in the late 1980s when sectors like the petroleum industry suffered a decline. The industry manufactures about 55

426

million tons of pulp each year in the U.S. It makes approximately 70 million tons of paper and paperboard products annually. This country's production of paper products is more than half the world's production. Per capita consumption of pulp has risen sharply in recent years. In 1940 it was 255 lb of pulp per person in the U.S.; in recent years it is near 600 lb. Compare this to 31 lb per person annually in Asia and 11 lb in Africa. There are about 200 pulp mills and 600 paper and paperboard mills in operation now. Some familiar names of companies in this industry are International Paper Co., Kimberly-Clark Corp., Champion International Co., Scott Paper Co., Crown Zellerbach Corp., Weyerhaeuser Co., Mead Corp., Boise Cascade Corp., James River Corp., and Georgia-Pacific Corp.

THE CHEMISTRY OF WOOD

Woody plants are made of strong, relatively thick-walled long cells that make good fibers. The cell wall in these types of plants is a complex mixture of polymers that varies in composition. But it can be roughly divided into 70% polysaccharides and 28% lignin.

The polysaccharides in wood are called holocellulose, or total cellulose carbohydrates. They can be subdivided into (1) cellulose (40%), a high molecular weight linear polymer composed of glucose units with high chemical resistance, and (2) hemicellulose (30%), other polysaccharides besides cellulose that are of lower molecular weight and have lower chemical resistance to acids and alkalies. The sugars in the hemicellulose are mostly xylose, galactose, arabinose, mannose, and glucose.

Lignin has been described as "the adhesive material of wood" because it cements the fibers together for strength. It is a complex cross linked polymer of condensed phenylpropane units joined together by various ether and carbon linkages. A representative structure of the phenylpropane units in lignin is given in Fig. 22.1. Lignin can be considered to be a polymer of coniferyl alcohol. About 50% of the linkages are β-aryl ethers. Lignin can be degraded with strong alkali, with an acid sulfite solution, and with various oxidizing agents. It is therefore removed from the wood to leave cellulose fibers, commonly called pulp. Although there are many differences between hardwood and softwood, the hardwoods always have less lignin and more hemicellulose (high in xylose), whereas the softwoods have more lignin and less hemicellulose (which is high in galactose, glucose, and mannose units).

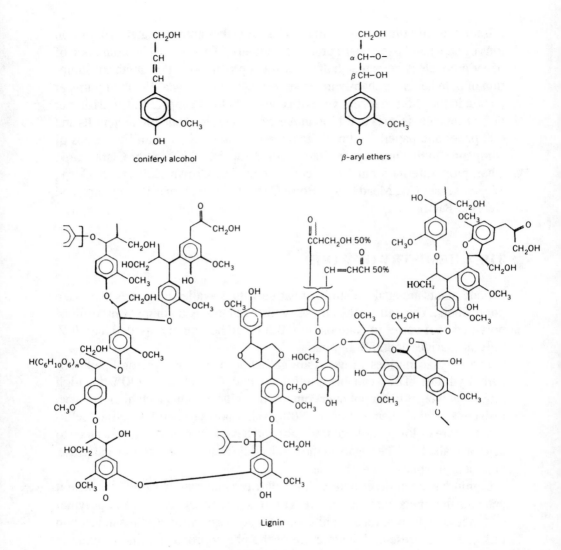

Figure 22.1 (*Source:* KO.)

Besides the holocellulose and lignin of the cell wall, wood contains about 2% extractives. These can be separated by steam distillation or solvent extraction and will be discussed later.

The average composition of most woods is summarized as follows:

Holocellulose	70%
Cellulose	40
Hemicellulose	30
Lignin	28
Extractives	2

PULP MANUFACTURE

The process of pulping, degrading the lignin to a more soluble form so the cellulose fibers can be separated from it, involves some interesting chemistry. The kraft or alkaline sulfate process dominates this part of the industry. Approximately 78% of all pulp is made by the kraft process, 3% by the acid sulfite process, 7% by the neutral sulfite semichemical (NSSC) process, 10% by a nonchemical, mechanical method called groundwood, and 2% by other methods.

The Kraft Process

Recalling that *kraft* is the German word for *strong* helps remind us that the strongest pulp fibers can be made by this method. Any pulping process lowers the molecular weight of the hemicellulose, depolymerizes the lignin, and gives a much larger percentage of cellulose fibers. The following table gives the percentage of these different components before and after this alkaline sulfate treatment.

Component	*Percent Before*	*Percent After*
Cellulose	40	36
Hemicellulose	30	7
Lignin	27	4
Extractives	3	0.5
Total	100	47.5

Inorganic Kraft Chemistry. The important chemistry in the kraft method is divided into inorganic and organic parts. Figure 22.2 summarizes the inorganic chemistry. The inorganic loop is a closed system with the exception of sodium

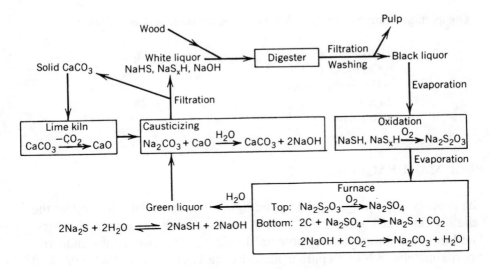

Figure 22.2 Kraft inorganic chemistry.

sulfate being added periodically. Only wood enters the loop and only pulp leaves.

Although the digester is shown as containing NaSH, NaOH, and NaS_xH, the typical entering white liquor is analyzed at about 59% NaOH, 27% Na_2S, and 13% Na_2CO_3—Na_2SO_4—Na_2SO_3. Digestion of the wood-white liquor mixture occurs at 170-175°C and 100-135 psi for 2-5 hr. A typical digester is 40 ft high with a diameter of 20 ft and can hold up to 35 tons of wood chips at a 1:4 wood: white liquor weight ratio. The organic chemistry of this digestion process is covered subsequently.

The resulting pulp is separated from the black liquor (colored with organics), which is then oxidized to $Na_2S_2O_3$ ($S^{-2} \rightarrow S^{+2}$) and further oxidized in the furnace to Na_2SO_4 ($S^{+2} \rightarrow S^{+6}$). The organic material from the digestion process, which we may simplify here as carbon, is oxidized in the furnace to CO_2 ($C° \rightarrow C^{+4}$) whereas the Na_2SO_4 is reduced back to Na_2S ($S^{+6} \rightarrow S^{-2}$), the original oxidation state of sulfur in the process. The CO_2 is absorbed by NaOH to form Na_2CO_3. Water is added to the material from the furnace, forming a green liquor containing NaSH and NaOH. The Na_2CO_3 is reacted with CaO and water to give more NaOH (causticizing) and $CaCO_3$, which is usually filtered and transformed on site back into CaO by a lime kiln.

Organic Kraft Chemistry. The organic chemistry of the alkali cleavage of lignin is summarized here. The phenoxide ion expels an alkoxide ion to form a quinonemethide intermediate, which then is attacked by hydroxide ions to eventually form an epoxide ring. Although this is somewhat simplified, it does

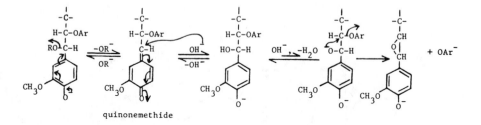

give an idea of how the degradation begins. The bisulfide ion present in the kraft process is even a better nucleophile than hydroxide, so when it is present it attacks the quinonemethide intermediate. An episulfide is formed that then hydrolyzes to a thiol-alcohol.

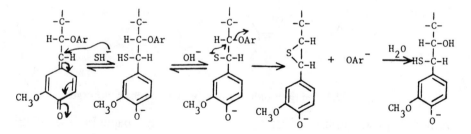

Basic hydrolysis also lowers the molecular weight of the polysaccharides in wood. Two types of hydrolysis occur, peeling and chain cleavage (see Fig. 22.3). Some organic materials from the black liquor can be isolated as useful side products. Much of it eventually is oxidized to carbon dioxide in the furnace.

Finishing Kraft Paper. After the crude pulp is obtained from the alkaline sulfate process, it must be bleached in stages with elemental chlorine, extracted with sodium hydroxide, and oxidized with calcium hypochlorite, chlorine dioxide, and hydrogen peroxide. This lightens it from a brown to a light brown or even white (difficult) color. Chlorination of the aromatic rings of residual lignin is probably what is occurring although this has not been completely

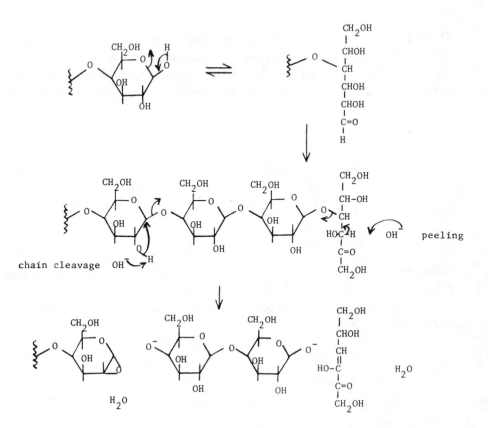

Figure 22.3 Two types of hydrolysis: peeling and chain cleavage.

studied. Typical end-uses of kraft pulp are brown bags, paper boxes, and milk cartons. A list of the major advantages and disadvantages of the kraft process versus other pulping methods follow:

Kraft Advantages	*Kraft Disadvantages*
Excellent paper strength	Poor pulp color
Low energy requirements	Low yield—43% after bleaching
Chemical recycling	High capital investment
Little pollution	High bleaching costs
Low chemical cost	Nonrecyclable bleaching effluent
Variety of wood species usable	Strong odors

Anyone having approached a kraft mill will be familiar with the last-named disadvantage. This odor is caused by methyl mercaptan and dimethyl sulfide, both of which are formed by bisulfide cleavage of methoxy groups in lignin.

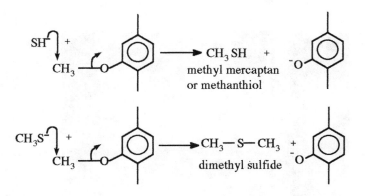

Much of the methyl mercaptan and dimethyl sulfide can be oxidized to dimethylsulfoxide, a useful side product that is a common polar, aprotic solvent in the chemical industry. This is in fact the primary method of its manufacture, as a kraft by-product. Recent reports that DMSO is a cure for common body aches and pains, including arthritis, have little scientific foundation and the chemical does not have FDA approval for most uses. Caution must be used when handling it because of its extremely high rate of skin penetration.

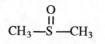

dimethylsulfoxide (DMSO)

Two other important side products of the kraft process are sulfate turpentine and tall oil. The turpentine is obtained from the gases formed in the digestion process. From 2-10 gal of turpentine can be obtained per ton of pulp. Tall oil soap is a black viscous liquid of rosin and fatty acids that can be separated from the black liquor by centrifuging. Acidification gives tall oil. These side products will be discussed later.

Other Pulp Processes

The acid sulfite process is used to obtain a higher quality paper. It is also more water polluting. Digestion occurs in a mixture of sulfur dioxide and calcium or magnesium bisulfite. The magnesium bisulfite process is better for pollution but still not so good as the kraft process. Sulfite pulp is used for bond paper and high-grade book paper.

Figure 22.4 Elaborate machinery for the drying of the pulp after the digesting process has been completed. (Courtesy of Georgia-Pacific Corporation, Bellingham, WA.)

In the NSSC process sodium sulfite is buffered with sodium carbonate, bicarbonate, and hydroxide to maintain a slightly alkaline pH during the cook. NSSC hardwood pulp is the premier pulp for corrugating medium and cannot be matched by any other process.

PAPER MANUFACTURE

Less chemistry is involved in the manufacture of paper once the pulp has been made, but it is a complex process that can be summarized in the following steps:

1. *Beating and refining* the pulp to make the fibers stronger, more uniform, more dense, more opaque, and less porous.
2. *Coagulating and coating* the fibers with aluminum sulfate, papermaker's alum.
3. Adding *fillers* to occupy the spaces between the fibers. These fillers are usually inorganic clays, calcium carbonate, or titanium dioxide.
4. Adding *sizing* to impart resistance to penetration by liquids. Most sizing is a soap or wax emulsion precipitated by the alum. This produces a gelatinous film on the fiber and a hardened surface.
5. Adding *wet strength resins* to increase the strength of the paper when wet. Urea-formaldehyde resins are typical.
6. *Dyeing.*
7. *Recycling.* This is on the increase, with 25% of the paper being recycled in recent years. It is estimated that 40% of paper is potentially recoverable. Of the 600 paper mills operating in the U.S., 140 of them depend on waster paper recycling.

There are many chemicals that are important in the manufacture of paper. These paper additives include pigments and dyes, wet-strength resins, sizes, thickeners, biocides, defoamers, etc. A good estimate of the total commerical valure of these additives is nearly $1 billion.

GENERAL USES OF PAPER PRODUCTS

Paper (50%)

Newsprint, books, tissue, corrugated boxes, bags, cigarette paper, food containers, plates, wallpaper, disposable clothing.

Paperboard (50%)

1. Fiberboard (fibers with added phenolics): paneling, furniture, insulation
2. Particleboard (waste wood chips or dust plus a resin): paneling, subflooring, general plywood and lumber replacement
3. Paper-base laminates (plies of wood plus a phenolic, urea or melamine resin): structural and machine parts

MISCELLANEOUS CHEMICALS USED ON WOOD

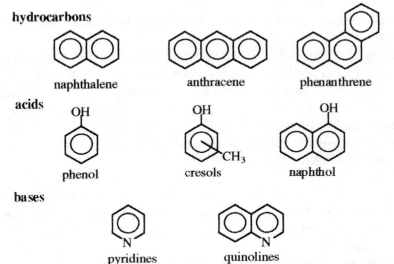

An interesting industry that has developed out of the necessity for preserving wood is now the second largest wood-related industry. Preservation against fungi, insects, borers, and mildew is accomplished by using one of three important types of preservatives. The first type is creosotes, which are mixtures of aromatic hydrocarbons with organic acids and bases.

Secondly, chlorinated phenols, especially pentachlorophenol, is used as a preservative.

Thirdly, inorganic salts of copper, chromium, arsenic, and tin have been used

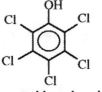

pentachlorophenol

as preservatives. The use of wood preservatives is about 1 billion lb, with 60% being creosotes, 20% pentachlorophenol, and 20% copper compounds. But these are cheap chemicals, so commercial value is under $1 billion. Creosotes from coal tar are the most widely used because, in addition to their cheapness, they are highly toxic to wood-destroying organisms, have a high degree of permanence because of low water solubility and volatility, and are easy to apply and penetrate deeply. In 1984 the EPA ruled that only people holding state pesticide licenses would be able to buy and use all three types of preservatives.

Flame retardants for wood have also been developed. They include inorganic compounds such as diammonium phosphate, ammonium sulfate, borax ($Na_2B_4O_7 \cdot 10H_2O$), boric acid, and zinc chloride. The mechanism of flame retardance in wood has no single explanation. It probably includes the following: (1) The fusing of the chemical at high temperatures to form a nonconbustible film that excludes oxygen, (2) the evolution of noncombustible gases, and (3) the catalytic promotion of charcoal formation instead of volatile combustible gas.

CHEMICALS OBTAINED FROM WOOD HYDROLYSIS AND FERMENTATION

Hydrolysis of the polysaccharides in wood to sugars and fermentation of the sugar to ethyl alcohol is no longer an economical process in this country. It cannot compete with synthetic alcohol made from ethylene or fermentation of corn.

A number of lower volume chemicals can be obtained from wood hydrolysis. Furfural is formed from the hydrolysis of some polysaccharides to pentoses,

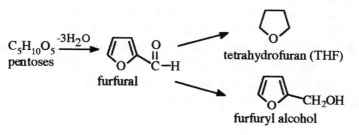

followed by dehydration. This process is still used in the former Soviet Union. Furfural is used in small amounts in some phenol plastics; it is a small minor pesticide and an important commercial solvent. It can be converted into the common solvent THF and an important solvent and intermediate in organic synthesis, furfuryl alcohol.

Vanillin is obtained in the United States from sulfite waste liquor by further alkaline hydrolysis of lignin. It is the same substance that can be obtained from vanilla bean extract and is the common flavoring in foods and drinks. Interestingly, natural and synthetic vanillin can be distinguished from each other by a slight difference in the amount of ^{13}C in their structure since one is biosynthetic in the bean and the other is isolated from a second natural product, wood, by hydrolysis of the lignin.

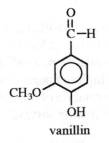

vanillin

CHEMICALS OBTAINED FROM WOOD CARBONIZATION AND DISTILLATION

Wood distillation was used previously in the United States to make methanol, acetic acid, and acetone. Up to 1-2% per wood weight of methanol, 4-5% acetic acid, and 0.5% acetone can be obtained. Formerly this was the only source of these compounds. It is no longer competitive with the synthetic processes. Some phenols can be obtained, as well as common gases such as carbon dioxide, carbon monoxide, methane, and hydrogen.

The manufacture of charcoal, especially briquettes, has been increasing in demand. It is the residue after combustion of the volatiles from a hardwood distillation. It consists of elemental carbon and incompletely decomposed organic material and many adsorbed chemicals. Carbonization is usually performed at about 400-500°C. The charcoal has a volatile content of 15-25% and can be made in about 37-46% yield by weight from wood.

NAVAL STORES INDUSTRY

These still important products, produced from softwood pines, were once used by the U.S. Navy in the days of wooden ships and were governed by the 1923 Federal Naval Stores Act.

Turpentine is a mixture of $C_{10}H_{16}$ volatile terpenes (hydrocarbons made of isoprene units). There are actually four different types and methods of making turpentine, including steam distillation of wood. The two pinenes, α and β, are major components of turpentine. Other compounds found in abundant amounts are camphene, dipentene, terpinolene, and Δ^3-carene. Although having been replaced by petroleum hydrocarbons as paint thinners (lower price, less odor), turpentine is still a good solvent and thinner in many specialty applications. The use pattern for turpentine is as follows: solvent, 11%; synthetic pine oil, 48%; polyterpene resins as adhesives, 16%; toxophene insecticides, 16%; and flavor and fragrance essential oils, 9%.

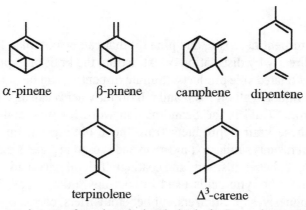

α-pinene   β-pinene   camphene   dipentene

terpinolene Δ^3-carene

Pine oil is a mixture of terpine-derived alcohols. It can be extracted from pine but is also synthetically made from turpentine, especially the a-pinene fraction, by reaction with aqueous acid. It is used in many household cleaners as a bactericide, odorant, and solvent. The major constituents of pine oil are shown here.

α-terpineol   β-terpineol   ϒ-terpineol   α-fenchone

borneol   octahydro-α-terpineol

Rosin, a brittle solid, mp 80°C, is obtained from the gum of trees and tree stumps as a residue after steam distillation of the turpentine. It is made up of 90% resin acids and 10% neutral matter. Resin acids are tricyclic monocarboxylic acids of formula $C_{20}H_{30}O_2$. The common isomer is l-abietic acid. About 38% of rosin is used as paper size (its sodium salt), in synthetic rubber as an emulsifier in polymerization (13%), and in adhesives (12%), coatings (8%), and inks (8%).

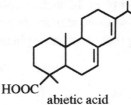

HOOC   abietic acid

Besides the turpentine, rosin, and pine oil that can be obtained from pines, directly or indirectly by distillation or extraction, the kraft pulp process now furnishes many related side products. Sulfate turpentine can be obtained from the black kraft liquor. Tall oil rosin and tall oil fatty acids can also be isolated from this liquor. "Tall" is the Scandinavian word for pine and is used to differentiate these kraft byproducts from those obtained from pine more directly. Tall oil rosin is similar to pine rosin and is used in paper sizing, printing inks, adhesives, rubber emulsifiers, and coatings. Tall oil fatty acids are C_{16} and C_{18} long-chain carboxylic acids used in coatings, inks, soaps, detergents, disinfectants, adhesives, plasticizers, rubber emulsifiers, corrosion inhibitors, and mining flotation reagents. The tall oil obtained from kraft liquor gives about 37% rosin and 36% fatty acids. The market for tall oil is expanding.

The last navel stores chemical that we will mention is tannin, an extract from the wood, bark, or leaves of many trees and plants. This is a mixture of complex, dark-colored sugar esters of polyhydroxy phenolic compounds related to catechol, pyrogallol, gallic acid, and ellagic acid. There is much variation with the species. Tannin has the ability to combine with proteins of animal skins to produce leather. This tanning process, probably involving hydrogen-bonding to the proteins, keeps the skin soft and pliable so it may be used in many leather products. Almost all tannin used by the United States is imported, especially from Argentina and Paraguay.

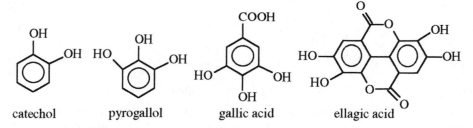

catechol   pyrogallol   gallic acid   ellagic acid

In 1991 it was reported that certain specific phenols and polyphenols, such as ellagic acid, commonly found in vegetables, fruits, and tea (especially green tea), have anticancer properties. Ellagic acid was effective in inhibiting the development of liver tumors. Other phenols, such as epigallocatechin-3-gallate, chlorogenic acid, and quercetin, also show anticarcinogenic properties. Studies suggest that a proper diet including these sources of phenols may help reduce the incidence of some cancers.

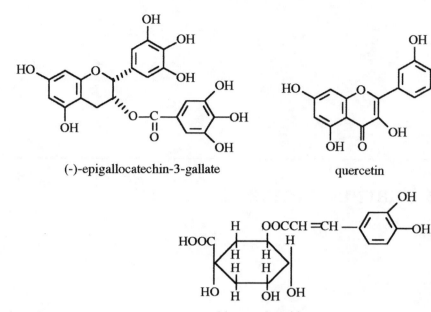

(-)-epigallocatechin-3-gallate

quercetin

chlorogenic acid

In Chemicals and Allied Products under Industrial Organic Chemicals (SIC 286) there is a subsector entitled Gum and Wood Chemicals (SIC 2861) that covers many of the miscellaneous chemicals that we have discussed here, including charcoal, tall oil, rosin, turpentine, and pine oil. The 1988 value of shipments was $566 million, down from a record $750 million in 1981. This is only 1% of industrial organic chemicals. However, the pulp and paper industry is large and is a prime user of chemicals and chemical processes, so it is good to know some of the basics of this industry.

23

The Pharmaceutical Industry

References

Kent, pp. 718-746

W & R II, pp. 213-279

C. O. Wilson, O. Gisvold, and R. F. Doerge, *Textbook of Organic Medicinal and Pharmaceutical Chemistry,* 7th ed. J. B. Lippincott Co.: Philadelphia, 1977.

D. Lednicer and L. A. Mitscher, *The Organic Chemistry of Drug Synthesis,* John Wiley & Sons: New York, 1977.

S. C. Stinson, "Bulk Drug Output Moves Outside U.S.," *C & News,* 9-16-85, pp. 25-59.

S. C. Stinson, "Innovation Fuels Growing Market for Antibacterial Drugs," *C & E News,* 9-29-86, pp. 33-67.

"AIDS," *C & E News* special issue, 11-23-87.

S. C. Stinson, "Better Understanding of Arthritis Leading to New Drugs to Treat It, " *C & E News,* 10-16-89, pp. 37-70.

INTRODUCTION

The pharmaceutical industry is an important segment of the chemical industry not because of its volume of chemicals, which is usually small, but because these chemicals are high priced per volume and because it employs about 30% of all technical personnel in Chemicals and Allied Products. The pharmaceutical industry is a technologically intensive industry; it is not uncommon for drug companies to spend 10% of their sales on research expenditures.

Drugs (SIC 283) is a very large segment of the chemical industry at a 1988 level of $44 billion, which is 17% of Chemicals and Allied Products. Fig. 23.1 gives

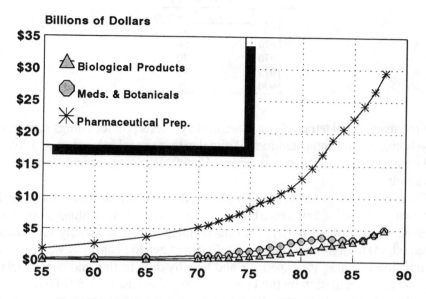

Figure 23.1 U.S. shipments of pharmaceuticals. (*Source*: AS.)

the trend since 1955 for the three major subsectors, Biological Products (SIC 2835, 2836), Medicinals and Botanicals (SIC 2833), and Pharmaceutical Preparations (SIC 2834). Percentages by shipments are 12% biological products, 13% medicinals and botanicals, and 75% pharmaceutical preparations.

Examples of biological products are bacterial and virus vaccines, serums, plasmas, and other blood derivatives. Medicinals and botanicals include bulk organic and inorganic medicinal chemicals and bulk botanical drugs and herbs. Examples are alkaloids, anesthetics, barbituric acid and derivatives, caffeine, hormones, insulin, morphine, penicillin, quinine, aspirin, sulfa drugs, and vitamins. Pharmaceutical preparations are drugs formulated and fabricated into their final form for direct consumption (tablets, capsules, etc.). The industry has grown rapidly in the past 20 years especially in pharmaceutical production.

The effect of the modern drug industry on the life expectancy in the United States can be seen in Table 23.1. In 1900 infectious diseases accounted for 500 deaths per 100,000 Americans; today the figure has dropped to 50. But many problems still face the industry, including most forms of cancer, arthritis, diabetes, senility, and viral diseases, even including the common cold.

There are five basic sources of pharmaceuticals. By dollar value of products, fermentation is probably the most important, whereas by tonnage, chemical synthesis is dominant. Fermentation is used for antibiotics such as penicillins,

TABLE 23.1

Year	Life Expectancy
1900	47
1920	56
1970	70
1975	75
1990	75

streptomycins, and tetracyclines. Chemical synthesis provides drugs such as the psychotropics and antihistamines. Animal extracts provide hormones. Biological sources lead to vaccines and serums. Vegetable extracts provide steroids and alkaloids.

Other important general characteristics of the pharmaceutical industry are the following: (1) use of multistage batch processes rather than continuous flow, (2) high level of product purity, (3) management which is usually technically oriented, (4) high promotional costs, (5) use of generic names as well as brand or trademark names, (6) expensive and lengthy drug testing and clinical trials, and (7) strict regulation by the Food and Drug Administration (FDA).

Although some efficacious drugs have been known for centuries, such as the antimalarial quinine first used in 1639, most important discoveries are of more recent origin. Smallpox vaccine was discovered around 1800, morphine in 1820, aspirin in 1894, and phenobarbital in 1912. But the discovery of the antibacterial activity of sulfur drugs in 1932 and penicillin in 1940 started the golden era of rapid expansion and discovery in the industry. Nearly all important drugs today have been discovered since 1940, some very recently.

What are the properties of an "ideal drug"? It should be nontoxic and without side effects. The fatal dose should be many times the therapeutic dose (it should have a maximum therapeutic index). The necessary dose should not require too frequent administration over too long a period (four tablets at 3-hr intervals is hard to remember). The efficiency of the drug should not be seriously reduced by changes in body fluids or by tissue enzymes. A drug should be stable and storable for long periods, even in extreme climates. Finally, it should be possible to dispense it in a variety of forms—pills, soft gelatin capsules, hard capsules, liquids, syrups for children, suppositories, ointments, and solutions for intravenous or intramuscular injection to name a few. Even implantation, time release capsules, and controlled release through membranes are now possible.

TYPES OF DRUGS

The pharmaceutical industry is so complex and diverse in its chemistry that it is difficult to know where to start, but actually 100 drugs account for 60% of

TABLE 23.2 U.S. Production of Medicinal Chemicals

Type	Production Million lb		
Antibiotics	28.8		
Penicillins		6.3	
Other Antibiotics		22.5	
Antihistamines	0.5		
Anti-infective Agents Except Antibiotics	19.1		
Autonomic Drugs	1.0		
Central Depressants and Stimulants	49.2		
Analgesics, Antipyretics, and Anti-inflammatory Agents		42.8	
Aspirin			23.7
Other Analgesics, etc.			19.1
Antidepressants		0.1	
Antitussives		0.3	
Other Central Depressants and Stimulants		6.0	
Dermatological Agents	15.0		
Expectorants and Mucolytic Agents	1.1		
Gastrointestinal Agents and Therapeutic Nutrients	99.7		
Vitamins	38.5		
Miscellaneous Medicinal Chemicals	5.3		
Total	258.2		

Source: SOC.

all prescriptions filled. Table 23.2 gives a breakdown of the 1988 production of different types of drugs to give some indication of relative importance. There are many different ways of subdividing drugs. Drugs that can be bought without a prescription from a doctor are called *over-the-counter* or *proprietary drugs*. Those that require a prescription are called *ethical* or *prescription drugs*, the purity of which are rigidly defined in the U.S. Pharmacopoeia (a drug or chemical this pure is described as USP grade). Drugs may also be divided by either structure or physiological activity. Many different types of chemical structures may still be useful in combating a certain type of illness. We will concentrate now on ethical drugs ranking high in pharmaceuticals used and will divide the discussion by physiological use rather than by chemical type: cardiovascular drugs, central nervous system depressants and stimulants, antibacterials, steroids, analgesics, and antihistamines.

According to the National Center for Health Statistics, the ten largest selling drugs in the United States in 1985 were the following:

1. Inderal (propranalol)
2. Lasix (furosemide)
3. Dyazide (triamterene, hydrochlorothiazide)

4. Ampicillin
5. Penicillin
6. Aspirin
7. Lanoxin (digoxin)
8. Tetracycline
9. Diphtheria tetanus toxoids pertussis
10. Polio vaccine

CARDIOVASCULAR AGENTS

Cardiovascular agents are used for their action on the heart or on other parts of the vascular system. They modify the total output of the heart or the distribution of blood to certain parts of the circulatory system.

Antihypertensive Agents

Antihypertensive agents, substances that lower high blood pressure, are an important subclass of cardiovascular agents. Reserpine, an indole alkaloid obtained from the Rauwolfia plant, was the first successful drug to treat high

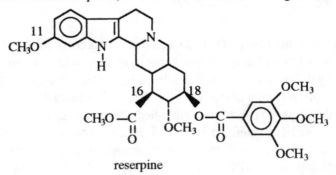

reserpine

blood pressure and was discovered in 1953. The plant extracts were first used in India to alleviate toothaches. They were brought to the United States in 1940. Reserpine was isolated in 1952. Its structure was determined in 1954 and it was proven by total synthesis in 1958.

In common with other indole derivatives, reserpine is susceptible to decomposition by light and oxidation, so it must be stabilized. Modifying the trimethoxyphenyl portion of the molecule gives other antihypertensive drugs with various potency and rapidity of action. Two other agents are syrosingopine and rescinnamine with the variations shown here. Another antihypertensive agent, deserpidine, differs from reserpine only in the absence of the methoxy group at C-11. The reserpine molecule is easily hydrolyzed at C-16 and C-18.

The mode of action of reserpine involves the release of norepinephrine (noradrenaline), responsible for heart contraction, which in turn is destroyed by normal processes to expand the heart and lower the blood pressure.

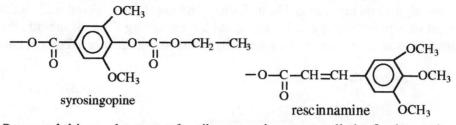

syrosingopine

rescinnamine

Propranolol is another type of antihypertensive agent called a β-adrenergic blocking agent because it competes with epinephrine (adrenaline) and norepinephrine at their receptor sites and protects the heart against undue stimulation. Propranalol is easily synthesized in two S_N2 substitution reactions from α-naphthol, epichlorohydrin, and isopropylamine. α-Naphthol attacks the better leaving group chlorine in the first S_N2 at a primary position of epichlorohydrin. Isopropylamine picks the primary and strained three-membered ring carbon in the second step. Notice that for propranalol, although it blocks the receptor site of epinephrine and norepinephrine, it bears only a vague chemical similarity to these substances. Propanalol was the largest selling drug in the U.S. in 1985. Sir James Black of the U.K. won the Nobel Prize in Medicine in 1988 for his discovery of propranolol in 1964, as well as other research.

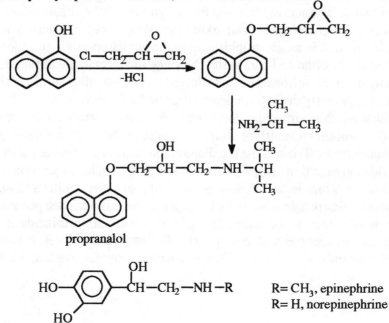

propranalol

R= CH$_3$, epinephrine
R= H, norepinephrine

Diuretics

Diuretics are drugs that increase the excretion of urine by the kidney, thereby decreasing body fluids. This alleviates the swelling of tissues that sometimes cause high blood pressure and heart, kidney, and liver failure. Furosemide is the most effective diuretic and is the second largest selling drug. It inhibits the readsorption of sodium in the kidney and promotes potassium excretion, two

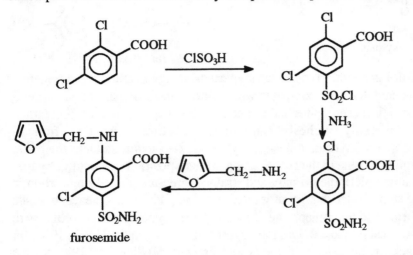

furosemide

ions intimately involved in water retention for the body. It lowers blood pressure as well. The starting material for its synthesis is 2,4-dichlorobenzoic acid (formed by chlorination and oxidation of toluene). Reaction with chlorosulfonic acid is an electrophilic aromatic substitution via the species $\oplus SO_2Cl$ attacking ortho and para to the chlorines and meta to the carboxylate. Ammonolysis to the sulfonamide is followed by nucleophilic aromatic substitution of apparently the less hindered chlorine by furfurylamine (obtained from furfural which is a major product from hydrolysis of many carbohydrates).

Dyazide®, another important diuretic, contains both triamterene and hydrochlorothiazide. Triamterene is a diuretic and is known to increase sodium and chloride ion excretion but not potassium ion. It is used in conjunction with a hydrothiazide, which is an excellent diuretic but also gives significant loss of potassium and bicarbonate ions. If the triamterene were not included potassium chloride would have to be added to the diet. Hydrochlorothiazide is an antihypertensive agent as well but, unlike other antihypertensives, it lowers blood pressure only when it is too high, and not in normotensive individuals.

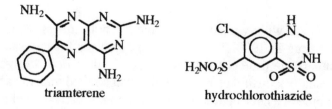

triamterene   hydrochlorothiazide

A number of thiazides can be synthesized from appropriate sulfonamideds by cyclization with dehydration. Conversion to hydrothiazides increases their activity by a factor of ten.

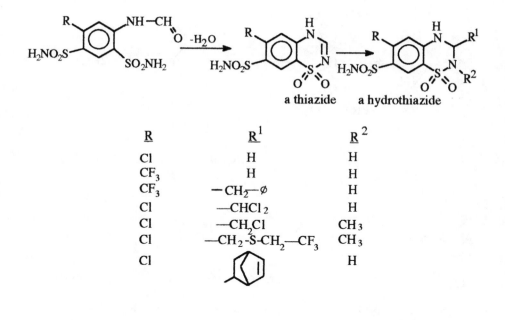

a thiazide   a hydrothiazide

R	R^1	R^2
Cl	H	H
CF$_3$	H	H
CF$_3$	—CH$_2$—∅	H
Cl	—CHCl$_2$	H
Cl	—CH$_2$Cl	CH$_3$
Cl	—CH$_2$-S-CH$_2$—CF$_3$	CH$_3$
Cl		H

CENTRAL NERVOUS SYSTEM PHARMACEUTICALS

Although this type of drug has various subclasses based on physiological response, that is, tranquilizers, stimulants, depressants, etc., we will subdivide and treat a few of them on the basis of their chemical classes.

Barbiturates

The barbiturates are widely used as sedative-hypnotic drugs. Barbital was introduced as a drug in 1903. The method of synthesis for thousands of its analogs has undergone little change. Urea reacts with various derivatives of malonic acid, usually a diethyl ester of a dialkyl substituted malonic acid. This is a classic example of a nucleophilic acyl substitution. A derivative of ammonia reacts with esters to form an amide, only in this case a cyclization to a strainless six-membered ring results because of the proximity of the bifunctionality. A partial mechanism is given here.

Reaction:

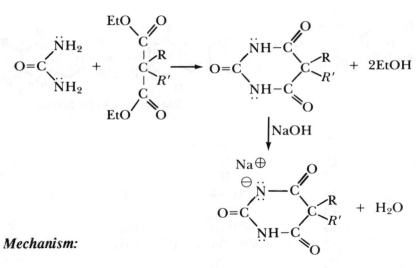

Mechanism:

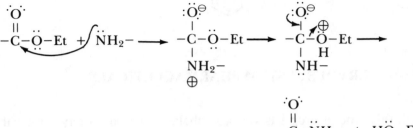

The barbiturates are usually administered as the sodium salts. The N—H bonds are acidic because the anion is resonance stabilized. Although barbituric acid is inactive, a range of activities is obtained that varies with the groups at *R* and *R'*. Some of the more important ones are listed below.

Name	R	R'
Barbituric acid	H	H
Barbital	Et	Et
Phenobarbital	Et	Phenyl
Butabarbital	Allyl	Isobutyl

Activity and toxicity both increase with the size of the groups. Branching and unsaturation decrease the duration of action. Phenobarbital and butabarbital are among the top prescription drugs. The maximum therapeutic index (tolerated dose/minimum effective dose) is highest when the two groups have a total of six to ten carbons. The mechanism of action is not completely understood, but they in some way reduce the number of nerve impulses ascending to the brain. Major drawbacks of their use are their habit formation and their high toxicity when alcohol is present in the bloodstream.

Benzodiazepines

A recently developed series of tranquilizers, drugs that relieve anxiety and nervous tension without impairing consciousness, have a benzene ring fused to a seven-membered ring containing two nitrogens. As a group they are called benzodiazepines. The two most successful are diazepam (Valium®) and chlordiazepoxide (Librium®) introduced in 1964 and 1960, respectively. Flurazepam (Dalmane®), first used in 1970, is a hypnotic. Diazepam is presently the most widely prescribed tranquilizer and chlordiazepoxide is not far behind.

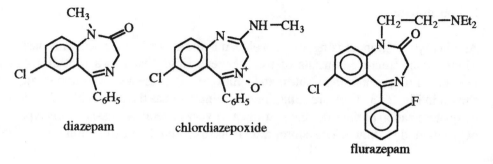

diazepam   chlordiazepoxide

flurazepam

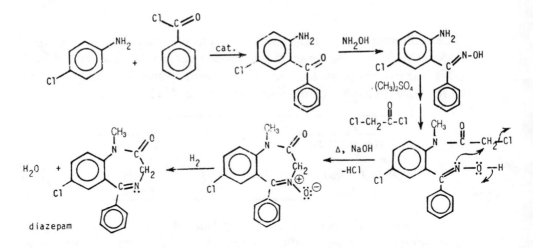

Figure 23.2 Synthesis of diazepam.

The synthesis of diazepam is outlined in Fig. 23.2. *p*-Chloroaniline (prepared from benzene by nitration, reduction of the nitro group to an amine, and chlorination of the *o,p*-directing aniline) is reacted with benzoyl chloride (from toluene by oxidation to benzoic acid, followed by acid chloride formation) in a Friedel-Crafts acylation. Since the position para to the amino group is taken, acylation occurs ortho. Formation of the oxime derivative is followed by methylation and then acylation of the amino group with chloroacetyl chloride. Heating in base splits out HCl as shown and forms the ring. Reduction of the amine oxide with hydrogen gives diazepam.

Diazepam is used for the control of anxiety and tension, the relief of muscle spasms, and the management of acute agitation during alcohol withdrawal, but it itself may be habit-forming. Chlordiazepoxide has similar uses and its synthesis is somewhat analogous to diazepam. Flurazepam is a hypnotic, useful for insomnia treatment. It is reported to provide 7-8 hr of restful sleep.

Phenothiazines

Another type of tranquilizing drug developed in the 1950s is the phenothiazines. Their basic structure consists of two benzene rings fused to a central six-membered ring containing a sulfur and a nitrogen. Representative examples are shown in Fig. 23.3. They are sometimes administered as the hydrochloride salt by quaternization of the side chain nitrogen. A very general synthesis of this type of compound is shown. The appropriate metasubstituted diphenylamine (made

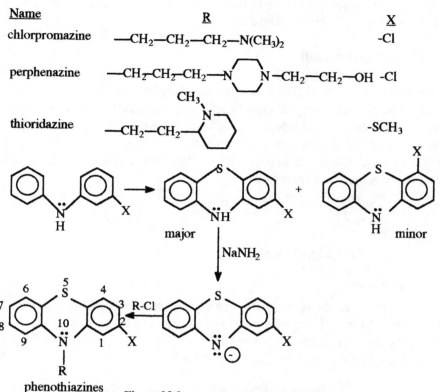

Figure 23.3 Some important phenothiazines.

by heating aniline with a substituted aniline hydrochloride) is reacted with sulfur and an iodine catalyst to close the ring. Usually only a small amount of the second isomer is formed. Treatment with the strong base sodium amide gives the anion on the ring nitrogen, which then displaces the chlorine of the appropriate second reactant by a simple S_N2 reaction.

A large number of structure-activity relationships have been performed on the phenothiazines. A chlorine at position 2 increases activity, as does a thioalkyl or acyl group. A substituent at positions 1,3,4, or simultaneous substitution in both aromatic rings results in a loss of the tranquilizing activity. A three-carbon side chain between the ring nitrogen and the more basic side chain nitrogen is optimal, but compounds with two carbons still possess some activity. Branching at the β-position of the side chain reduces tranquilizing activity. A piperazine ring (as in perphenazine) increases the drug potency.

There are three main therapeutic applications of these drugs. They have an antiemetic effect (stop vomitting). They are used with anesthetics, potent analgesics (pain relievers), and sedatives to permit their use in smaller doses. Finally, they are used most widely to relieve anxiety and tension in various severe mental and emotional disorders.

Tricyclic Antidepressants

Following the discovery of the phenothiazines many compounds with structures similar to them were studied. Other tricyclic compounds were found to be powerful stimulants, or antidepressants, to the central nervous system. Depressed individuals may respond with an elevation of mood, increased physical activity, mental alertness, and an improved appetite. Imipramine and amitriptyline hydrochlorides are good examples.

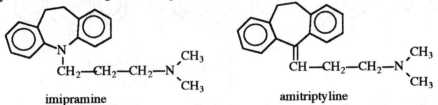

imipramine   amitriptyline

The synthesis of amitriptyline starts from the key intermediate dibenzosuberone (which comes from phthalic anhydride) and can proceed by two pathways (Fig. 23.4). Treatment of dibenzosuberone with cyclopropyl Grignard gives the tertiary alcohol after hydrolysis. Reaction of the alcohol with hydrochloric acid

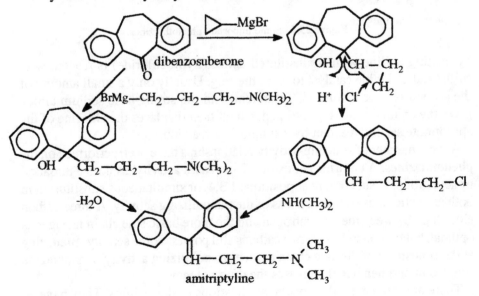

Figure 23.4 Synthesis of amitriptyline.

proceeds with rearrangement and opening of the strained cyclopropane to give a chloride. S_N2 displacement of the chloride with dimethylamine forms amitriptyline. Alternatively, dibenzosuberone can be reacted with dimethyl-aminopropyl Grignard to form an alcohol, which upon dehydration forms amitriptyline.

Activity in these tricyclic compounds is restricted to compounds having a two- or three-carbon side chain and methyl-substituted or unsubstituted amino groups in the side chain. Some compounds with substituents on the aromatic ring are active. Finally, the two-carbon bridge linking the aromatic rings may be —CH_2—CH_2— or —CH=CH—.

Amitriptyline is recommended for the treatment of mental depression, with improvement in mood seen in two to three weeks after the start of medication. Imipramine is used in similar cases.

ANTIBACTERIAL AGENTS

Before the 1930s bacterial diseases were a major cause of death. Pneumonia and tuberculosis were major killers. Since the advent of the sulfa drugs and penicillins many bacterial diseases have been controlled. The antibacterial drugs will be discussed by type. We will see that some of these are antibiotics, that is, an antibacterial substance produced by a living organism such as a bacterium or fungus rather than synthesized in the laboratory.

There are four general properties for a good antibacterial agent. It must be selective. Eliminating all species from the body may leave the patient prone to superinfection. It should kill bacteria rather than just prevent their multiplication; it should be bactericidal rather than bacteriostatic. Bacteria should not develop resistance to the drug. Lastly, absorption of the drug into the body should be rapid and the desired level maintained in the body for long periods.

Sulfa Drugs (Sulfonamides)

In 1935 Gerhard Domagk observed that prontosil, an azo dye, was effective against streptococcus bacteria. He won the 1939 Nobel Prize in Medicine for this discovery. Actually hydrolysis of the dye in the body forms the active ingredient

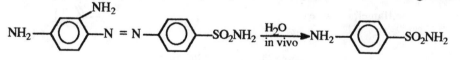

prontosil   sulfanilamide

p-aminobenzenesulfonamide, or sulfanilamide. Over 5000 sulfonamides have been synthesized and tested. The physiologically active ones are known collectively as sulfa drugs. Most involve variations of groups in place of the hydrogens of the sulfonamide moiety. A general synthesis of these compounds is outlined in Fig. 23.5. Aniline is protected by acetylation to acetanilide to limit the chloro-sulfonylation to the para position. Acetylation deactivates the ring toward multi-electrophilic attack. Various amines react with sulfonyl chloride

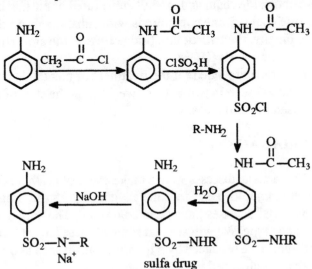

Figure 23.5 Synthesis of sulfonamides.

Name		R
sulfanilamide		-H
sufadiazine		
sulfisoxazole		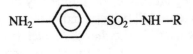
sulfamethoxazole		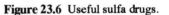

Figure 23.6 Useful sulfa drugs.

to give acetylated sulfonamides. Hydrolysis then removes the acetyl group to give the active drug. Sometimes the drug is administered as its sodium salt, which is soluble in water.

Some common sulfa drugs are pictured in Fig. 23.6 with the appropriate *R* group designated. Sulfadiazine is probably the best for routine use. It is eight times as active as sulfanilamide and exhibits fewer toxic reactions than most of the sulfonamides. Most of the common derivatives have an *R* group that is heterocyclic. These groups cause greater absorption into the body, yet they are easily hydrolyzed to the active sulfanilamide.

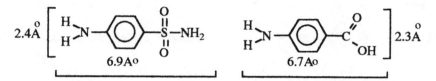

Unlike many drugs the mode of action of sulfonamides *is* understood. They are bacteriostatic. Sulfanilamide mimics *p*-aminobenzoic acid (PABA), essential for incorporation into enzymes regulating bacterial growth but nonessential for human growth. The bacteria mistake sulfur for carbon, form inactive enzymes, and cannot grow. Note that the molecular geometry of sulfanilamide and PABA are similar.

If an alkyl, alkoxy, or other functional group is substituted for the *p*-amino group all activity is lost. Groups at the ortho and meta positions also cause inactivity.

Sulfonamides are historically important but have been largely replaced by other newer antibacterials. They are still used in urinary infections and in the treatment of bronchitis. The danger of crystal formation in the kidneys is circumvented by administering a mixture of sulfonamides. This changes the solubility characteristics but still has an effect on the bacteria.

Penicillins

In 1929 Fleming discovered that certain molds contained antibiotics. This initial report was studied in detail by Chain and Florey. All three won the Nobel Prize in Medicine for 1945 because of their discovery of the penicillins. Examination of their general structure shows them to contain a fused ring system of unusual design. A four-membered ring amide, or b-lactam, structure is bonded to a five-membered thiazolidine ring. Over 30 penicillins have been isolated from various fermentation mixtures and over 2000 different *R* groups have been made synthetically. The most important pharmaceutically are shown in Fig. 23.7. They

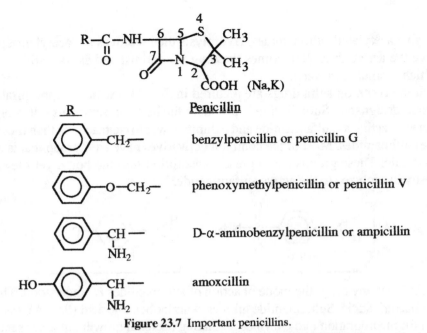

Figure 23.7 Important penicillins.

work by inactivating enzymes that are essential for cell wall development. As a result, the bacteria are enclosed only by a fragile cell membrane and they do not survive.

The free carboxylic acid is not suitable for oral administration, but the sodium or potassium carboxylates of most penicillins are soluble in water and are readily absorbed orally. Salts of penicillins with organic bases have limited water solubility but provide effective blood levels over a long period. Although total syntheses of the penicillins have been reported, they are not yet a viable alternative to large-scale fermentation. Large tanks from 5,000-30,000 gal capacities are used. The penicillin is separated by solvent extraction. The mold grows best at 23-25°C, pH 4.5-5.0. The fermentation broth is made from corn steep liquor with lactose and inorganic materials added. Sterile air permits growth of the mold over a 50-90-hr period.

The strong acid in the stomach leads to hydrolysis of the amide side chain and a β-lactam opening. An electron-attracting group at the α-position of the amide side chain inhibits the electron displacement involving the carbonyl group and the β-lactam ring, thus making such modifications as penicillin V, ampicillin and amoxicillin more acid-stable so they can be taken orally.

For years the most popular penicillin was a natural one, penicillin G. It remains the agent of choice for the treatment of more different kinds of bacteria than any other antibiotic. It is usually given by injection because it is not acid stable and is absorbed poorly through the intestine. Penicillin G procaine, made by reacting the free acid with procaine to form the amine salt, is long acting.

Penicillin G can be hydrolyzed in the laboratory to 6-aminopenicillanic acid which can be acylated to the more acid-resistant penicillin V and ampicillin, both of which can be taken orally. Ampicillin has a broader spectrum of antibacterial activity than G or V. Amoxicillin gives more complete absorption through the intestines and causes less diarrhea. There is little or no effect of food on its absorption rate.

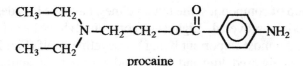

procaine

Cephalosporins

The cephalosporins (Fig. 23.8) are β-lactams like the penicillins, but instead of a five-membered thiozolidine ring, they contain a six-membered dihydrothiazine ring. They are otherwise similar in general structure to the penicillins and inactivate enzymes that are responsible for bacterial cell wall formation. Cephalosporin C, which itself is not antibacterial, is obtained from a species of

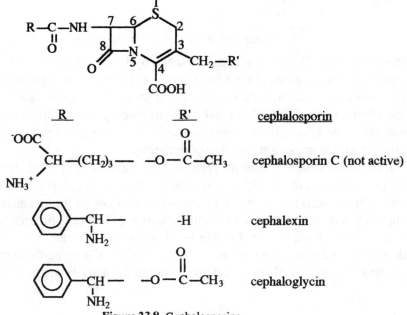

Figure 23.8 Cephalosporins.

fungus. Chemical modification of this structure to 7-aminocephalosporanic acid by removal of the R - C =O allows the preparation of the active cephalosporins such as cephalexin and cephaloglycin. These are orally active because they have an α-amino-containing R group that is stable to the gastric acid in the stomach. Other cephalosporins are easily made by acylation of the 7-amino group with different acids or nucleophilic substitution or reduction of the 3-acetoxy group.

Tetracyclines

Another group of compounds, the tetracyclines, have the broadest spectrum of any antibacterial discovered. Eight such compounds have been introduced into medical use, the most important being tetracycline itself (Fig. 23.9). It is the eighth most widely used drug and the second most popular antibacterial behind

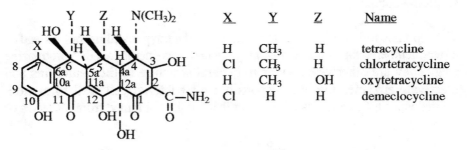

X	Y	Z	Name
H	CH$_3$	H	tetracycline
Cl	CH$_3$	H	chlortetracycline
H	CH$_3$	OH	oxytetracycline
Cl	H	H	demeclocycline

Figure 23.9 Some important tetracyclines.

the penicillins. The tetracyclines are made by fermentation procedures or by chemical modifications of the natural product. The hydrochloride salts are used most commonly for oral administration and are usually encapsulated because of their bitter taste. Controlled catalytic hydrogenolysis of chlortetracycline, a natural product, selectively removes the 7-chloro atom and produces tetracycline, the most important member of the group.

Interesting structure-activity relationships have been studied. Besides the variations shown in Fig. 23.9, hydroxyl substitution at C-6 is not essential for antibacterial activity. Removal of the 4-dimethylamino group causes loss of about 75 percent of the activity. If epimerization of C-4 or C-5a occurs, activity is decreased dramatically. Double bond formation between C-5a and C-lla decreases the activity. An aromatic C ring is inactive. Changing the carboxamide group at C-2 to a nitrile or acetyl group makes the activity negligible.

The Macrolides

At present more than 30 compounds with large rings have been isolated from fermentation processes. Only erythromycin and related structures have antibacterial properties. The macrolide antibacterials have three common chemical

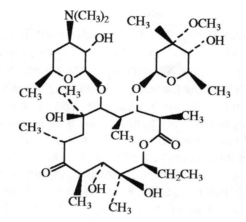

erthromycin A

characteristics: (1) a large lactone ring of 12-16 atoms, (2) a ketone group, and (3) an amino sugar. A neutral sugar moiety may also be present. Erythromycin B differs only in one less hydroxyl group. They appear in some way to inhibit protein synthesis in the bacteria. Their activity was reported in 1952, their structure determined in 1957, and their complex stereochemistry found in 1965. Since then they have been the challenge of many chemists who are interested in total synthesis. The natural erythromycin is effective against a number of organisms that have developed resistance to penicillin, tetracycline, and streptomycin.

STEROIDS

Although we are discussing most drugs by groups in their biological activity, it is convenient to study as a group steroids that are all related chemically but that cause a variety of physiological responses. Steroid drugs include anti-inflammatory agents, sex hormones, and synthetic oral contraceptives. A *steroid* is a general term for a large number of naturally occurring materials found in many plants and animals. Their general structure includes a fused set of three cyclohexanes and one cyclopentane.

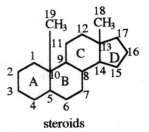

steroids

Examples are shown in Fig. 23.10. Stereochemistry is indicated by dotted lines (α-bonds, behind the plane) and solid lines (β-bonds, in front of the plane) of any substituents on the rings. Since the sex hormones are the molecules mainly responsible for differentiating the sexes, it is amazing how similar the male and female hormones are in chemical structure. The only difference between testosterone and progesterone is a hydroxy versus an acetyl group. But what a result! The natural sex hormones are used to treat prostate cancer, to alleviate menopausal distress, and to correct menstrual disorders. Other common natural steroids are estradiol, cholesterol, and cortisone. Infamous cholesterol causes deposits in the gall bladder and arteries that result in gallstones and some heart attacks. Cortisone is found in the adrenal gland, which is concerned with electrolyte balance and carbohydrate metabolism.

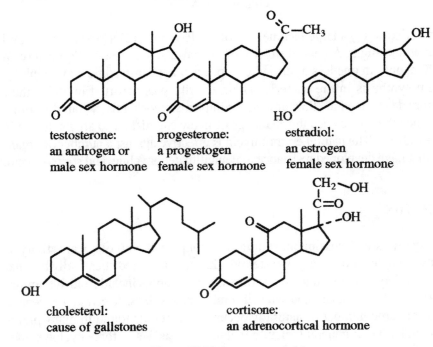

testosterone:
an androgen or
male sex hormone

progesterone:
a progestogen
female sex hormone

estradiol:
an estrogen
female sex hormone

cholesterol:
cause of gallstones

cortisone:
an adrenocortical hormone

Figure 23.10 Important steroids.

Oral Contraceptives

The oral contraceptives are synthetic drugs used to mimic the action of the natural progestogens and estrogens. They are combinations of these two types of synthetic derivatives. Note the differences in structure of norethindrone to progesterone and of mestranol to estradiol. The triple bond and other changes at C-17, plus removal of the methyl between rings A and B, allow them to be taken orally by easing the passage of the compound into the blood stream. There is evidence of a relationship between their use and blood clotting, breast cancer and heart disease. Some have also been found to actually be abortifacient, preventing pregnancy after conception, rather than before. They induce abortion rather than prevent conception.

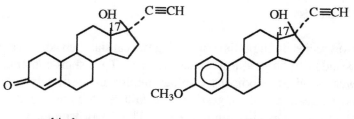

norethindrone   mestranol

Adrenal Cortex Hormones

The adrenal glands secrete over 50 different steroids, the most important of which are aldosterone and hydrocortisone. Aldosterone causes salt retention in the body. It is not commercially available. Hydrocortisone is useful for its anti-inflammatory and antiallergic activity. Cortisone and its derivatives have similar activity and it is reduced in vivo to hydrocortisone. The two substances are used to treat rheumatoid arthritis. The 11-β-hydroxyl of hydrocortisone is believed

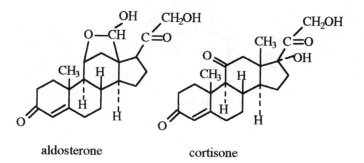

aldosterone   cortisone

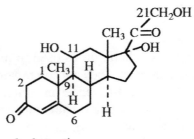

hydrocortisone

to be of major importance in binding to the receptors of enzymes. Anti-inflammatory activity is significantly increased by various substituents: 6α-fluoro, 9α-fluoro, 21-hydroxy, 2α-methyl, 9α-chloro, and a double bond at C-1.

Cardiac Steroids

Plants (two species of Digitalis) containing the cardiac steroids have been used as poisons and heart drugs at least since 1500 B.C. Toad skins containing cardiac steroids were good arrow poisons. Cardiac steroids are absolutely indispensable in the modern treatment of congestive heart failure. A commercially available important cardiac steroid is digitoxigenin. The 5β,14βstereochemistry is an important prerequisite for most cardiac activity. The 3β-hydroxy group is essential. The 17α,β-unsaturated carbonyl system is necessary. The cardiac steroids inhibit sodium- and potassium-dependent ATPase, an enzyme responsible for maintaining the unequal distribution of sodium and potassium ions across cell membranes in the heart.

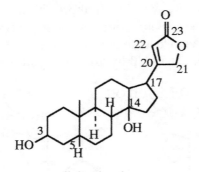

digitoxigenin

Steroid Semisynthesis

The availability of various steroids as drugs is dependent on a combination of three things: isolation of certain steroids economically from natural sources in acceptable yields, conversion into other steroids with the aid of microbial oxidation reactions, and modification with organic synthetic reactions. To sample this fascinating area of research, we will focus on cortisone. Russel E. Marker was the "founding father" of modern steroid chemistry. His synthesis of progesterone from diosgenin in the 1930s is still used commercially today. The bulk of the world's supply of steroid starting material is derived from two species of plants, the Mexican yam and the humble soybean. Diosgenin is

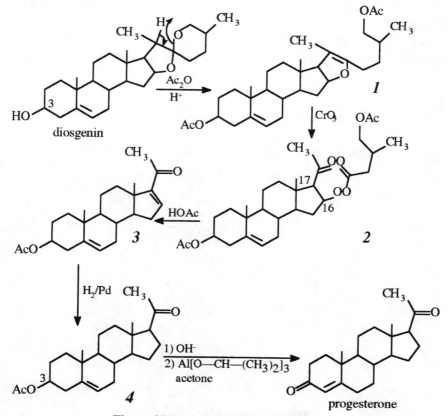

Figure 23.11   Synthesis of progesterone.

isolated from the yam in large amounts. Treatment with acetic anhydride opens the spiran ring as shown in Fig. 23.11. It also acetylates the C-3 hydroxyl to give *1*. Oxidation of the newly formed double bond with chromium trioxide makes the desired acetyl group at C-17 of compound *2*. Treatment with acetic acid hydrolyzes the ester to a hydroxyl at C-16, which then dehydrates to the double bond of compound *3*, called 16-dehydropregnenolone acetate. Selective catalytic hydrogenation of the new double bond follows to give *4*, pregnenolone acetate. The acetate at C-3 is removed by basic hydrolysis to a hydroxy group, which is then oxidized with aluminum isopropoxide (the Oppenauer reaction) to a keto group. The basic reaction conditions isomerize the double bond so that a conjugated α,β-unsaturated ketone is formed, namely, progesterone. The various intermediates shown in this synthesis are currently turned out in tonnage quantity. Other routes to progesterone are commercially used, but this is representative.

Large-scale commercial production of cortisone (Fig. 23.12) from progesterone (Fig. 23.13) starts with a microbiologic oxidation with a soil organism,

Figure 23.12 A Glatt mixer used in the manufacture of drugs.

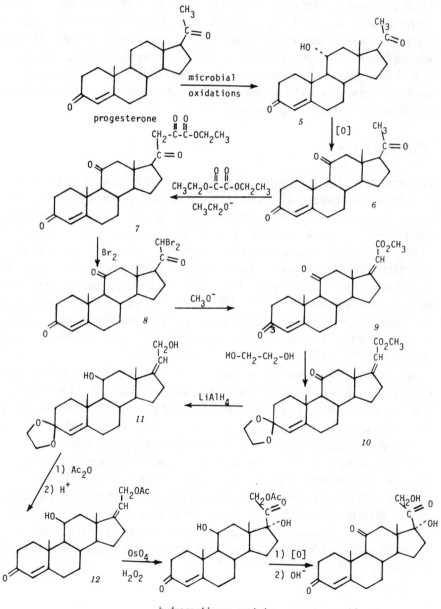

Figure 23.13 Manufacture of cortisone.

Rhizopus arrhizus, to convert progesterone into 11α-hydroxyprogesterone *(5)* in 50% yield. Oxidation of the alcohol with a number of reagents leads to the trione *6*. Condensation with ethyl oxalate gives *7*, which activates the appropriate carbon toward selective bromination to form *8*. A Favorskii rearrangement followed by dehydrohalogenation gives *9*. After the ketone at C-3 is protected as its ketal *10*, reaction with lithium aluminum hydride reduces the ester and the C-11 ketone to the alcohol *11*. Acetylation of one of the alcohol groups (the less hindered primary alcohol) and removal of the protecting group at C-3 then gives *12*. Osmium tetroxide and hydrogen peroxide oxidize the double bond to give hydrocortisone acetate. Oxidation of the alcohol group and hydrolysis of the acetate gives cortisone.

ANALGESICS

An important class of compounds that have members from both ethical and over-the-counter drugs are those that relieve pain. Aspirin is by far the most common type of analgesic and it is the sixth largest selling drug. It is also an antipyretic, that is, it lowers abnormally high body temperatures. A third use is in reducing inflammation caused by rheumatic fever and rheumatoid arthritis. Salicylic acid has been known for its analgesic properties since the early 1800s. Kolbe and Lautermann prepared it synthetically from phenol in 1860. Acetylsalicylic acid was first prepared in 1853 by Gerhardt but remained obscure until Hoffmann discovered its pharmacologic activities in 1899. It was first used in medicine by Dreser, who named it aspirin by taking the "a" from acetyl and adding it to "spirin," the old name for salicylic acid.

The industrial synthesis of aspirin is still based on the original synthesis of salicylic acid from phenol by Kolbe. Reaction of carbon dioxide with sodium phenoxide is an electrophilic aromatic substitution ($O=C\delta+=O\delta-$) on the ortho, para-directing phenoxy ring. The ortho isomer is steam-distilled away from the para isomer. Salicylic acid reacts easily with acetic anhydride to give aspirin. Aspirin is the only drug manufactured on the scale of other industrial chemicals. Over 20 million lb/yr are produced in the U.S. and it sells for $2.50/lb. In the last step a 500-gal glass-lined reactor is needed to heat the salicylic acid and acetic anhydride for 2-3 hr. The mixture is transfered to a crystallizing kettle and cooled to 3°C. Centrifuging and drying of the crystals yields the bulk aspirin. The excess solution is stored and the acetic acid is recovered to make more acetic anhydride.

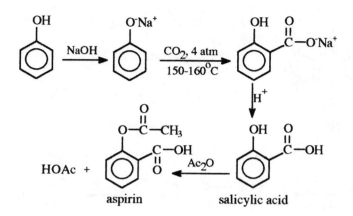

The antipyretic and analgesic actions of aspirin are believed to occur in a certain area of the brain. It is also thought by some that the salicylates exert their analgesia by their effect on water balance. Aspirin is anti-inflammatory because it inhibits the biosynthesis of chemicals called prostaglandins.

The irritation of the stomach lining caused by aspirin can be alleviated with the use of mild bases such as sodium bicarbonate, aluminum glycinate, sodium citrate, aluminum hydroxide, or magnesium trisilicate (a common trademark for this type of aspirin is Bufferin®). A widely used combination *a*spirin with *p*henacetin and *c*affeine is called APC. Caffeine is just a mild alkaloid stimulant. Both phenacetin and the newer replacement acetaminophen are derivatives of *p*-aminophenol. Although these latter two are analgesics and antipyretics, the aniline-phenol derivatives show little if any anti-inflammatory activity. *p*-Aminophenol itself is toxic but acylation of the amino group makes it a convenient drug. Since phenacetin is de-ethylated in the body, acetaminophen is believed to be responsible for the drug activity and is now considered safer to use, but it should not be taken for over ten days unless directed by a physician. A common trademark for acetaminophen is Tylenol®. Excedrin® is acetaminophen, aspirin, and caffeine. Acetaminophen is easily synthesized from phenol.

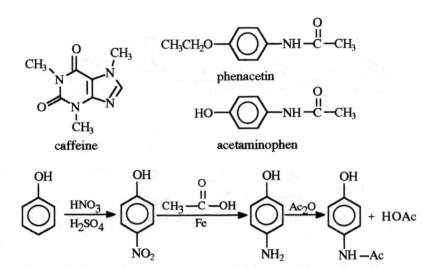

caffeine

phenacetin

acetaminophen

A very popular alternative to aspirin and acetaminophen is ibuprofen, which has tradenames such as Motrin® and Advil®. It can be synthesized from isobutylbenzene by a Friedel-Crafts acylation with acetyl chloride, followed by formation of a cyanohydrin. Treatment with HI and phosphorus reduces the benzylic hydroxyl to a hydrogen and hydrolyzes the nitrile to a carboxylic acid. Ibuprofen has good analgesic and anti-inflammatory action.

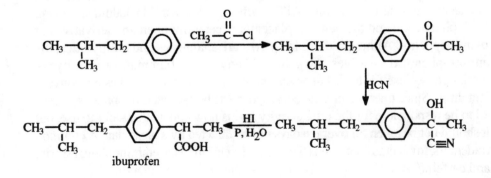

ibuprofen

The U.S. pain reliever market is estimated at $1.8 billion. Aspirin's share has recently decreased from 60% in 1983 to 44%. Acetaminophen has 32% of the market and ibuprofen 20%. But aspirin may be back on the upswing. After earlier reports that it was linked to Reye's syndrome in children, the recent news is that a single aspirin tablet taken every other day halves the risk of heart attacks among healthy men.

Propoxyphene (Darvon®) is a stronger analgesic than aspirin but has no antipyretic effects. It is sometimes taken in combination with aspirin and acetaminophen. It has widespread use for dental pain since aspirin is relatively ineffective, but it is not useful for deep pain. It must be prescribed. The starting material for propoxyphene is propiophenone, made from benzene and propionyl chloride by a Friedel-Crafts acylation . It undergoes a Mannich reaction with formaldehyde and dimethylamine. A Grignard reaction with benzyl magnesium

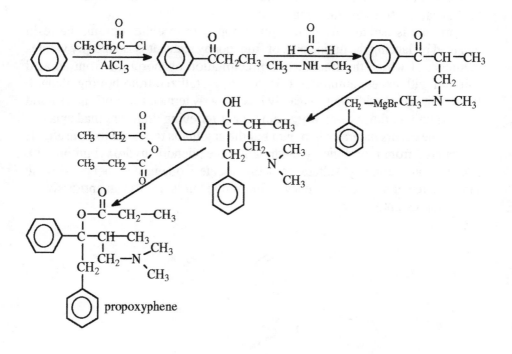

bromide follows. Esterification with propionic anhydride gives propoxyphene (4-dimethylamino-1,2-diphenyl-3-methyl-2-butyl propionate). Only one stereoisomer is active physiologically. It is administered as the amine hydrochloride salt.

The discovery of morphine's analgesic activity by Serturner in 1806 started a long series of studies of the alkaloids from the opium poppy, including morphine's first correctly postulated structure in 1925 and its total synthesis in 1952. Codeine is the methyl ether of morphine. The depressant action of the morphine group is the most useful property, resulting in an increased tolerance to pain, a sleepy feeling, a lessened perception to external stimuli, and a feeling of well-being. Respiratory depression and addiction are its serious drawbacks. The important structure-activity relationships that have been defined are: (1) a tertiary nitrogen, the group on the nitrogen being small; (2) a central carbon atom, of which none of the valences is connected to hydrogen; (3) a phenyl group connected to the central carbon; and (4) a two-carbon chain separating the central carbon from the nitrogen.

Morphine is isolated from the opium poppy from either opium, the resin obtained by lancing the unripe pod, or from poppy straw. It is isolated by various methods, of which the final step is precipitation of morphine from an acid solution with excess ammonia. It is then recrystallized from boiling alcohol. Because it causes addiction so readily it is properly termed a narcotic and should be used only in those cases where other pain-relieving drugs are inadequate.

Codeine occurs naturally in opium but the amount is too small to be useful. It is prepared from morphine by methylating the phenolic hydroxyl group with diazomethane, dimethyl sulfate, or methyl iodide. Codeine does not possess the same degree of analgesic potency as morphine but is used as an antitussive, a cough repressant.

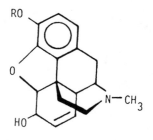

morphine, R=H
codeine, R=CH$_3$

ANTIHISTAMINES

These drugs alleviate allergic conditions such as rashes and runny eyes and nose. Antihistamines are decongestants which are used for swelled sinuses and nasal passages during the common cold. These symptoms are caused by histamine and hence the drugs that get rid of them are antihistamines. They are also sleep inducers. All the most popular antihistamines, sold under such tradenames as Dimetapp®, Actifed®, and Benadryl®, have a structure including the group *R*—*X*—C—C—N< , where *X* can be nitrogen, oxygen, or carbon. The mode of action may be considered to be a competition, in tissue, between the antihistaminic agent and histamine for a receptor site. The combining of the antihistaminic agent with the receptive substance at the site of action prevents the histamine from exerting its characteristic effect on the tissue. Figure 23.14 gives the structures of histamine and the three most prescribed antihistamines.

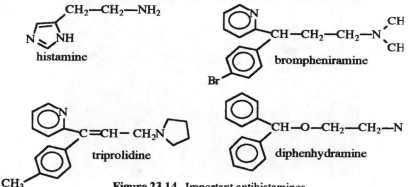

Figure 23.14 Important antihistamines.

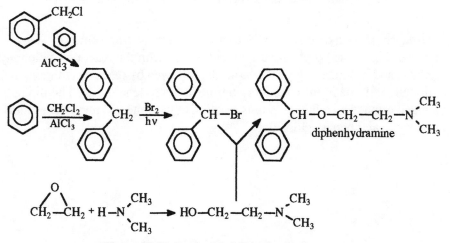

Figure 23.15 Synthesis of diphenhydramine.

Figure 23.16 Tablet printing occurs with a high degree of automation.

These complex molecules can be easily synthesized with some key steps. For instance (Fig. 23.15) diphenhydramine can start with the reaction of diphenyl-methane and bromine to give the bromide, followed by reaction with dimethyl-aminoethanol, made from dimethylamine and ethylene oxide. The diphenyl-methane is made by a Friedel-Crafts reaction of benzene and methylene chloride or benzene and benzyl chloride.

OTHER DRUGS

This chapter is hopefully a good introduction into the fascinating chemistry and pharmacology of the drug industry. We have tried to summarize representative important types of drugs to give an overall view of the industry. Although many other areas and chemical types could have been included, we mainly emphasized only the most used ethical drugs. A more complete description of these and other pharmaceuticals cannot be justified in a survey course.

24

Surfactants, Soaps, and Detergents

References

W & R II, pp. 182-212
B. F. Greek, "Sales of Detergents Growing Despite Recession," *C & E News*, 1-28-91,
   pp. 25-52.

INTRODUCTION TO THE INDUSTRY

A general area of the chemical industry that manufactures most of the surfactants, soaps, and detergents is called Soaps, Cleaners, and Toilet Goods (SIC 284). Divisions of this category are given in Table 24.1. We concentrate primarily on SIC 2841, although all divisions of soaps, cleaners, and toilet goods use surface active (surfactant) chemicals, which are further modified into finished products. Some 7 billion lb of surfactant chemicals serve all these sectors. Soaps, cleaners, and toilet goods contribute 14.6% of total shipments for Chemicals and Allied Products. Fig. 24.1 shows the trend in U.S. shipments for Soap and Other Detergents (SIC 2841) and its subsectors of Soap and Detergents, Nonhousehold (SIC 28411), Household Detergents (SIC 28412), and Household Soaps (SIC 28413). Notice the rapid increase for soaps and detergents in the 1970s from $2.5 billion to $8 billion in 1981. Slower growth occurred in the 1980s.

TABLE 24.1 Shipments for Soaps, Cleaners, and Toilet Goods

SIC	Industry	Shipments ($ billion)
2841	Soap and Other Detergents	12.306
2842	Polishes and Sanitation Goods	5.858
2843	Surface Active Agents	3.399
2844	Toilet Preparations	16.294
284	Soaps, Cleaners, Toilet Goods	37.856

Source: AS.

In general, surfactants are chemicals that, when dissolved in water or another solvent, orient themselves at the interface between the liquid and a second solid, liquid, or gaseous phase and modify the properties of the interface. Surfactants are not only important as the active constituent of soaps and detergents but are also vital in the stabilization of emulsions, in fabric softening, in oil well drilling, etc. Surfactants are the most widely applied group of compounds in the Chemicals and Allied Products industries. We will concentrate on their use in cleaning, that is, in soaps and detergents.

All surfactants have a common molecular similarity. A portion of the molecule has a long nonpolar chain, frequently a hydrocarbon chain, that promotes oil solubility but water insolubility (the hydrophobic portion—water hating). Another part of the molecule promotes oil insoluble and water soluble properties (the hydrophilic portion—water loving).

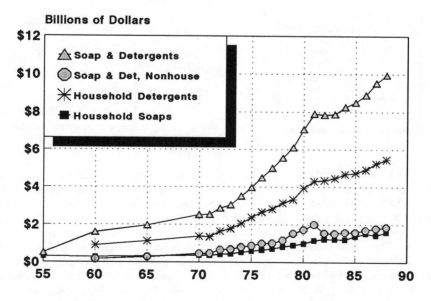

Figure 24.1 U.S. shipments of soaps and detergents. (*Source*: AS.)

Surface layer of surfactant →
molecules
[• Hydrophillic end]
[| Hydrophobic end]
Micelles in body of liquid →

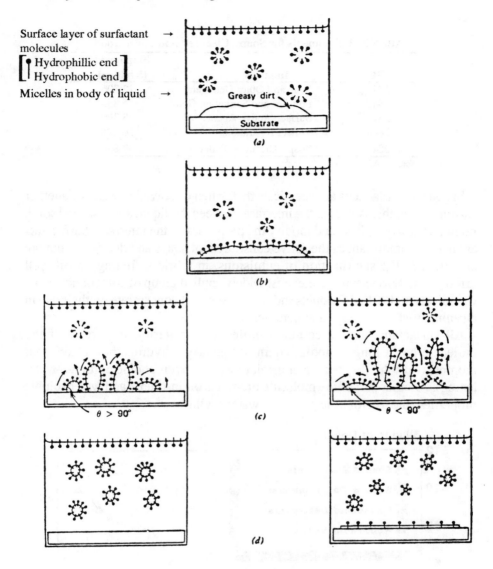

Figure 24.2 "Solubilizing" effect of surface active agents. *(a)* Greasy dirt comes into contact with surfactant solution. *(b)* Hydrophobic ends of surfactant molecules dissolve in the grease. *(c)* The surfactant affects the contact angle θ between the dirt and the substrate. If θ < 90°, total removal of the grease is impossible. *(d)* Further agitation displaces the greasy dirt as macroscopic particles. These form an emulsion if agitation is sufficient. The particles form the center of micelle-like structures. Removal of grease is seldom complete (θ < 90° as in the diagrams on the right rather than the simple "rollback" mechanism on the left). Usually the main body of grease is removed from a strongly adsorbed monomolecular or duplex layer of grease. Agitation is an essential part of the process. (*Source:* W & R II. Reprinted by permission of John Wiley & Sons, Inc.)

Figure 24.2 summarizes the cleaning action of surfactants nicely. The surfactant lines up at the interface and also forms micelles, or circular clusters of molecules. In both cases the hydrophobic end of the molecule gets away from water molecules and the hydrophilic end stays next to the water molecules (like dissolves like). When grease or dirt come along (primarily hydrophobic in nature) the surfactants surround it until it is dislodged from the substrate. The grease molecules are suspended in the emulsion by the surfactant until they can be washed away with freshwater.

Surfactants can be divided into four general areas. These will be discussed separately. Their approximate percentages of production are given in parentheses: cationics (10%), anionics (62%), nonionics (28%), and amphoterics (<1%). Major anionics are soaps (17%), linear alcohol sulfates (AS) (7%), linear alcohol ethoxysulfates (AES) (7%), and linear alkylbenzenesulfonates (LAS) (31%). Table 24.2 gives the 1988 surfactant production in millions of pounds.

TABLE 24.2 Surfactant Production

Type	Production (Million lb)
Amphoteric	**41.1**
Anionic	**4,559.5**
Carboxylic acids and salts (soaps)	1,269.3
Phosphoric and polyphosphoric acid esters and salts	63.1
Sulfonic acids and salts (LAS)	2,285.7
Sulfuric acid esters and salts (AS/AES)	941.5
Cationic	**703.2**
Amines and amine oxides	466.2
Quaternary ammonium salts	224.7
Nonionic	**2,012.2**
Carboxylic acid amides	186.7
Carboxylic acid esters	421.6
Ethers	1,382.1
Total	**7,316.1**

Source: SOC.

CATIONIC SURFACTANTS

In cationics the long hydrophobic alkyl chain is in the cationic portion of the molecule. Another way of saying this is that the organic part is positive. Practically all industrially important cationics are fatty nitrogen compounds and many are quaternary nitrogen compounds such as tallow fatty acid trimethylammonium chloride. In the more general structure $R^1R^2R^3R^4N^+X^-$, R^1 is a long alkyl chain, the other R's may be alkyl or hydrogen, and X^- is halogen or sulfate. The long hydrocarbon chain is derived from naturally occurring fats or triglycerides, that is, triesters of glycerol having long chain acids with an even number of carbons, being of animal or vegetable origin. A common fat source for cationics is inedible tallow from meat packing plants. If the fatty acid is desired the ester is hydrolyzed at high temperature and pressure, or with a catalyst such as zinc oxide or sulfuric and sulfonic acid mixtures. The fatty acid is then converted into the "quat" by the following sequence of reactions.

Structure:

$$CH_3(CH_2)_n\overset{\overset{\displaystyle CH_3}{|+}}{\underset{\underset{\displaystyle CH_3}{|}}{N}}{-}CH_3 \quad Cl^-$$

$$n = 15 \text{ or } 17$$

Source of R:

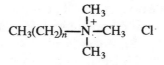

fat or triglyceride   glycerol (glycerin)   fatty acid

Conversion to quat:

$$\underset{\text{fatty acid}}{R-\overset{\overset{\displaystyle O}{\|}}{C}-OH} \xrightarrow{NH_3} \underset{\text{amide}}{R-\overset{\overset{\displaystyle O}{\|}}{C}-NH_2} \xrightarrow{-H_2O} \underset{\text{nitrile}}{R-C{\equiv}N} \xrightarrow{H_2}$$

$$\underset{\text{primary amine}}{R-CH_2-NH_2} \xrightarrow{CH_3Cl} \underset{\underset{\underset{\displaystyle CH_3Cl^-}{|+}}{}}{R-CH_2\overset{\overset{\displaystyle CH_3}{|}}{N}-CH_3}$$

"quat"

Cationic surfactants are not very good for cleaning because most surfaces carry a negative charge and the cationic portion adsorbs on the surface instead of dissolving the grease. But they do have other important surfactant applications. They inhibit the growth of bacteria, are corrosion inhibitors, are used in ore flotation processes (separating phosphate ore from silica and potassium chloride from sodium chloride), and are good fabric softeners and antistatic agents. They also find use in hair conditioners and other personal care applications.

ANIONIC SURFACTANTS

This is by far the most important type of surfactant and will be discussed under separate subtypes. In anionics the long hydrophobic alkyl chain is in the anionic part of the molecule. The organic part is negative.

Soaps

The first type of cleansing agent, used by humankind for centuries, was soap. Although it has now been supplemented by various synthetic detergents in advanced countries for laundry and household use, it is still preferred for personal hygiene. In less-developed countries it is still preferred for laundry use.

$$\text{fats} \xrightarrow[\Delta]{\text{NaOH}} \text{glycerol} + \quad R-\overset{\overset{\displaystyle O}{\|}}{C}-O^-Na^+$$

Soaps are the sodium or potassium salts of certain fatty acids obtained from the hydrolysis of triglycerides. The potassium salts form the "soft soaps" that have become popular recently. The fats used in soap manufacture come from diverse natural sources. Animal tallows and coconut oil are the favored sources of the triglycerides, and quite often mixtures from different sources are used to vary hardness, water solubility, and cleansing action of the final product. Palm, olive, cottonseed, castor, and tall oil are other sources. The side chains are usually C_{12}-C_{18} in length. Manufacturing processes are both batch and continuous. Sometimes the triglyceride is steam-hydrolyzed to the fatty acid without strong caustic and then in a separate step it is converted into the sodium salt. Either way gives a similar result. Soaps have some disadvantages compared to synthetic detergents: they are more expensive, they compete with food uses for fats and oils, and their calcium and magnesium salts formed in hard water are very insoluble and precipitate onto the clothing being washed. They also tend to clog automatic washers. They deteriorate on storage and are unstable in acid solutions. This is why in 1940 only 1% of cleaning agents were detergents; in 1970 they were 85%. Fig. 24.3 shows this historical replacement of soaps by detergents from 1940-1970. However, there will always be that small market

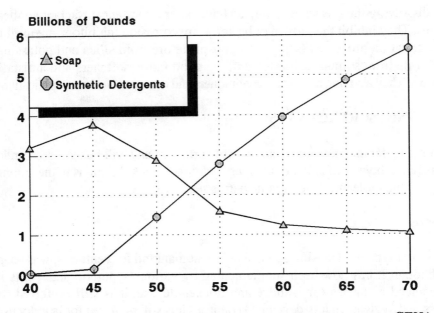

Figure 24.3 U.S. consumption of soaps vs. total synthetic detergents. (*Source*: CEH.)

for soaps, presently still 17% of all surfactants and still at the 1970 level of about 1 billion lb. Advantages of soaps include greater biodegradability, less toxicity, and less phosphate builders.

Straight Chain Detergent Intermediates

It is necessary for any soap or detergent to have a high degree of linearity for it to be biodegradable by bacteria. Many synthetic anionic detergents are based on straight chain primary alcohols and α-olefins. New technology allows these materials to be manufactured from ethylene (using Ziegler polymerization catalysts) or from linear alkanes (paraffins), followed by conversion into linear alkyl chlorides by chlorination or linear α-olefins by dehydrogenation, in addition to being formed from naturally occurring straight chains. These processes will be discussed under LAS detergents, but it is important to realize that synthetic long straight-chain compounds are now available.

α-Olefin Sulfonates (AOS)

These materials are made by the reaction of C_{12}-C_{18} α-olefins with sulfur trioxide followed by reaction with caustic. The product is a complex mixture of compounds, the most important of which are the two pictured here. Note the shift in the double bond. Disulfonates are also formed.

$$R-(CH_2)_n-CH=CH_2 \xrightarrow[\text{2) NaOH}]{\text{1) SO}_3}$$

$$R-CH=CH-(CH_2)_n-\overset{\overset{O}{\|}}{\underset{\underset{O}{\|}}{S}}-O^-Na^+ \quad + \quad R-CH-CH_2-(CH_2)_n-\overset{\overset{O}{\|}}{\underset{\underset{O}{\|}}{S}}-O^-Na^+$$
$$\underset{OH}{|}$$

AOS have solubility characteristics particularly suitable for liquid formulations. They are relatively new as commercial surfactants but may be the surfactant of the future. They do not presently have a large share of the market.

Secondary Alkanesulfonates (SAS)

Like AOS, secondary alkanesulfonates are new on the market. They are made by the action of sulfur dioxide and air directly on C_{14}-C_{18} n-paraffins (a sulfoxidation reaction). The sulfonate group can appear in most positions on the chain. They are used only in liquid formulations and are manufactured only in Europe.

$$C_{16}H_{34} + SO_2 + O_2 \longrightarrow \overset{C_8H_{17}}{\underset{C_7H_{15}}{\diagdown}} CH-\overset{\overset{O}{\|}}{\underset{\underset{O}{\|}}{S}}-OH \xrightarrow{\text{NaOH}}$$

+ isomers

$$\overset{C_8H_{17}}{\underset{C_7H_{15}}{\diagdown}} CH-\overset{\overset{O}{\|}}{\underset{\underset{O}{\|}}{S}}-O^-Na^+$$

+ isomers

Linear Alcohol Sulfates (AS)

Alcohol sulfates are usually manufactured by the reaction of a primary alcohol with sulfur trioxide or chlorosulfonic acid followed by neutralization with a base. These are high foam surfactants but they are sensitive to water hardness and high levels of phosphates are required. This latter requirement is expected to harm the market for this type of detergent in the future, but they still have 7% of the total surfactant market. Sodium lauryl sulfate ($R = C_{11}$) is a constituent of shampoos to take advantage of its high-foaming properties.

$$R-CH_2-OH \xrightarrow{SO_3 \text{ or } ClSO_3H}$$

$$R = C_{11}-C_{17}$$

$$R-CH_2-O-\underset{\underset{O}{\overset{\|}{\|}}}{\overset{O}{\overset{\|}{S}}}-OH \xrightarrow[\text{or } NH_4OH]{NaOH} R-CH_2-O-\underset{\underset{O}{\overset{\|}{\|}}}{\overset{O}{\overset{\|}{S}}}-O^-Na^+$$

$$\text{(or triethanolamine)}$$

The linear alcohols can be made from other long-chain linear materials, but a new process with a triethylaluminum catalyst allows their formation directly from ethylene and oxygen.

$$n CH_2{=}CH_2 + O_2 \xrightarrow{Et_3Al} R-CH_2-OH$$

Linear Alcohol Ethoxysulfates (AES)

Alcohol ethoxysulfates are made by reaction of 3-7 mol of ethylene oxide with a linear C_{12}-C_{14} primary alcohol to give a low molecular weight ethoxylate.

$$C_{14}H_{29}OH + n CH_2{-}CH_2 \longrightarrow C_{14}H_{29}O-(CH_2-CH_2-O)_{\overline{n}}H$$

$$\downarrow SO_3 \text{ or } ClSO_3H$$

$$C_{14}H_{29}O-(CH_2-CH_2-O)_{\overline{n}}\underset{\underset{O}{\overset{\|}{\|}}}{\overset{O}{\overset{\|}{S}}}-OH$$

$$\xrightarrow{NaOH}$$

$$C_{14}H_{29}O-(CH_2-CH_2-O)_{\overline{n}}\underset{\underset{O}{\overset{\|}{\|}}}{\overset{O}{\overset{\|}{S}}}-O^-Na^+$$

They have high foam for shampoos and are "kind to the skin." They are also used in light duty products such as dishwashing detergents. It is the least sensitive of the anionics to water hardness and will therefore benefit in the trend away from phosphates. They have 7% of the total surfactant market.

Linear Alkylbenzenesulfonates (LAS)

By far the most important of all synthetic detergents are the alkylbenzene-sulfonates, having captured 31% of the market. Originally the cheap tetramer of propylene was used as the source of the alkyl group. This tetramer is not a single compound but a mixture. However, they are all highly branched. For example:

$$4CH_3-CH=CH_2 \xrightarrow[\Delta]{H^+} CH_3-\overset{\overset{\displaystyle CH_3}{|}}{CH}-CH_2-\overset{\overset{\displaystyle CH_3}{|}}{CH}-CH_2-\overset{\overset{\displaystyle CH_3}{|}}{CH}-CH=CH-CH_3$$

nonlinear $C_{12}H_{24}$

The nonlinear dodecene was then used to alkylate benzene by the Friedel-Crafts procedure. Sulfonation and treatment with caustic completed the process. This nonlinear alkylbenzenesulfonate formed the basis for the heavy duty household washing powders of the 1950s and early 1960s with excellent cleaning ability. But rivers and lakes soon began foaming since enzymes present in bacteria could not degrade these "hard" detergents because of the highly branched side chain. They were banned in 1965.

Thus *n*-alkanes (C_{10}-C_{14}) separated from the kerosene fraction of petroleum (by urea complexation or absorption with molecular sieves) are now used as one source of the alkyl group. Chlorination takes place anywhere along the chain at any secondary carbon. Friedel-Crafts alkylation followed by sulfonation and caustic treatment gives a more linear alkylbenzenesulfonate (LAS) which is "soft" or biodegradeable. The chlorination process is now the source of about 40% of the alykyl group used for the manufacture of LAS detergent.

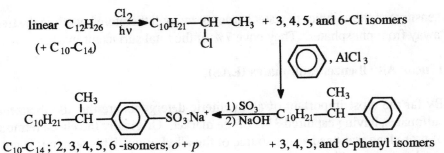

The other 60% of the alkyl groups for LAS detergents are made through linear α-olefins. *n*-Alkanes can be dehydrogenated to α-olefins, which then can undergo a Friedel-Crafts reaction with benzene as described above for the nonlinear olefins. Sulfonation and basification gives the LAS detergent.

$$\text{linear } C_{12}H_{26} \xrightarrow{-H_2} \text{linear } C_{10}H_{21}CH=CH_2 + \underset{\text{(linear } C_{12}H_{24})}{\bigcirc} \xrightarrow{HF} C_{10}H_{21}-\overset{\overset{\displaystyle CH_3}{|}}{CH}-\bigcirc$$

Alternatively, linear α-olefins can be made from ethylene using Ziegler catalysts to give the ethylene oligomer with a double-bonded end group.

$$6CH_2{=}CH_2 \xrightarrow{\Delta} C_{10}H_{21}-CH=CH_2$$
$$\text{(linear } C_{12}H_{24})$$

LAS detergents made from the chlorination route have lower amounts of 2-phenyl product. Use of the α-olefins gives greater 2-phenyl content, which in turn changes the surfactant action somewhat.

NONIONICS

In nonionics the molecule has a nonpolar hydrophobic portion and a more polar, but not ionic, hydrophilic part capable of hydrogen bonding with water. For

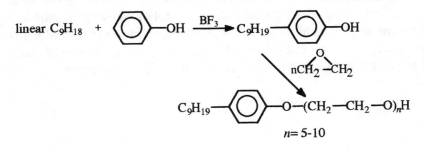

some years the major nonionics have been the reaction products of ethylene oxide and nonylphenol. Dehydrogenation of *n*-alkanes from petroleum (C_9H_{20}) is the source of the linear nonene.

They are now being replaced by the polyoxyethylene derivative of straight-chain primary or secondary alcohols with C_{10}-C_{18}. These linear alcohol ethoxylate nonionics (AEO) are more biodegradable than nonylphenol derivatives and have better detergent properties than LAS. They are expected to be the fastest growing for household detergents because they require less phosphates and work well in energy-saving cooler wash water. Nonionics are the fastest growing type of surfactant, holding 28% of the market, up 7% in just the last 6 years.

$$\text{linear} \quad C_{14}H_{29}\text{—OH} + n\overset{\displaystyle O}{\overset{\displaystyle \diagup\diagdown}{CH_2\text{—}CH_2}} \longrightarrow C_{14}H_{29}\text{—O—}(CH_2\text{—}CH_2O)_n H$$

AMPHOTERICS

These surfactants carry both a positive and a negative charge in the organic part of the molecule. They still have a long hydrocarbon chain as the hydrophobic tail. They may behave as anionics or cationics depending on the pH. They are

$$
\begin{array}{cc}
\underset{CH_3}{\overset{CH_3}{R\text{—}N^+\text{—}CH_2\text{—}\overset{\displaystyle O}{\overset{\|}{C}}\text{—}O^-}} &
\underset{\underset{CH_3}{|}}{\overset{\displaystyle O}{R\text{—}CH\text{—}\overset{\|}{C}\text{—}O^-}}\\
& CH_3\text{—}\overset{+}{N}\text{—}CH_3
\end{array}
$$

derivatives of amino acids and may have one of two general formulas, where *R* is linear C_{12}-C_{18}. Amphoterics are used in shampoos. Near pH = 7 they are less irritating and "milder." They can be used with alkalies for greasy surfaces as well as in acids for rusty surfaces.

DETERGENT BUILDERS

The finished household soap or detergent is more than just a surfactant. It is a complex formulation that includes bleaches, fillers, processing aids, fabric softeners, fragrances, optical brighteners, foam stabilizers, soil-suspending agents, enzymes, and opacifying agents. We will not discuss all of these. But one of the most important and controversial materials in a detergent is the builder, phosphate being a common one. Although the calcium and magnesium salts of docedecylbenzenesulfonic acid are more soluble than those of fatty acids, these ions in solution interfere with the dislodging of dirt from the substrate, and the dirt-suspending power is also affected because of their double positive charge. A chemical must be added to the detergent to sequester or complex the ions. These are called *builders*. Builders are actually 58% by weight of the chemicals found in a detergent, with the surfactants only 36%, and other additives making up the remaining 6%.

The first important commerical builder was sodium tripolyphosphate, $Na_5P_3O_{10}$, first used with Tide® detergent in 1947. Besides sequestering polyvalent metal ions, it prevents redeposition of dirt, buffers the solution to pH=9-10, kills bacteria, and controls corrosion and deposits in the lines of automatic washers.

$$O=\overset{\overset{\displaystyle O^-Na^+}{|}}{\underset{\underset{\displaystyle O^-Na^+}{|}}{P}}-O-\overset{\overset{\displaystyle O}{\|}}{\underset{\underset{\displaystyle O^-Na^+}{|}}{P}}-O-\overset{\overset{\displaystyle O^-Na^+}{|}}{\underset{\underset{\displaystyle O^-Na^+}{|}}{P}}=O$$

In the late 1960s phosphate builders came to be seen as an environmental problem. Phosphates pass unchanged through sewage works and into rivers and lakes. Since they are plant nutrients they cause blue-green algae to grow at a very fast rate on the surface, causing oxygen depletion. This is called *eutrophication*. The search for phosphate substitutes began, and some states banned their use. Figure 24.4 lists some of the compounds tried as phosphate substitutes.

Ethylenediaminetetracetic acid (EDTA) is a good sequestering agent but its cost is excessive. Nitrilotriacetate is effective but has been suggested to be teratogenic and carcinogenic so it is not used in the U.S. Sodium citrate is harmless but does not work well. Benzene polycarboxylates are expensive and are not biodegradable. Sodium carbonate is not successful in hard water areas. Commerical use of zeolites and poly-α-hydroxyacrylate is just beginning. Sodium sulfate occurs as a byproduct of any sulfate or sulfonate detergent, but has limited use as a builder, as does sodium silicate. The present breakdown of builders used in detergents is sodium carbonate, 40%; sodium tripolyphosphate, 31%; sodium silicate, 9%; zeolites, 7%; sodium citrate, 3%; and other, 10%.

Phosphate detergents are no longer criticized as much as in the early 1970s since we now know only about 50% of phosphate comes from detergents, with 33% from household wastes and 17% from fertilizer runoff contributing to eutrophication. Tertiary treatment of sewage to precipitate phosphate as iron phosphate may be the answer. The average level of phosphates in detergents has already dropped in the U.S., but the search for a phosphate substitute will probably continue.

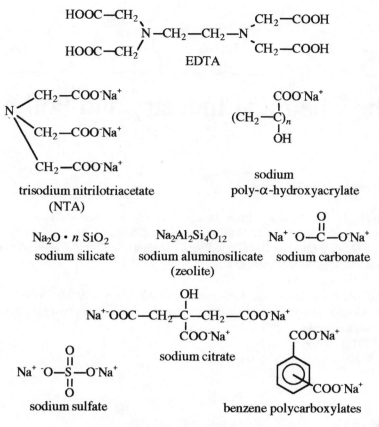

Figure 24.4 Phosphate substitutes.

25

The Chemical Industry and Pollution

References

G. W. Ingle, Ed., "TSCA's Impact on Society and the Chemical Industry," ACS Symposium Series, ACS, Washington, DC, 1983

"The Chemical Balance: Benefiting People, Minimizing Risks," Chemical Manufacturers Association, Washington, D.C.

W & R I, pp. 255-268

L. N. Davis, *The Corporate Alchemists,* William Morrow & Co.: New York, 1984

M. E. Gleiter, Department of Chemistry, University of Wisconsin-Eau Claire, personal communication

IO, 1991, pp. 12-1 to 12-6.

C & E News, selected articles, 1980-1991.

INTRODUCTION

There are many areas of the chemical industry that must be controlled to avoid ill effects on health and the environment. Throughout this book we have tried to stress individual pollution-related problems. The details of these topics can be found in various chapters. A list of these subjects already discussed is given in Table 25.1. You may wish to review these topics at this time. Other pollution problems are covered in the optional studies that are listed in the Appendix.

The purpose of this chapter is to summarize and generalize the various pollution, health, and environmental problems especially specific to the chemical industry and to place in perspective government laws and regulations as well as industry efforts to control these problems. A brief survey of air and water pollution problems will be given, but these are characteristic of all industry and the topics are too vast to be covered adequately in this course. We will be more concerned with toxic chemical pollution and its control and will spend some time on the Toxic Substances Control Act (TSCA, TOSCA) of 1976.

TABLE 25.1 Environmental Problems Discussed Previously

Subject	Chapter
SO_2 in the atmosphere from H_2SO_4 plants	2
Phosphate and eutrophication of lakes	2, 24
Road deicing and its effect on local plant life	5
Electrolysis of brine in mercury cells	6
Combustion of petroleum containing sulfur and nitrogen	7
Tetraethyllead and contamination of the air	7
Disadvantages of burning unleaded gasoline	7
Sulfur extraction from petroleum and natural gas	7
Chlorofluorocarbons and ozone depletion	12
Threshold limit values of organic chemicals	8-13
Known and suspected organic carcinogens	8-13
Possible health problems with use of PVC and formaldehyde polymers	16
Coatings solvents and air pollution	19
Toxicity and persistency of chlorinated pesticides	20
Polychlorinated biphenyls	20
Toxicity of other types of pesticides	20
Dioxin toxicity and teratogenicity	20
Relative pollution problems of various pulping processes	22
Health risks and side effects of some drugs	23
Biodegradable vs. nonbiodegradable detergents	24
Phosphate substitutes in detergents	24

GENERAL POLLUTION PROBLEMS

Air

Since the advent of the Industrial Revolution there has been an air pollution problem. For years the control of air pollutants was nonexistent. Many industries and governments suggested that "the solution to pollution was dilution," that is, build larger and higher smokestacks to dilute and disperse the fumes before health or the environment are affected. This can no longer be diluted to insignificant concentrations, especially in large metropolitan areas. In 1873 several thousand people died in London because of air pollution. In 1952 another acute air problem in London killed 4000 people. In 1909 approximately 1000 deaths were attributed to "smog" in Glasgow, Scotland, the first time this word was used. We are all familiar with the continuing battle of the large cities in the U. S. to alleviate the air pollution problem. Recently, the four most important challenges in air pollution control facing all of us (for we are all polluters) are the following:

1. Acid rain. Lakes in some areas of the world are now registering very low pH's because of excess acidity in rain. This was first noticed in

Scandinavia and is now prevalent in eastern Canada and the northeastern U. S. Normal rainfall has a pH of 5.6 (because of CO_2 in the air forming H_2CO_3). However, excessive use of fossil fuels (especially coal) with high sulfur and nitrogen content cause sulfuric and nitric acids in the atmosphere from the sulfur dioxide and nitrogen oxide products of combustion. Some rain in the Adirondack mountains of upper New York State has been measured with a pH of 3.0. This problem is not specific to the chemical industry but should be of concern to all of us.

$$SO_2 \xrightarrow[H_2O]{O_2} H_2SO_4$$

$$NO_x \xrightarrow[H_2O]{O_2} HNO_3$$

2. *Carbon dioxide content.* The increased burning of fossil fuels in the last couple hundred years is slowly increasing the concentration of carbon dioxide in the atmosphere, which absorbs more infrared radiation than oxygen and nitrogen. Atmospheric carbon dioxide is up more than 10% since 1960. As a result, there is a so-called "greenhouse effect" and the average temperature of the earth may be increasing. The polar ice caps may be melting, oceans may be rising, and more desert areas may be forming.

3. *Lead.* The use of unleaded gasoline is rapidly allowing a solution to this problem. But is the increasing use of aromatic hydocarbons, necessary for acceptable octane ratings in unleaded gasoline, causing possible increases in polynuclear aromatic hydrocarbons to be added to our air? Compounds such as benzopyrene are known carcinogens.

4. *CFCs.* This is adequately discussed in Chapter 12, but certainly it deserves to be listed here. Chlorine atoms from photodissociation of CFCs in the stratosphere has led to depletion of the ozone layer protecting us from ultraviolet rays. These substances have been outlawed in some applications, but not all. Substitutes have been hard to find. Do we change our way of living by giving up certain products, or do we increase R&D spending to find substitutes more quickly? These questions must be answered.

Things do appear to be improving. In the last few years the level of dust, soot, and other solid particulates in air has decreased. The SO_2 concentration in urban areas has dropped. The CO concentration in urban areas has fallen. In 1970 the Clean Air Amendment allowed the Environmental Protection Agency (EPA) to establish air quality standards and provisions for their implementation and enforcement. This has gone a long way to controlling multi-industrial pollution. This law was strengthened in 1977. The Clean Air Act Amendment of 1990 imposed many new standards. Controls for industrial sources of 41 pollutants must be in place by 1995; for 148 other pollutants, the deadline is 2003.

Pollution from these 189 chemicals will have to decrease by 90%. Production of CFCs and carbon tetrachloride is to be halted by 2000, methyl chloroform by 2002, HCFCs by 2030. Automakers will have to begin in 1994 to phase in new cars emitting 60% less nitrogen oxides and 40% less hydrocarbons. The bill will be costly. For the government administration, $11 billion/yr will be spent by 1995, $22-25 billion/yr by 2005. For industry, $35-50 billion/yr will be the price tag by 2005. The EPA will be spending $32 billion/yr for air pollution controls. But what have we as individuals done to help? When was the last time we walked instead of drove a car? How many families now get by with *only* one car? How much of our total energy needs are real?

Water

A number of critical water problems face us today. We have already discussed information relating to areas of the chemical industry specifically, such as phosphate in detergents and the now banned nonbiodegradable detergents. Certainly efforts by general industry must be continued to eliminate local contamination that may occur, whether it be from oil spills or the typical manufacturing plant down the street. Mercury concentration must be continually monitored in the fish population, since heavy metals tend to concentrate in the fatty tissues of animals and increase in concentration as they go up the food chain. General sewage problems face every municipal sewage treatment facility regardless of size.

Although primary treatment (solid settling and removal) is required and secondary treatment (use of bacteria and aeration to enhance organic degradation) is becoming more routine, tertiary treatment (filtration through activated carbon, applications of ozone, chlorination, etc.) should be set as the ideal for all large urban areas. The 1972 Clean Water Act has done some good at improving water quality. It allowed federal funding for sewage treatment, established effluent standards for water quality, and required permits for point-source discharges. The 1987 Clean Water Act will do more to guarantee continued progress. Though details of the implementation of this act are still to be decided, guidelines limiting effluents from chemical plants have been developed. Chemical industry facilities would be required to sample and monitor stormwater runoff. Wastewater pretreaters that discharge water into sewer systems have new requirements. Pollutant standards for sewage sludge have been set. States must identify toxic "hot spots" in their water and develop plans to alleviate the problems. Finally, states are establishing water-quality standards for 126 priority toxic pollutants.

In recent years the EPA has found many examples of clear-cut improvements in water quality in the United States, but much more needs to be done.

A CHRONOLOGY OF POLLUTION BY THE CHEMICAL INDUSTRY AND ITS CONTROL

Although general air and water pollution and controls are affected by the chemical industry, these problems and solutions are not unique to our industry. Certainly the area of toxic chemicals is unique. Because this problem is so diverse it is by itself a series of complex pollution problems. We have discussed some of these already (see Table 25.1). We now attempt to summarize these and other toxic chemical problems in chronological order. This includes many examples of pollution caused more specifically by the chemical industry or a closely allied industry as well as the laws and controls that have been enacted by governments and the industry to solve some of these unique problems.

Before 1700: Pollution has been with us a long time. There was copper pollution near Jericho on the west bank of the Jordan river due to *copper smelting* for tool manufacture thousands of years ago. *Deforestation* of many areas near the Mediterranean Sea for the building of ships was a norm. Poor agricultural methods led to soil erosion. In 2500 B.C. Sumerians used sulfur compounds to manage insects and in 1500 B.C. the Chinese used natural products to fumigate crops. Pesticides began polluting the environment hundreds of years ago.

1773: The *LeBlanc* process for making *soda ash* for the growing glass, soap, and paper industries of Europe was discovered in this year. Salt and sulfuric acid were heated to give the salt cake needed to react with limestone in the process. But in addition to the salt cake, large amounts of HCl were also released as a gas into the atmosphere or as hydrochloric acid into water. In 1864 the *Alkali Act* in England became one of the first milestones in pollution control, when the discharge of HCl was forbidden.

$$2NaCl + H_2SO_4 \longrightarrow 2HCl + Na_2SO_4$$
$$CaCO_3 + Na_2SO_4 \longrightarrow Na_2CO_3 + CaSO_4$$

Late 1800s: The *lead chamber process* for manufacturing sulfuric acid was prevalent in this period. *Arsenic* was a common contaminant in the pyrites used as a source of sulfur for this process. Now the cleaner contact process is used and most of the raw material is elemental sulfur.

1906: The Pure Food and Drug Act established the *Food and Drug Administration (FDA)* that now oversees the manufacture and use of all foods, food additives, and drugs. The law was toughened considerably in 1938, 1958, and 1962.

1917: During World War I munition factory workers making *trinitrotoluene (TNT)* developed *jaundice* from inhaling the dust. Also in 1917 a TNT explosion near Manchester, England killed 41 and wounded 130 additional.

1921: An Oppau, Germany, *nitrate plant exploded* killing over 600 people. This was probably the worst ever chemical explosion up to 1984.

1924-25: Illnesses and 15 deaths were recorded at Ethyl Corporation among workers developing *lead gasoline additives*.

1935: The *Chemical Manufacturers Association (CMA),* a private group of people working in the chemical industry and especially involved in the manufacture and selling of chemicals, established a *Water Resources Committee* to study the effects of their products on water quality.

1947: A French freighter docked in Texas City, Texas, caught fire and exploded with 2500 tons of *ammonium nitrate* on board. Then the nearby Monsanto chemical plant blew up, followed by oil refineries, tin smelters, and tanks filled with chlorine, sulfur, and nitrate. The exposions were more powerful at ground level than the atomic blasts of Hiroshima and Nagasaki. The final toll was 462 dead, 50 missing, 3000 injured, and $55 million in property damages. This was the worst chemical disaster ever in this country.

1948: CMA established an *Air Quality Committee* to study methods of improving the air that could be implemented by chemical manufacturers.

1958: The *Delaney Amendment* to the Food and Drug Act defined and controlled *food additives*. After this passed, any additives showing an increase in cancer tumors in rats, even if extremely large doses were used in the animal studies, had to be outlawed in foods. This controversial law is still being debated today and has been used to ban a number of additives including the artificial sweetener cyclamate.

1959: Just before Thanksgiving the government announced that it had destroyed *cranberries* contaminated with a chemical, *aminotriazole,* that produced

aminotriazole

cancer in rats. The cranberries were from a lot frozen from two years earlier when the chemical was still an approved weed killer. The animal studies were not completed until 1959. Even though there was no evidence that the 1959 crop was contaminated, cranberry sales dropped precipitously and public fears about dangerous chemicals in food lingered.

1960: *Diethylstilbestrol (DES),* taken in the late 1950s and early 1960s to prevent miscarriages and also used as an animal fattener, was reported to cause vaginal *cancer* in the daughters of these women and caused premature deliveries, miscarriages, and infertility in the daughters.

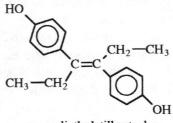

diethylstilbestrol

1962: *Thalidomide,* a prescription drug used as a tranquilizer and flu medicine for pregnant women in Europe to replace dangerous barbiturates that cause 2000-3000 deaths per year by overdoses, was found to cause *birth defects.* Thalidomide had been kept off the market in America because a government scientist insisted on more safety data for the drug, but 8000-10,000 deformities were reported, especially in Germany. In 1962 the *Kefauver-Harris* Amendment to the Food and Drug Act began requiring that drugs be *proven safe* before put on the market.

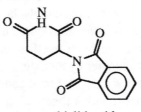

thialidomide

1962: A marine biologist by the name of *Rachel Carson* published her book entitled *Silent Spring* outlining many environmental problems associated with *chlorinated pesticides,* especially DDT. Introduced during World War II, DDT was found to be very effective against insects, was cheap, convenient to use, and had lasting pesticidal action. Its acute toxicity to humans in normal exposure was low. But it was found to accumulate in the body's fatty deposits, had side effects on wildlife, and was very persistent in the environment. Carson's book set off an extensive debate about safety of many different types of toxic chemicals, a debate that is still going on today. Despite some shortcomings *Silent Spring* forced industry to take a hard look at the way their products were affecting the environment. DDT was banned in the U. S. in 1972.

1965: *Nonlinear, nonbiodegradable synthetic detergents* made from propylene tetramer were banned after these materials were found in large amounts in rivers, so much as to cause soapy foam in many locations. *Phosphates* in detergents were also being investigated for their *eutrophication* effect on lakes. They were later banned in detergents by many states in the 1970s.

1965: A strange disease was reported in the area around *Minamata Bay* in Japan. Forty-six people died and many more became ill. The illness was due to *mercury poisoning* from a plastics factory. The Chisso Corporation used mercury as a catalyst in making acetaldehyde. Dimethyl mercury becomes concentrated up the food chain and the heavy reliance on food from the sea life in the bay caused the epidemic. Mercury also became a source of worry for many U.S. rivers and has been monitored closely since then. Chisso was finally found guilty in 1973, but 300 people had died by 1980.

1966: *Polychlorinated biphenyls (PCBs)* were first found in the environment and in contaminated fish. They were banned in 1978 except in closed systems.

1969: The artificial sweetener *cyclamate* was banned because of a study linking it to *bladder cancer* in rats when large doses were fed. At least 20 subsequent studies have failed to confirm this result but cyclamate remains banned. In 1977 *saccharin* was found to cause cancer in rats. It was banned by the FDA temporarily, but Congress placed a moratorium on this ban because of public pressure. Saccharin is still available today.

calcium cyclamate   sodium saccharin

1970: A nationwide celebration, called *Earth Day,* especially on college campuses across the United States, emphasized respect for the environment and an increased awareness by industry and the public about the effects of many substances on the fragile environment.

1970: The *Clean Air Amendment* was passed. This is described earlier in this chapter. It was strengthened in 1977 and 1990.

1971: After *TCDD (dioxin)* had been found to be a contaminant in the herbicide 2,4,5-T and was tested as a teratogen in rats in 1968, the herbicide was outlawed by the EPA on most food crops. A complete discussion of this chemical and its history is given in Chapter 20.

1971: CMA established the *Chemical Emergency Transportation System (CHEMTREC)* to provide immediate information on chemical transportation emergencies. In 1980 this was recognized by the Department of Transportation as the central service for such emergencies.

1972: The *Clean Water Act* was passed. This is discussed previously. It was strengthened in 1987.

1974: A *nylon 6* plant in *Flixborough, England,* exploded during the oxidation of cyclohexane to cyclohexanone. Twenty-eight people were killed. Although it was not a pollution problem, it certainly increased public concern about the chemical industry at a time when it was undergoing vigorous scrutiny.

1974: Three workers in a Goodrich *poly (vinyl chloride)* plant in Louisville developed a rare *angiosarcoma of the liver.* This started the investigation of vinyl chloride as a possible carcinogen.

1975: The state of Virginia closed a *kepone* pesticide plant because 70 of the 150 employees were suffering from kepone poisoning. The James River, which furnishes one fifth of all U.S. oysters, was contaminated. Kepone is made by dimerizing hexachlorocyclopentadiene and hydrolyzing to a ketone. Kepone is now banned. The plant was first owned by Life Science, then by Allied.

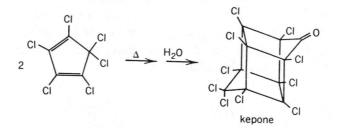

kepone

1976: A plant manufacturing 2,4,5-trichlorophenol in *Seveso, Italy,* exploded and liberated substantial amounts of *TCDD (dioxin).* Although it caused quite a scare and the town was evacuated, there were no known deaths and no increase in birth defects reported. Some chloracne (a skin disease) occurred and one liver cancer was diagnosed.

1976: *The Toxic Substances Control Act (TSCA or TOSCA)* was initiated. It has far-reaching effects specifically for the chemical industry and will be discussed in detail in the next section. An immediate effect that it had was to direct the EPA to develop rules to limit manufacture and use of *PCBs.*

1976: *The Resource Conservation and Recovery Act (RCRA)* was passed. It became effective in 1980, and governs in detail how generators of chemical wastes manage their hazardous wastes. This includes the generation, handling, transportation, and *disposal of hazardous wastes.*

1977: *Polyacrylonitrile* plastic bottles for soft drinks and beer were taken off the market as possible carcinogens because of migration of acrylonitrile into the drink. Now most plastic food containers of this type are poly (ethylene terephthalate).

1977: Employees in an Occidental Petroleum plant manufacturing *dibromochloropropane (DBCP)* became *sterile.* DBCP was used as a soil fumigant and nematocide. It is now banned.

$$CH_2—CH—CH_2—Cl$$
$$\quad\; | \qquad |$$
$$\quad\; Br \quad\; Br$$

DBCP

1977: *Benzene* was linked to an abnormally high rate of *leukemia* at a Goodyear plant. This further increased the concern with benzene use in industry.

1977: *Tris(2,3-dibromopropyl) phosphate (Tris),* used to treat children's sleepwear to reduce flammability, was banned from use. The chemical was linked to *kidney cancer* in mice and rats and was mutagenic in bacteria. At the time it was used on 40-60% of children's sleepwear, mostly polyester, to enable it to meet federal requirements for flame retardancy.

$$(CH_2—CH—CH_2—O)_3—P{=}O$$
$$\quad\;\; | \qquad |$$
$$\quad\;\; Br \quad\; Br$$

Tris

1978: There was a ban on *chlorofluorocarbons (CFCs) as aerosol propellants* because they may react with ozone in the stratosphere, increase the penetration of ultraviolet sunlight, alter the weather, and increase the risk of skin cancer. CFCs are discussed in Chapter 12.

1978: An old chemical dump in Niagara Falls, New York, near *Love Canal,* began leaking into the environment. A state of emergency and an evacuation of the neighboring area resulted.

1980: *Asbestos* dust had been known for years in industry to cause a rare form of *lung cancer* when inhaled. A rule in 1980 caused regulation of asbestos use and repair in school buildings.

1980: The *Comprehensive Environmental Response, Compensation, and Liability Act* established a $1.6 billion, five-year *"Superfund"* to clean up landfills. This was funded by the chemical industry (87.5%) and general government revenues (12.5%). It will be expanded in dollar amount and in number of landfills affected later in the 1980s. In 1982 of some 14,000 abandoned hazardous waste sites across the nation, the EPA proposed 418 of them for priority cleanup. New Jersey had the most priority sites (65) followed by Michigan (46), Pennsylvania (30), and New York (26). Alaska, The District of Columbia, Georgia, Hawaii, Nevada, the Virgin Islands, and Wisconsin had no sites. The notorious Love Canal ranked 116th worst.

1982: Bottles of *Tylenol®,* a common pain reliever, were found to contain *sodium cyanide* that had purposely been placed there. Seven deaths occurred in Chicago. The murderer has never been found. This incident caused stricter packaging guidelines for the pharmaceutical industry. Most drugs now are sealed into their containers with a plastic or metal wrapping that cannot be removed without it being noticed.

1983: Over a two-year period 600 people in Spain died from so-called *"olive oil"* bought from door-to-door salespersons. It actually was oil contaminated with toxic chemicals that was to be used industrially.

1984: On December 3 the *worst* chemical and general *industrial accident* in history occurred in *Bhopal, India.* A Union Carbide plant making carbamate insecticide developed a leak in one of their underground storage tanks for the very toxic chemical *methyl isocyanate,* MIC, CH_3—N=C=O, used in the manufacture of their largest selling carbamate, carbaryl or Sevin®. Isocyanates react exothermicly with water and methyl isocyanate has a low bp, 39°C. As the tank heated up some of the isocyanate was hydrolyzed with a caustic safety tank, but a large amount escaped into the atmosphere. At least 2500 people died that night in the neighborhood next to the plant. As much as 45 tons of MIC may have escaped. The shock of this disaster is still being felt and the final consequences remain to be seen. A detailed reassessment of safety and environmental standards for many chemical plants has resulted. A West Virginia MIC plant in the U. S. is similarly designed and makes 30-35 million lb of MIC per year. Suits against Union Carbide have been filed. Even arson has not yet been ruled out.

1985: Three employees of a silver recovery firm near Chicago were convicted of *murder.* Film Recovey Systems *recovered silver* from used X-ray films *using sodium cyanide.* In 1983 a worker became ill and died. Cyanide level in the blood was a lethal dose. The president and part owner, the plant manager, and the plant foreman were responsible. Plant safety conditions were completely inadequate and much different from that found with any other company, but it may become a landmark decision because it was the first time murder was leveled against corporate officials.

1986: The *Safe Drinking Water Act Amendments* require EPA to set standards for 83 chemicals.

1986: The *Emergency Planning and Community-Right-to-Know Act* was signed into law. Companies involved in the production and handling of hazardous materials have had to submit *safety data sheets* or lists of chemicals kept on site. Companies must also report inventories of specific chemicals kept in the workplace and annual release of hazardous materials into the environment.

1986: The *Superfund Amendments and Reauthorization Act* of 1986 established a $9 billion, five-year fund to pay for continued cleanup of 375 hazardous waste sites. This is scheduled for reauthorization in 1991. Funding for the programs comes from the following: $2.75 billion from a tax on crude oil; $2.5 billion from a broad-based tax on corporations; $1.4 billion from a tax on 42 chemical feedstocks; $1.25 billion from general federal revenues; $0.60 billion in funds recovered from responsible parties; and $0.50 billion from a motor-fuels tax. At the time of this writing over 30,000 sites have been inventoried, over 1,100 are on the National Priority List for cleanup, hundreds have had short-term cleanups, but only 50 have had complete cleanups.

1988: The *Chemical Diversion and Trafficking Act* contains three key provisions to address the problem of diverting chemicals to make illegal drugs: (1) the seller of chemicals must keep detailed records; (2) sellers must report suspicious purchases and unusual or excessive losses; and (3) the Drug Enforcement Administration is authorized to control export and import transactions.

1989: The *Great Apple Scare* occurred. Alar®, or daminozide, was found in apples and apple products as a residue. It is a growth regulator that keeps apples longer on trees and helps yield more perfectly shaped, redder, firmer fruit. It also maintains firmness in stored apples by reducing ethylene production. Concern about Alar®'s carcinogenicity focussed not on the compound itself, but on a breakdown product, unsymmetrical dimethylhydrazine (UDMH). Heat treatment in apple processing can cause Alar® to break

$$HO{-}\overset{\overset{\textstyle O}{\|}}{C}{-}CH_2{-}CH_2{-}\overset{\overset{\textstyle O}{\|}}{C}{-}NH{-}N\!\!\begin{smallmatrix}CH_3\\[4pt]CH_3\end{smallmatrix} \xrightarrow{\;H_2O\;} HO{-}\overset{\overset{\textstyle O}{\|}}{C}{-}CH_2{-}CH_2{-}\overset{\overset{\textstyle O}{\|}}{C}{-}OH$$

daminozide, Alar® succinic acid

$$+$$

$$\begin{smallmatrix}H\\[4pt]H\end{smallmatrix}\!\!N{-}N\!\!\begin{smallmatrix}CH_3\\[4pt]CH_3\end{smallmatrix}$$

UDMH

down. Uniroyal, its producer, voluntarily halted sales in the U.S. for food group uses.

1989-91: A recent series of four serious explosions over these three years, though unrelated to each other, has caused renewed concern over plant safety. In 1989 one of the world's largest HDPE plants, owned by Phillips 66 in Pasadena, TX, exploded when ethylene and isobutane leaked from a pipeline. Twenty were killed. In 1990 an Arco Chemical Co. plant in Channelview, TX had an explosion in the petrochemicals complex which killed 17. A treatment tank of wastewater and chemicals blew up. These two accidents in the Houston area caused more deaths than the previous ten years combined. A 1991 explosion in Sterlington, LA of an Argus Chemical nitroparaffin plant resulted in eight deaths. Nitroparaffins are used in pharmaceuticals, fine chemicals, cosmetics, and agrochemicals. Also in 1991 an explosion of an Albright & Wilson Americas plant in Charleston, SC killed six. Ironically Antiblaze 19®, a phosphonate ester and flame retardant used in textiles and polyurethane foam, was being manufactured from trimethyl phosphite, dimethyl methylphosphonate, and trimethyl phosphate.

THE TOXIC SUBSTANCES CONTROL ACT (TSCA)

Probably the law that has specifically affected the chemical industry the most is TSCA. Since it was signed on October 11, 1976 and became effective on January 1, 1977, it has caused many changes in the industry and will create further modifications in the years to come. The basic thrust of the law is threefold: (1) to develop data on the effects of chemicals on our health and environment, (2) to grant authority to the EPA to regulate substances presenting an unreasonable risk, and (3) to assure that this authority is exercised so as not to impede technological innovation.

TSCA is a "balancing-type law." It is concerned with unreasonable risks. It attempts to balance risks versus benefits for all chemicals and uses. The EPA administrator must consider (1) effects on health, (2) effects on the environment, (3) benefits and availability of substitutes, and (4) economic consequences.

Specific bans on chemicals or uses have not been the most important outcome of TSCA. Only one type of chemical, PCBs, was specifically targeted in the original law and it is now outlawed in some of its uses. EPA administration of the law has led to a ban of chlorofluorocarbons as aerosol propellants, restrictions on dioxin waste disposal, rules on asbestos use, and testing rules on chlorinated solvents. It has led to a central bank of information on existing commercial chemicals, procedures for further testing of hazardous chemicals, and detailed permit requirements for submission of proposed new commercial chemicals.

After TSCA was passed the EPA began a comprehensive study of all commercial chemicals. Of the 55,000 chemicals made only 9.9% of them are made at the 1 million lb/yr or more level. These account for 99.9% of all chemical production. Detailed records are now available on all chemicals made at the time TSCA was initiated. Some are lacking in full health and environmental testing. The consensus is that this should be done, but the resources available for such testing are limited in private and government laboratories.

Next, the EPA began requiring companies to submit premanufacturing notices (PMNs) 90 days before a chemical's manufacture is started. EPA may stop the manufacture or prohibit certain uses. The PMN must include detailed information. They were initiated in 1979. Some 1100 were filed from 1979-1981 and the number now averages 700 new PMNs per year. About 1-7% are chemicals of concern. The PMN system has been criticized by many in the industry. There has been a 54% decline in new chemical introductions since PMNs have been initiated. Eventually it may concentrate new product development into large companies that can afford the extra testing and administrative costs.

Finally, a number of existing chemicals have been designated by EPA as being hazardous and requiring extra precautions and further testing. The list is revised periodically. The 1990 list of acute hazardous waste published by EPA contained 203 chemicals. These are listed in Table 25.2, which is given at the end of the chapter. A second list about twice as long is considered toxic waste (nonacute but still hazardous). The EPA has also tested a number of chemicals as potential carcinogens. The current ones are listed in Table 25.3 at the end of this chapter, which includes 194 possible carcinogens with a hazard ranking of low, medium, or high.

The cost of TSCA administration is high. Over 600 people are now employed in the Office of Toxic Substances. In the 1990s the cost to the government will be over $100 million/yr. Direct public and private costs total millions of dollars and there may be other indirect costs that cannot be estimated. Certainly some things could be done to get more for our money. The PMN system could be modified to spend less time on low-risk chemicals. The ability to regulate existing chemicals should be increased. Voluntary compliance by industry should be stressed because it is cheaper and more efficient, but this must obviously be backed up by the possibility of regulatory action by the government.

Perhaps it is appropriate that we end this topic and this book the way we began, with the top 50 chemicals. How does TSCA relate to them. The physical and chemical properties of all 50 are easily cataloged. Human and mammalian toxicology are known for 45 of them. The other 5 are not expected to be of concern (nitrogen, oxygen, ethylene, propylene, and ammonium nitrate). Ecotoxicology data are available on all but 8 (carbon dioxide, nitrogen, oxygen, soda ash, sulfuric acid, ammonium sulfate, terephthalic acid, and sodium silicate). Four compounds have been recommended by EPA for further priority testing: ethylene oxide, propylene oxide, toluene, and xylene.

The final verdict is still out on whether TSCA is sufficient to maintain adequate control of toxic chemicals. The years ahead may show that further regulation, legislation, and enforcement are necessary or that less is optimum to avoid restrictions on innovation. No doubt, just as now, a variety of opinions will exist.

ARE THINGS BETTER TODAY?

Since the beginning of the environmental movement in the 1960s many people have asked repeatedly if we are better off environmentally today and, if so, can we do even more than we have done. In some respects statistics show a bad side. The chemical industry (SIC 28) leads all industries in toxic chemical emissions into the air with 0.89 billion lb annually, about one third of the total of 2.4 billion lb. Primary metals (SIC 33) is next with 0.22 billion lb, and paper (SIC 26) is third with 0.21 billion lb. Air emissions are the largest source of toxic emissions, some 39%, with underground (19%) , off-site (18%), public sewage (9%), land (9%), and surface water (6%) making up the rest. Pollution control is getting more expensive. In 1991 it is a little over $100 billion, but by 2000 it should be about $150 billion. The private sector will be absorbing 60% of this cost, local governments 22%, the EPA 7%, non-EPA federal government 8%, and the state governments 3%.

The volume of hazardous waste material is getting very large and estimates as high as a trillion lb of hazardous waste material that must be processed have been quoted by EPA for all industries. Private companies will be doing about 90% of this processing in the future, with the government, mostly federal, doing the rest. Hazardous waste treatment alone will be a $40 billion business by the year 2000.

On the brighter side, the chemical industry is spending much more on pollution control that it once did. In 1989 it is estimated that the chemical industry spent $1.45 billion on capital expenditures for pollution abatement; annual operating expenditures for pollution abatement was another $3.35

billion. The Council on Environmental Quality estimates that air pollution control benefits amount to over $20 billion/yr and thousands of lives are saved. Water pollution control benefits top $10 billion/yr. Since 1961 employees working on environmental problems have tripled. The chemical industry now accounts for 22% of all expenditures for pollution control in the United States. A chemical plant worker is four times safer than the average American industrial employee. The National Safety Council ranked the chemical industry first in safety of all American industries in 1980 as compared to sixth place in 1975. As an example, the biggest chemical company, Du Pont, has about 3000 employees in environmental protection and spends nearly 7% of its total sales on pollution control. The EPA found in 1988 that U.S. industry recorded an 11% drop in total toxic emissions compared to 1987. Total toxic releases were 6.24 billion lb. U.S. chemical companies reported emitting 0.36 billion lb of toxic chemicals into water, 2.43 billion lb into the air, 0.56 billion lb into landfills, and 1.22 billion lb into underground storage wells for a total of 4.57 billion lb or 73% of the total. Texas and Louisiana accounted for 24% of all releases. But the chemical industry reported the greatest amount of pollution reduction, 0.16 billion lb, of any single industry. In 1991 many chemical companies joined EPA's call for a voluntary reduction plan named the 33-50 Program. EPA has targeted 17 chemicals (Table 25.4) accounting for 1.4 billion lb of toxic waste. The goal is to cut environmental emissions of these chemicals to 33% below the 1988 levels by 1992, and 50% below 1988 levels by 1995. Companies that have subscribed are Amoco Chemical, BASF, Bayer, Dow Chemical, Du Pont, Monsanto, Occidental Chemical, and Union Carbide.

In summary, it does appear that dramatic improvements have been made in pollution control by the chemical industry. Perhaps the place most in need of further work is toxic waste disposal, where concentrated efforts could improve many hazardous waste sites by 2000. Only time will tell whether or not we can complete the job and meet the challenge.

TABLE 25.2 EPA's Acute Hazardous Wastes

Acetaldehyde, chloro-
Acetamide, N-(aminothioxomethyl)-
Acetamide, 2-fluoro-
Acetic acid, fluoro-, sodium salt
Acetyl-2-thiourea
Acrolein
Aldicarb
Aldrin
Allyl alcohol
Aluminum phosphide
5-(Aminomethyl)-3-isoxazolol
4-Aminopyridine
Ammonium picrate
Ammonium vanadate
Argentate(1-), bis(cyano-C)-, potassium
Arsenic acid H_2AsO_4
Arsenic oxide As_2O_5
Arsenic oxide As_2O_3
Arsenic pentoxide
Arsenic trioxide
Arsine diethyl-
Arsonous dichloride, phenyl-
Aziridine
Aziridine, 2-methyl-
Barium cyanide
Benzenamine, 4-chloro-
Benzenamine, 4-nitro-
Benzene,(chloromethyl)-
1,2 Benzenediol, 4-[1-hydroxy-2(methylamino)ethyl]-
Benzeneethanamine, alpha, alpha-dimethyl
Benzenethiol
2H-1Benzopyran-2-one,4-hydroxy-3-(3-oxo-1-phenyl-butyl)-,& salts when present at concentrations greater than 0.3%
Benzyl chloride
Beryllium
Bromoacetone
Brucine
2-Butanone, 3,3-dimethyl-1(methylthio), O-[methylamino)carbonyl]oxime
Calcium cyanide
Calcium cyanide $Ca(CN)_2$
Carbon disulfide
Carbonic dichloride
Choloroacetaldehyde
p-Chloroaniline
1-(o-Chlorophenyl) thiourea
3-Chloropropionitrile
Copper cyanide
Copper cyanide Cu(CN)
Cyanide (soluble cyanide salts), not otherwise specified
Cyanogen chloride
Cyanogen chloride (CN)Cl
2-Cyclohexyl-4,6-dinitrophenol
Dichloromethyl ether
Dichlorophenylarsine

Diedrin
Diethylarsine
Diethyl-p-nitrophenyl phosphate
O,O-Diethyl O-pyrazinyl phosphorothioate
Diisopropylfluorophosphate (DFP)
1,4,5,8-Dimethanonaphthalene 1,2,3,4, 10,10-hexachloro-1,4,4a ,5, 8, 8a-hexahydro-, (1alpha, 4 alpha, 4abeta,5alpha, 8alpha, 8abeta)
1,4,5,8-Dimethanonaphthalene 1,2,3,4, 10,10-hexachloro-1,4,4a,5,8, 8a-hexahydro-, (1alpha,4alpha, 4abeta,5beta, 8beta, 8abeta)
2,7:3,6-Dimethanonaphthalene [2,3-b]oxirene, 3,4, 5,6,9,9-hexachloro-1a,2,2a,3,6,6a,7,7a-octa-hydro-, (1aalpha, 2beta, 2aalpha,3beta,6beta, 2aalpha,3beta,6beta,7beta,7aalpha) 7beta, 7aalpha)-
2,7:3,6-Dimethanonaphthalene [2,3-b]oxirene, 3,4, 5,6,9,9-hexachloro-1a,2,2a,3,6,6a,7,7a-octa-hydro-, (1aalpha, 2beta, 2abeta,3 alpha, 6alpha, 6abeta, 7beta,7aalpha)-, & metabolites
Dimethoate
alpha, alpha-Dimethylphenethylamine
4,6-Dinitro-o-cresol, & salts
2,4-Dinitrophenol
Dinoseb
Diphosphoramide, octamethyl-
Diphosphoric acid, tetraethyl ester
Disulfoton
Dithiobiuret
Endosulfan
Endothall
Endrin
Endrin, & metabolites
Epinephrine
Ethanedinitrile
Ethanimidothioic acid, N-[[(methylamino)carbonyl]oxy]-, methyl ester
Ethyl cyanide
Ethyleneimine
Famphur
Fluorine
Fluoroacetamide
Fluoroacetic acid, sodium salt
Fulminic acid, mercury (2+) salt
Heptachlor
Hexaethyl tetraphosphate
Hydrazinecarbothioamide
Hydrazine, methyl
Hydrocyanic acid
Hydrogen cyanide
Hydrogen phosphide
Isodrin
3(2H)-Isoxazolone, 5-(aminomethyl)-
Mercury, (acetato-O)phenyl-
Mercury fulminate
Methanamine, N-methyl-N-nitroso-
Methane, isocyanato-

TABLE 25.2 (continued)

Methane, oxibis[chloro-

Methane, tetranitro-

Methanethiol, trichloro-

6,9-Methano-2,4,3-
benzodioxathiepin, 6,7,8,9,10,
10-hexachloro-1,5,5a,6,9,9a, hexahydro-, 3-oxide

4,7 Methano-1H-indene, 1,4,5,6.7,8,8-heptachloro-
3a,4,7,7a-tetrahydro

Methomyl

Methyl hydrazine

Methyl isocyanate

2-Methyllactonitrile

Methyl parathion

alpha-Naphthylthiourea

Nickel carbonyl

Nickel carbonyl Ni(CO)$_4$

Nickel cyanide

Nickel cyanide Ni(CN)$_2$

Nicotine, & salts

Nitric oxide

p-Nitroaniline

Nitrogen dioxide

Nitrogen oxide NO

Nitrogen oxide NO$_2$

Nitroglycerine

N-Nitrosodimethylamine

Nitrosomethylvinylamine

Octamethylpyrophosphoramide

Osmium oxide OsO$_4$

Osmium tetroxide

7-Oxabicyclo[2.2.1]heptane-2,3-dicarboxylic acid

Parathion

Phenol, 2-cyclohexyl-4,6-dinitro-

Phenol, 2,4-dinitro-

Phenol, 2-methyl-4,6-dinitro-, & salts

Phenol, 2-(1-methylpropyl)-4,6-dinitro-

Phenol, 2,4,6-trinitro-, ammonium salt

Phenylmercury acetate

Phenylthiourea

Phorate

Phosgene

Phosphine

Phosphoric acid, diethyl 4-nitrophenyl ester

Phosphorodithioic acid, O,O-diethyl
S-[2-(ethylthio)ethyl] ester

Phosphorodithioic acid, O,O-diethyl-S-
[(ethylthio)methyl] ester

Phosphorodithioic acid. O,O-dimethyl S-[2-
(methylamino)-2-oxoethyl] ester

Phosphorofluoridic acid, bis(1-methylethyl) ester

Phosphorothioic acid, O,O-diethyl O-(4-nitro-
phenyl) ester

Phosphorothioic acid, O,O-diethyl O-pyrazinyl
ester

Phosphorothioic acid,
O-[4-[(dimethylamino)sulfonyl]phenyl] O,O-
dimethyl ester

Phosphorothioic acid, O,O,-dimethyl O-(4-
nitrophenyl) ester

Plumbane, tetraethyl-

Potassium cyanide

Potassium cyanide K(CN)

Potassium silver cyanide

Propanal, 2-methyl-2-(methylthio-),O[(methylamino)carbonyl]oxime

Propanenitrile

Propanenitrile, 3-chloro-

Propanenitrile, 2-hydroxy-2-methyl-

1,2,3-Propanetriol, trinitrate

2-Propanone, 1-bromo-

Propargyl alcohol

2-Propenal

2-Propen-1-ol

1 ,2-Propylenimine

2-Propyn-1-ol

4-Pyridinamine

Pyridine, 3-(1-methyl-2-pyrrolidinyl)-, (S)-, & salts

Selenious acid, dithallium(1+) salt

Selenourea

Silver cyanide

Silver cyanide Ag(CN)

Sodium azide

Sodium cyanide

Sodium cyanide Na(CN)

Strontium sulfide SrS

Strychnidin-10-one, +a salts

Strychnidin-10-one, 2,3-dimethoxy-

Strychnine, & salts

Sulfuric acid, dithallium(1 +) salt

Tetraethyldithiopyrophosphate

Tetraethyl lead

Tetraethyl pyrophosphate

Tetranitromethane

Tetraphosphoric acid, hexaethyl ester

Thallic oxide

Thallium oxide Tl$_2$O$_3$

Thallium(I) selenite

Thallium(I) sulfate

Thiodiphosphoric acid, tetraethyl ester

Thiofanox

Thioimidodicarbonic diamide [(H$_2$N)C(S)]$_2$NH

Thiophenol

Thiosemicarbazide

Thiourea (2-chlorophenyl)-

Thiourea 1-naphthalenyl-

Thiourea, phenyl-

Toxaphene

Trichloromethanethiol

Vanadic acid, ammomium salt

Vanadium oxide V$_2$O$_5$

Vanadium pentoxide

Vinylamine, N-methyl-N-nitroso-

Warfarin, & salts,when present at concentrations
greater than 0.3%

Zinc cyanide

Zinc cyanide Zn(CN)$_2$

Zinc phosphide Zn$_3$P$_2$, when present at concentrations greater than 10%

TABLE 25.3 EPA's Possible Carcinogens

Acetamide, N-fluoren-2-yl	HIGH	Chlorambucil	MED
Acrylonitrile	MED	Chlordane	MED
Aldrin	HIGH	Chlornaphazine	LOW
Amitrole	MED	Chloroform	MED
Arsenic	HIGH	Chloromethyl methyl ether (technical grade)	HIGH
Arsenic acid	HIGH	4-Chloro-*o*-toluidine, hydrochloride	LOW
Arsenic disulfide	HIGH	Chromium	None
Arsenic pentoxide	HIGH	Ammonium bichromate	HIGH
Arsenic trichloride	HIGH	Ammonium chromate	HIGH
Arsenic trioxide	HIGH	Calcium chromate	HIGH
Arsenic trisulfide	HIGH	Chromic acid	HIGH
Cacodylic acid	None	Lithium chromate	HIGH
Calcium arsenate	HIGH	Potassium bichromate	HIGH
Calcium arsenite	HIGH	Potassium chromate	HIGH
Cupric acetoarsenite	HIGH	Sodium bichromate	HIGH
Dichlorophenylarsine	None	Sodium chromate	HIGH
Diethylarsine	None	Strontium chromate	HIGH
Lead arsenate	HIGH	Chrysene	LOW
Potassium arsenate	HIGH	Coke Oven Emissions	HIGH
Potassium arsenite	HIGH	Creosote	HIGH
Sodium arsenate	HIGH	Cyclophosphamide	MED
Sodium arsenite	HIGH	Daunomycin	MED
Asbestos	HIGH	DDD	MED
Auramine	LOW	DDE	MED
Azaserine	HIGH	DDI	MED
Aziridine	HIGH	Diallate	LOW
Benz(c) acridine	LOW	Diaminotoluene (mixed)	MED
Benz(a) anthracene	MED	Dibenz(a,h) anthracene	HIGH
Benzene	MED	1,2:7,8-Dibenzopyrene	MED
Benzidine and its salts	HIGH	1,2-Dibromo-3-chloropropane	HIGH
Benzo (b)fluoranthene	HIGH	3,3-Dichlorobenzidine	MED
Benzo(k)fluoranthene	None	1,2-Dichloroethane	LOW
Benzo(a)pyrene	HIGH	1,1-Dichloroethylene (Vinylidene chloride)	LOW
Benzotrichloride	MED	Dieldrin	HIGH
Benzyl chloride	LOW	1,2:3,4-Diepoxybutane	MED
Beryllium	MED	1,2-Diethylhydrazine	MED
Beryllium chloride	HIGH	Diethylstilbestrol	HIGH
Beryllium fluoride	HIGH	Dihydrosafrole	MED
Beryllium nitrate	HIGH	3,3'-Dimethoxybenzidine	MED
alpha - BHC	MED	Dimethyl sulfate	MED
beta - BHC	LOW	Dimethylaminobenzene	MED
gamma BHC (Lindane)	MED	7,12-Dimethylbenz(a)anthracene	HIGH
Bis (2-chloroethyl) ether	MED	3,3'-Dimethylbenzidine	MED
Bis (chloromethyl) ether	HIGH	Dimethylcarbamoyl chloride	HIGH
Bis (2-ethylhexyl) phthalate	LOW	1,1-Dimethylhydrazine	MED
Cadmium	MED	1,2-Dimethylhydrazine	HIGH
Cadmium acetate	MED	Dinitrotoluene (mixed)	MED
Cadmium bromide	MED	2,4-Dinitrotoluene	MED
Cadmium chloride	MED	2,6-Dinitrotoluene	LOW
Carbon tetrachloride	MED	1,4-Dioxane	LOW

Under review.

TABLE 25.3 EPA's Possible Carcinogens

1,2-Diphenythydrazine	MED	N-Nitrosodiethylamine	HIGH
Epichlorohydrin	LOW	N-Nitrosodimethylamine	MED
Ethyl carbamate (urethane)	LOW	N-Nitrosodi-*n* -propylamine	MED
Ethyl 4,4' -dichlorobenzilate	MED	N-Nitroso-N-ethylurea	HIGH
Ethylene dibromide	HIGH	N-Nitroso-N-methylurea	HIGH
Ethylene oxide	MED	N-Nitroso-N-methylurethane	HIGH
Ethylenethiourea	MED	N-Nitrosomethylvinylamine	MED
Ethyl methanesulfonate	HIGH	N-Nitrosopiperidine	MED
Formaldehyde	MED	N-Nitrosopyrrolidine	HIGH
Glycidylaldehyde	MED	5-Nitro-*o*-toluidine	LOW
Heptachlor	HIGH	Pentachloroethane	LOW
Heptachlorepoxide	HIGH	Pentachloronitrobenzene	LOW
Hexachlorobenzene	MED	Pentachlorophenol	None
Hexachlorobutadiene	##	Phenacetin	LOW
Hexachloroethane	##	Polychlorinated biphenyls (PCBs)	MED
Hydrazine	HIGH	Aroclor 1016	MED
Indeno[1,2,3-cd]pyrene	LOW	Aroclor 1221	MED
Isosafrole	LOW	Aroclor 1232	MED
Kepone	MED	Aroclor 1242	MED
Lasiocarpine	MED	Aroclor 1248	MED
Lead	##	Aroclor 1254	MED
Lead acetate	##	Aroclor 1260	MED
Lead phosphate	##	1,3-Propane sultone	MED
Lead stearate	##	1,2-Propylenimine	HIGH
Lead subacetate	##	Saccharin	LOW
Lead sulfide	##	Safrole	LOW
Melphalan	HIGH	Selenium sulfide	LOW
Methyl chloride	LOW	Streptozotocin	HIGH
3-Methylcholanthrene	MED	2,3,7,8-Tetrachlorodibenzo-*p*-dioxin (TCDD)	HIGH
4,4'-Methylenebis(2-chloroaniline)	MED	1,1,1,2-Tetrachloroethane	LOW
Methyl iodide	LOW	1,1,2,2-Tetrachloroethane	LOW
N-Methyl-N'-nitro-N-nitrosoquanidine	MED	Tetrachloroethylene	LOW
Methylthiouracil	MED	Thioacetamide	MED
Mitomycin C	MED	Thiourea	MED
1-Naphthylamine	LOW	*o*-Toluidine	LOW
2-Naphthylamine	HIGH	*p*-Toluidine	LOW
Nickel	LOW	*o*-Toluidine hydrochloride	LOW
Nickel ammonium sulfate	LOW	Toxaphene	MED
Nickel carbonyl	MED	1,1,2-Trichloroethane	LOW
Nickel chloride	LOW	Trichloroethylene	LOW
Nickell cyanide	LOW	Trichlorophenol (mixed)	LOW
Nickel hydroxide	LOW	2,4,5-Trichlorophenol	None
Nickel nitrate	LOW	2,4,6-Trichlorophenol	LOW
Nickel sulfate	LOW	Tris(2,3-dibromopropyl) phosphate	MED
2-Nitropropane	MED	Trypan blue	MED
N-Nitrosodi-*n*-butylamine	MED	Uracil mustard	MED
N-Nitrosodiethanalamine	HIGH	Vinyl chloride	HIGH

TABLE 25.4 EPA's Toxic Waste List Under the 33-50 Program

Benzene	Methyl ethyl ketone
Cadmium	Methyl isobutyl ketone
Carbon tetrachloride	Nickel
Chloroform	Tetrachloroethylene
Chromium	Toluene
Cyanide	Trichloroethane
Dichloromethane	Trichloroethylene
Lead	Xylene
Mercury	

Source: Reprinted with permission from *Chem. Eng. News* **1991**, 69(27), 7.
Copyright 1991 American Chemical Society.

Figure 25.1 Sunset on a chemical plant. (Courtesy of Amoco Chemicals Corporation, Alvin, TX.)

Appendix ─────────────
Possible Subjects
for Further Study

Although it might not be specifically mentioned CEH, KO, and Ullmann have information on all these topics and should be routinely consulted first for basic material. Then proceed to the other leading references listed if any. You should also consult with any of the references listed at the beginning of this text.

Subject	*Leading References*
1. Asbestos	*C & E News*, 3-4-85, pp. 28-41
2. Bhopal, India tragedy	*C & E News*, 2-11-85, special issue
3. Bromine, inorg. and org. bromides —manufacture and uses	Kent, pp. 232-233
4. Carbon black	Austin, pp. 130-148
5. Ceramics	Austin, pp. 149-169
6. Chemistry of steel manufacture	CEH, KO
7. Coal technology	Kent, pp. 66-129
8. Cosmetics	*C & E News*, 4-29-85, pp. 19-46; 4-4-88, pp. 21-41
9. Dyes and pigments	W & R II, pp. 399-464; *C & E News*, 7-25-88, pp. 7-14
10. Explosives	Kent, pp. 700-717; Austin, pp. 389-408
11. Flame retardants	*C & E News*, 4-24-78, pp. 22-36; 3-7-83, pp. 15-18
12. Flavors and fragrances	*C & E News*, 7-30-84, pp. 7-13
13. Food additives	W & R II, pp. 383-398

14. Glass industry	Austin, pp. 193-212
15. History of industrial chemistry and petrochemicals	Taylor, F.S. *A History of Industrial Chemistry*; Spitz, P.H. *Petrochemicals*
16. Industrial catalysts	W & R II, pp. 220-239; *C & E News*, 2-17-86, pp. 27-50; 5-29-89, pp. 29-56; *Chemical Week*, 6-29-88, pp. 20-64
17. Industrial fermentation	Kent, pp. 631-69; White, pp. 35-56
18. Industrial solvents	W & R II, pp. 280-303
19. Lubricating oils	W & R II, pp. 304-319
20. Metals and alloys	Austin, pp. 242-259
21. Misc. specialty fibers	Kent, pp. 414-422
22. Nuclear power	Kent, pp. 862-916
23. Oils, waxes, and fats	Kent, pp. 428-449
24. Photographic products	Austin, pp. 409-423
25. Plasticizers	W & R II, pp. 320-338
26. Plastics, specialty & engineering	W & R II, pp. 83-97; *C & E News*, 8-18-86, pp. 21-46
27. Portland cement	Austin, pp. 170-192
28. Sugars	Kent, pp. 577-606
29. Synthetic fuels	*C & E News*, 10-26-81, pp. 22-32; *Chemtech*, 12-82, pp. 744-750
30. Vitamins	W & R II, pp. 371-382
31. Zeolites	*C & E News*, 9-27-82, pp. 10-15

Index

513